INVERTEBRATE LEARNING

Volume 1
Protozoans Through Annelids

INVERTEBRATE LEARNING

INVERTEBRATE LEARNING

Volume 1
Protozoans Through Annelids

Edited by
W. C. Corning and J. A. Dyal

Department of Psychology
University of Waterloo
Waterloo, Ontario, Canada

and
A. O. D. Willows

Department of Zoology
University of Washington
Seattle, Washington

PLENUM PRESS · NEW YORK-LONDON · 1973

Library of Congress Catalog Card Number 72-90335

ISBN 0-306-37671-7

To Dan Mazilli

PREFACE

Since the publication of the second volume of *Comparative Psychology* by Warden, Warner, and Jenkins (1940), there has not been a comprehensive review of invertebrate learning capacities. Some high-quality reviews have appeared in various journals, texts, and symposia, but they have been, of necessity, incomplete and selective either in terms of the phyla covered or the phenomena which were reviewed. Although this lack has served as a stimulus for the present series, the primary justification is to be found in the resurgence of theoretical and empirical interests in learning capacities and mechanisms in simpler systems of widely different phylogenetic origin. Intensive research on the physiological basis of learning and memory clearly entails exploration of the correlations between levels of nervous system organization and behavioral plasticity. Furthermore, the presence of structural–functional differentiation in ganglionated systems, the existence of giant, easily identifiable cells, and the reduced complexity of structure and behavior repertoires are among the advantages of the "simple systems" strategy which have caused many neuroscientists to abandon their cats, rats, and monkeys in favor of mollusks, leeches, planaria, crayfish, protozoa, and other invertebrate preparations. Behavioral research continues to reveal remarkable capacities in these simple organisms and encourages us to believe that the confluence of the invertebrate learning data with the more voluminous vertebrate literature will contribute substantially to the enrichment of all of the neurobehavioral sciences.

While the emphasis of these three volumes is upon learning, each of the authors has included phylogenetic, anatomical, and physiological information which seems to be especially relevant to the observed capacities for behavioral plasticity. However, our concern for a comprehensive review of learning capacities and mechanisms has precluded discussion of many diverse and interesting reflexes and behavioral adaptations which would need to be discussed in a treatment of the complete behavior repertoires of invertebrates. Nonetheless, it is our hope that the present three volumes will prove to be a valuable reference source which will stimulate further exploration of invertebrate plasticity by neurobehavioral scientists.

We wish to thank Kathy Bernhardt, Marjorie Kohli, Betty Ledger, Madeline Bailey, and Gabrielle Schreiber for their tolerance and invaluable assistance in the preparation of the manuscripts.

W. C. CORNING
J. A. DYAL
A. O. D. WILLOWS

January 1973

CONTENTS OF VOLUME 1

Chapter 1
Invertebrate Learning and Behavior Taxonomies
J. A. Dyal and W. C. Corning

Chapter 2
Protozoa
W. C. Corning and R. Von Burg

Chapter 3
Behavioral Modifications in Coelenterates
N. D. Rushforth

Chapter 4
Platyhelminthes: The Turbellarians
W. C. Corning and S. Kelly

Chapter 5

Behavior Modification in Annelids

J. A. Dyal

CONTENTS OF VOLUME 2

Chapter 8
Learning in Insects Excepts Apoidea
Thomas M. Alloway

CONTENTS OF VOLUME 3

Chapter 1

INVERTEBRATE LEARNING AND BEHAVIOR TAXONOMIES[1]

J. A. DYAL

AND

W. C. CORNING

Department of Psychology
University of Waterloo
Waterloo, Ontario, Canada

I. PROLOGUE TO "LEARNING"

A. Inadequacy of Contemporary Theories of Learning

It is difficult to reconcile formal learning theories with the intricate, complex, and adaptive organism–environment transactions which we observe in invertebrates. While it may turn out that the delicate behavioral chiaroscuros are "nothing more than" the stark operational definitions found in Kimble (1961), when invertebrate transactions are watched for extended periods of time there is little rapport between the textbook and the observations. This is not a plea for naturalistic research so much as a comment on the lack of a conceptualization which could encompass the adaptive repertoires shown by invertebrates. This volume reviews behavior modification of invertebrates and permits itself rather broad latitude concerning what may or may not turn out to be "grade A certified" learning. It will be shown that behavior modification occurs at all levels of invertebrate phylogeny and that some quite complex paradigms are successfully performed by quite simple systems.

[1]Preparation of this chapter was facilitated by Grants A 0305 and A 0351 from the National Research Council of Canada.

The ability of a system to permanently alter its behavior as a consequence of "experience" is not unique to the molar behavior of higher organisms and thus is not the exclusive domain of psychologists and zoologists. Indeed, the immunological response is frequently cited as being a mnemonic analogue, and recent research in biochemical genetics indicates that cellular read-out mechanisms can be altered by the cell's transactions or "experience." Some attempts have been made to relate these changes to the sorts of phenomena in which psychologists are interested (Barondes, 1965; Gaito, 1963; Hyden, 1959; Smith, 1962; Mekler, 1967). It is clear at higher levels of organization, also, that proper structural and biochemical organization depend on experience. For example: (a) Retinal cells will undergo degeneration if deprived of light (Brattgärd, 1952; Rasch *et al.*, 1961). (b) Stimulation can increase the number of dendritic spines in cortical cells (Schapiro and Vukovich, 1970). (c) Deprivation of environmental interaction but maintenance of stimulus input will produce animals that are not perceptually competent (Held, 1965). (d) Complex environments can change brain chemistry and improve function (Krech *et al.*, 1966). (e) Brain polysome concentration is affected by stimulation (Appel *et al.*, 1967).

Does a retinal cell "learn" to perform better metabolically because of its "experience" with light? Does a cortical cell "learn" to process complex patterns because of the animal's "experience" with patterns? It would be difficult to exclude such ontogenetic considerations from any learning theory which aims at completeness, particularly when the adult brain may still be differentiating and demonstrating experientially induced plasticity in form and function. Structure and experience are obviously inseparable, and the organism's response to and processing of environmental input can never be the same from epoch to epoch, even if stimulus conditions are invariant. Child had stressed this many years ago in discussing organism–environment interaction:

> The nature of this equilibrium and the reciprocal relation between the structural and functional aspects of organisms are graphically illustrated in Child's analogy of the river and its channel . . . . The river receives tributary water, suspended solids, and energy, here erodes its banks and there deposits the same material as a bar, and throughout its course exhibits the closest adjustment of energy of flow to configuration of channel. The relation between structure and function in the organism is similar in character to the relation between the river as an energetic process and its banks and channel. From the moment that the river began to flow it began to produce structural configurations in its environment, the products of its activity accumulated in certain places and modified its flow, but just so long as the flow continues the process of equilibration goes on. The river is neither the water current nor the channel; it is both of these as they have developed together, and the only way the river can be fully understood is by the consideration of it as a process. (Herrick, 1924, pp. 16–17)

B. Toward a Definition of Learning

Why Try to Define Learning?

It seems likely that the amount of effort which has been expended by latter-day soul searchers (psychologists) in debating the criteria for demonstrating the existence of learning has been exceeded only by our philosophical brethren in their debates over the criteria for establishing the existence of God. To what end? Why all this concern over "the definition of learning"? Should we impatiently dismiss such efforts as "research-stifling semantics," as Agranoff seems to propose (Bullock and Quarton, 1966, p. 164)? Or are there sound reasons for continuing to engage ourselves in definitional matters? While recognizing that excessive concern with problems of conceptual classification can inhibit a free-swinging empiricism, there appear to be both logical and scientific reasons to continue to be challenged by such problems. The logic of the matter supports the importance of achieving as much precision as possible in our definitions in order to "lay bare the principal features or structure of a concept, partly in order to make it definite, to delimit it from other concepts, and partly in order to make possible a systematic exploration of the subject matter with which it deals" (Cohen and Nagel, 1934, p. 232). The scientific exploration of the subject matter is facilitated by increased explicitness of the critical operations by virtue of the increased precision of experimental procedures which such conceptual refinements afford. On the other hand, we must assure ourselves that our concepts are *expansive* enough, so as not to restrict penetration of new observational domains. The necessity of some trade-off between precision and expansivity is especially critical when we attempt to carry definitions of learning based on mammals over to the simple invertebrate systems which are the subjects of analysis of this book. It should also be kept in mind that the definition of any scientific concept is open-ended; it is modified and refined as new data and more illuminating conceptualizations emerge.

Some Definitions of Learning

Many of the complexities involved in the problem of defining learning have been well reviewed in several of the easily available texts (e.g., Kimble, 1961; Razran, 1971; Thorpe, 1956). An excellent discussion of the problem and many points of view are presented in the NRP Bulletin on "Simple Systems for the Study of Learning Mechanisms" (Bullock and Quarton, 1966). As Kimble notes, learning definitions range from the factual to the theoretical, with all points in between. Some definitions concentrate on the adaptive significance of learning: "We can define learning as that process which manifests itself by adaptive changes in individual behavior as a result of experience" (Thorpe, 1956, p. 55). Others specifically exclude this criterion as irrelevant: "We consider any systematic change in behavior to be learning whether or not the change is adaptive, desirable for certain purposes, or in accordance with any other such criteria" (Bush and Mosteller, 1955).

Some definitions are quite specific and exclusive, as that of Miller discussed below. Others, while less precise, are more inclusive and general—for example, Thorpe's second definition, in which he defines learning as "the organization of behaviour as a result of individual experience" (1956, p. 55).

All of these definitional instances seem to be strongly influenced by the experimental preparation and the general Weltanshauung of the definer. Thus we find Miller opting for the following restrictive definition of "grade A certified" learning based on mammalian preparations:

"Learning is a relatively permanent increase in response strength that is based on previous reinforcement and that can be made specific to one out of two or more arbitrarily selected stimulus situations."

Miller elaborates this definition to show that (a) "relatively permanent" means "days and months rather than seconds or minutes"; (b) "increase in response strength" of a particular response may be reflected in decrease of a given response when it is in conflict with one which is being strengthened; (c) "reinforcement" in both classical and instrumental paradigms involves association by contiguity; (d) the final criterion, regarded as the most fundamental, "means that our instrumental learning or classical conditioning procedures should be able to either cause S_1 to elicit R_1 and S_2 to elicit R_2, or to cause S_1 to elicit R_2 and S_2 to elicit R_1" (Miller, 1967).

The above criteria taken together provide a most stringent and conservative definition of learning and exclude most instances of habituation, sensitization, and pseudoconditioning. There is little doubt that there would be high agreement among authorities that any phenomenon exhibiting these characteristics could be a legitimate instance of learning, grade A and certified!

Eisenstein has made use of a preparation involving the single, isolated thoracic ganglion of a cockroach in his search for neural mechanisms of learning. It is thus not surprising that his definition of learning emphasizes the coding of a temporal sequence or pattern of events as "the most fundamental aspect of learning." His formal definition is as follows:

"A system is said to demonstrate learning when its output (Response) to a given test input pattern of (Stimulus) is a function of the total previous input–output pattern of which the test input was a part. That is, a system can be said to have learned if its output to a given test input is a function of the specific input–output pattern to which it has been exposed."
(Eisenstein, 1967)

This definition has two important implications: (a) Experimental and control groups are not seen as providing a comparison of procedures for exhibiting "learning" or "nonlearning" but as procedures for establishing that the system organizes its output in terms of the temporal pattern of the previous input; such a demonstration requires at least two patterns of input. (b) It permits the exploration of a much broader range of systems under a common organizing rubric. According to this view, it is no longer a conceptual strain to consider isolated simple systems and intact simple organisms as appropriate preparations which may contribute to our understanding of the mechanisms underlying (neurophysiologically, neurochemically, and evolutionarily) Miller's grade A certified learning.

Our point of view is consistent with that expressed by Chorover to the effect that "we are really searching for a paradigm to use in studying behavior, not searching for a permanent definition. It would be absurd to exclude by a narrow definition consideration of phenomena that have attracted real interest among workers in the field" (Bullock and Quarton, 1966, p. 165). We suggest that, at this point, our knowledge of the basic mechanisms of learning and memory is too limited to follow the traditional pattern of defining learning by exclusion. Indeed, were it not for the heavy heuristic load carried by the term *learning* we would prefer to drop it altogether and substitute such terms as *behavioral plasticity* and *plasticity of the nervous system* (Konorski, 1948; Livingston, 1966) or preferably the more neutral and descriptive phrase *behavioral modification based on experience* (bm/be). Thus our domain of interest is in

> *the characteristics of systems which exhibit (1) a change in behavior (response, output) as a function of characteristics of the events which have preceded or followed that response when it occurred on previous occasions and (2) which changes are presumed to be mediated by structural changes in the system.*

This "definition" recognizes that "we . . . simply cannot define by mechanism at present" (Bullock and Quarton, 1966, p. 168) and permits those of us who are interested in learning to cast a broader net to encompass such diverse behavioral modifications as immunological reactions, maturational effects, one-trial learning, habituation, sensitization, pseudoconditioning, aversive inhibitory conditioning, classical conditioning, avoidance conditioning, appetitive-reward conditioning, and insight learning.

Polythetic Operationism and the Problem of Levels

It is believed that the extension of our domain of interest to *behavior modification based on experience* will make it easier to define the subspecies of such behavioral modification on the basis of polythetic operational definitions which include both multidimensional and multilevel criteria. The polythetic approach to taxonomic problems within biology and psychology has been supported by Jensen (1967), following Sokal (1966). It is contrasted with the traditional "monothetic" approach, which attempts to define the "essence" of a concept by the use of a small number of critical key features. The polythetic approach requires that multiple features be specified and assigns degrees of affinity between phenomena or taxonomic categories on the basis of the degree of overlap among the sets of features rather than by congruence of small subsets of *critical* features. It is asserted that the advantage of the approach to taxonomy is that it is more "natural" and more "useful." Now the utility of a particular concept depends on its ability to organize a domain of observations and to lead to novel observations; thus *in part* the utility of a concept is to be judged by its integrative capacity. Furthermore, the integrative power of a particular concept or taxonomy is greater to the extent that it permits one to relate several levels of analysis (e.g., neuroelectrical, neurochemical, and behavioral). For example, quite in the spirit of polythetic operationism, Thompson and Spencer (1966) advance nine criteria as constituting an "operational definition of habituation." This definition is multidimensional and appears to be applicable to two levels of analysis, behavioral and electrophysiological. It is, of course, quite conceivable that all of the behavioral criteria could be met in two different preparations, yet the electrophysiological mechanisms or neurochemical mechanisms subserving the response decrement might be quite different.[2] The utility of the definition will thus depend on the perspective from which it is being evaluated. From the perspective of the individual investigator whose interest is confined to a single level of analysis (e.g., behavioral), a single-level polythetic definition is quite adequate. On the other hand, such a behavioral definition might be of limited utility to the investigator who is concerned with events at a quite different level (e.g., neurochemical) or who is interested in integrating two or more levels.[3] From the perspective of the scientific enterprise as a whole, it seems clear that viable theories of learning will require integration of several dimensions at several levels of analysis; i.e., they must be polythetic *and* multilevel. In this process, a generic term *(habituation)* may be elaborated by defining many subspecies of habituation (habituation 1,2,3,..., *n*), which may refer to polythetic definitions at various levels of analysis.[4]

[2]Indeed, there is reason to believe that several different neurophysiological mechanisms could mediate the response decrement at the behavioral level, e.g., presynaptic transmitter depletion, postsynaptic desensitization, or active involvement of an inhibitory system (Corning and Lahue, 1972).

[3]Problems of research tactics and strategy related to attempts to integrate multiple levels of analysis of invertebrate behavior are discussed by Abraham *et al.* (1972), and by Corning, Lahue, and Dyal (in preparation).

[4]It seems relevant to note at this point that several investigators who work with electro-

II. BASIC BEHAVIOR MODIFICATIONS

A. Habituation (Nonassociative Response Decrement)

The waning of the probability/amplitude of an unconditioned response as a function of repeated presentation of an effective stimulus is a phenomenon ubiquitous throughout the phylogenetic scale. It is observed at all levels of analysis from the behavior of intact organisms to the behavior of single nerve fibers and even the behavior of intact organisms not having nervous tissue (protozoa, fungi). Although the following polythetic operational definition of behavioral habituation provided by Thompson and Spencer (1966) is based exclusively on vertebrate (primarily mammalian) preparations, many of its features have been demonstrated in invertebrates.

1. *Repeated application of an effective stimulus typically results in a negatively accelerated decrease in responsiveness as a function of number of stimulus presentations.*

This characteristic function is found in all of the major invertebrate phyla which have been studied, including protozoans, planarians, annelids, coelenterates, echinoderms, arthropods, and mollusks.

2. *Spontaneous recovery of the response occurs over time when the habituating stimulus is no longer presented* (protozoans, planarians, annelids, coelenterates, echinoderms, arthropods, and mollusks).

3. *The rate and/or degree of spontaneous recovery is some inverse function of the number of stimulus presentations such that spontaneous recovery is reduced by continued stimulus repetition after the response has completely disappeared (annelids).*

4. *The response decrement becomes successively more rapid with repeated series of habituation and spontaneous recovery* (protozoans, annelids, arthropods, and mollusks).

5. *The rate and/or degree of habituation increases with the rate of stimulus presentation* (protozoans, planarians, and annelids).

6. *The rate and/or degree of response decrement depends on the intensity of the stimulus being fast for weaker stimuli and slower (or nonexistent) for strong stimuli* (protozoans, planarians, and annelids).

7. *Habituation of response to a particular stimulus generalizes to other stimuli.*

---

physiological events (e.g., Hoyle, Kandel, and Morrell) prefer to define the modifications which they obtain in their preparations as *learning analogues*. However, whether they are dealing with *analogues* or *homologues* will ultimately be decided by the addition of extensive sets of criteria dealing with mechanisms to the polythetic operations defining habituation 1, 2, or 3 or learning 1, 2, or 3, etc. The reader is referred to Jensen (1967) for an extensive discussion of the role of polythetic operational definition in helping to differentiate between homologues and analogues.

This is an obviously imprecise characterization. Thompson and Spencer cite four studies, three of which show within-modality generalization and one which demonstrates cross-modal generalization. Studies of stimulus generalization of habituation in annelids have revealed that the effect is complexly determined, being dependent on order of training and apparently other unknown variables (Clark, 1960a,b).

There are several types of generalized habituation which are logically possible. These are represented in Table I. In the case of reflex habituation, the standard paradigm is represented as requiring a stimulus (S^u) which innately elicits a response (R^u). The training involves the repeated presentation of S^u and a final test with S^u in which it is observed that there is a decrement ($\downarrow \varDelta$) in the capacity of S^u to elicit R^u. Generalization of the reflex habituation could be observed to occur to other stimuli (S^o) (type S), to other responses (R^x) (type R), and to other stimuli and their associated responses (S^o, R^o) (type SR). Only type S generalization of habituation has been studied, so that it is not now certain that the other two logical possibilities do in fact exist. Clearly, considerable effort needs to be expended with a wide variety of preparations, both vertebrate and invertebrate, in order to establish a more precise characterization of this parameter.

8. *Presentation of another (usually strong) stimulus results in recovery of the habituated response (dishabituation)* (protozoans, planarians, annelids, and mollusks).

Thompson and Spencer point out that the demonstration of dishabituation "has been perhaps the most important method of distinguishing between habituation and fatigue." However, it should be noted that the use of a stronger stimulus to dishabituate provides a less than convincing differentiation of "habituation" and "fatigue," since it rests on the assumption that recovery from fatigue (whether "effector" or "central") cannot be similarly facilitated by strong extraneous stimuli. It is clear that a conceptually more

Table I. Paradigms Illustrating Types of Nonassociative Response Decrement (Habituation)[a]

Pretraining test	Training stimulus	Test stimulus	Observe	Phenomenon
S^u–R^u	S^u	S^u	$\downarrow \Delta S^u$–R^u	Reflex habituation
S^o–R^u	S^u	S^o	$\downarrow \Delta S^o$–R^u	Generalization of habituation (type S)
S^u–R^x	S^u	S^u	$\downarrow \Delta S^u$–R^x	Generalization of habituation (type R)
S^o–R^o	S^u	S^o	$\downarrow \Delta S^o$–R^o	Generalization of habituation (type SR)

[a] Symbols: –, elicits; $\downarrow \Delta$, response decrement between pretraining baseline and post-training test.

convincing test would involve the use of a weaker stimulus as a dishabituating stimulus.

9. *The amount of dishabituation produced by the repeated application of the dishabituating stimulus decreases as a negative exponential function of the number of presentations.*

Extensive reviews of habituation phenomena, including studies of invertebrates and speculations regarding neurophysiological mechanisms which may be involved, may be found in Horn and Hinde's outstanding book *Short-Term Changes in Neural Activity and Behaviour* (1970), in Horn (1971), and in Razran's masterful synthesis of the Western and Russian literature *Mind in Evolution* (1971). Razran also provides a provocative interpretative comment regarding the relationship of habituation to other learning phenomena in an evolutionary context (pp. 47–54). With regard to the conceptual status and evolutionary significance of habituation, he comments as follows:

Plainly, collated East-and-West evidence accords neither with the exclusionist view of keeping habituation out of inner-club learning (Hilgard and Bower, 1966; McGeoch and Irion, 1952) nor with the overextended optimism that it is the master key to all learning (Thompson and Spencer, 1966). An intermediate position, which is also an evolutionary one, is ineluctable. Habituation is a prodromal stage: informing about some of learning's most significant properties, uninforming about others, and misinforming about still others—not unlike the early stage of any evolving life mechanism, including that of neural action. Or, habituation is a low—more exactly, the lowest—level of learning, functioning and prevailing in its functioning at the very dawn of life and under most drastic experimental restrictions of its neural substratum in higher animals. (Razran, 1971, 44–45)

B. Sensitization and Pseudoconditioning (Nonassociative Response Increments)

Few other behavioral concepts have lent themselves to such conceptual confusion as have the concepts of sensitization and pseudoconditioning. The myopic focusing of "learning psychologists" on conditioning procedures has relegated other behavioral modifications to the terminological purgatory of "pseudo" effects. Hilgard and Marquis née Kimble (1961) define these effects *in relation to the classical conditioning paradigm* as follows:

"Sensitization (Wendt). *The increase in the strength of a reflex originally evoked by a conditional stimulus through its conjunction with an unconditional stimulus and response. It differs from conventional conditioning in that the response which is strengthened is appropriate to the conditional stimulus not to the unconditional stimulus.* (Syn. alpha conditioning—Hull; *cf.* pseudoconditioning—Grether)."

"Pseudoconditioning (Grether). *The strengthening of a response to a previously neutral stimulus through the repeated elicitation of the response by another stimulus without paired presentation of the two stimuli.*"

Many researchers have been critical of these conceptualizations. They point out that their effect has been to relegate these phenomena to positions of secondary interest as "control problems" in defining classical conditioning. In fact, Kimble himself acknowledges that "Unfortunately, the amount of experimentation on pseudoconditioning is very small. It seems quite likely that the paucity of evidence is directly traceable to the connotations of the prefix pseudo, and that research on an important problem has been inhibited on grounds that are purely semantic" (1961, pp. 63–64).

Others have demurred on the grounds that the differentiations have been taken to imply different processes or mechanisms, a point they are not ready to concede. For example, Thorpe contends that pseudoconditioning "seems to be essentially a part of the process which we have just been considering. There seems to be no sufficient reason for giving it a new name" (1956, p. 85). Similarly, Wickens and Wickens (1942) have regarded pseudo-conditioning as a case of true conditioning, and Kimble suggests that "pseudoconditioning may be a part of all conditioning in which a noxious stimulus is employed" (1961, p. 62).

While the identification of sensitization and pseudoconditioning as legitimate phenomena in their own right has been retarded by semantic and conceptual confusion, the recent statements of Razran (1971) and Wells (1967) may well serve as powerful heuristics toward a more meaningful taxonomy and more intensive research. Razran defines sensitization at the most general level as a "more or less permanent increment in an innate reaction upon repeated stimulation." He makes a valuable distinction between two generic types of sensitization: *nonassociative sensitization* and *conditioned sensitization*. Nonassociative sensitization is the response increment in an innate reaction which is the consequence of repeated presentation of a *single* stimulus. Conditioned sensitization involves an increment in an innate reaction which is based on the pairing of (or contingency between) two stimuli.

He describes two species of nonassociative sensitization *(incremental sensitization* and *pseudoconditioning)* and two species of associative sensitization *(alpha conditioning* and *beta conditioning)*. In the case of nonassociative sensitization, the basic operation is to present repeatedly a stimulus (S^u) which innately elicits a response (R^u) and then to test to see if that response (R^u) (or some other response R^x) has increased in probability/amplitude when elicited by that stimulus (S^u) or some other stimulus (S^o). Although Razran specifies only two species of nonassociative sensitization, inspection of Table II will reveal that there are actually four basic operational paradigms which should be distinguished.

In the first instance, we have what Razran calls incremental sensitization. We prefer to designate this case as *reflex potentiation* (Kimble and Ray, 1965) in order to preserve the term *sensitization* for the more generic usage. In this

Table II. Paradigms Illustrating Types of Nonassociative Response Increment (Sensitization)[a]

Pretraining test	Training stimulus	Test stimulus	Observe	Phenomenon
S^u–R^u	S^u	S^u	$\uparrow\Delta S^u$–R^u	Reflex potentiation
S^o↛R^u	S^u	S^o	$\uparrow\Delta S^o$–R^u	Pseudoconditioning (type S)
S^u↛R^x	S^u	S^u	$\uparrow\Delta S^u$–R^x	Pseudoconditioning (type R)
S^o↛R^x	S^u	S^o	$\uparrow\Delta S^o$–R^x	Pseudoconditioning (type SR)

[a] Symbols: –, elicits; ↛, does not elicit, or elicits weakly; $\uparrow\Delta$, response increment between pretraining baseline and post-training test.

procedure, a stimulus (S^u) is presented repeatedly, and if an increment in its tendency to elicit R^u is observed, then reflex potentiation is demonstrated. It may be seen that the procedure for demonstrating reflex sensitization is exactly the same as that used to demonstrate reflex habituation, but the direction of the effect is opposite.

Nonassociative pseudoconditioning is the taxonomic term which designates the general procedure in which the sensitized response (R^u) is shown to be elicited by a stimulus (S^o) other than the sensitizing stimulus, or the sensitizing stimulus (S^u) is shown to now elicit a new response (R^x). In the former case, which we call *pseudoconditioning type S,* we find that following repeated presentations of S^u another stimulus S^o has increased in its capacity to elicit R^u. It is this paradigm which has usually been called "pseudoconditioning" when used as a control procedure in classical conditioning. In this case, S^u is the unconditioned stimulus, S^o is the conditioned stimulus, and R^u is the unconditioned response. *Pseudoconditioning type R* involves an increment in the capacity of S^u to elicit a different response (R^x) as a result of previous repeated training involving S^u–R^u. This type of pseudoconditioning appears to be analogous to response generalization in classical conditioning. In *pseudoconditioning type SR,* it is shown that prior to sensitization training S^o does not elicit (or elicits weakly) R^x, but following repeated presentations of S^u there is an increment in the capacity of S^o to elicit R^x.

Illustrations of several nonassociative phenomena are available in a series of experiments on sensitization in spinal frogs. Franzisket (1951, 1963) showed that repeated presentation of a tactual stimulus (S^u) to a specific reflexogenous zone resulted in an increment in the probability that a hind leg wiping response (R^u) would be elicited by the tactual stimulus. Similar demonstrations of reflex *potentiation* (S^u–R^u) have been made by Kimble and Ray (1965) and Corning and Lahue (1971). In addition, Kimble and Ray showed that if the stimulus (S^u) were precisely localized to a specific point within the reflexogenous zone, *reflex habituation* (S^u–R^u) occurred. Franzisket (1963) demonstrated that the sensitization procedure (repeated presentation of S^u) resulted in an increased capacity of another stimulus (mechanical shock) to elicit the wiping reflex. This is of course a demonstration of *pseudoconditioning type S*. Corning and Lahue (1971) also demonstrated pseudoconditioning type S by showing that the wiping response was not specific to the original training site but was elicited by applying the stimulus to another site in the reflexogenous zone. Furthermore, their experiment demonstrated the occurrence of *pseudoconditioning type SR* by showing that a stimulus (S^o) applied to another reflexogenous zone (e.g., a contralateral anterior zone) elicited the reflex specific to that zone (wiping of the *front* leg) at a level equivalent to the trained zone (i.e., S^o–R^x).

Reflex potentiation and pseudoconditioning type S have been demonstrated in a wide variety of vertebrates (*cf.* Razran, 1971) and in some invertebrate preparations, including annelids (Evans, 1966*a,b*; Roberts, 1966), coelenterates (Pantin, 1935*a,b*), crustaceans (Wiersma, 1949), and cephalopod mollusks (Bullock, 1948; Young, 1958, 1960).

Pseudoconditioning type R and pseudoconditioning type SR have rarely been investigated. The one experiment which seems to have demonstrated pseudoconditioning type R was conducted by Pressman and Tveritskaya (1969) with dogs. As reviewed by Razran (1971), Pressman and Tveritskaya administered air puffs to the eye and shock to the left hind leg *on alternate days*. They found that the air puff came to elicit leg flexion and the shock came to elicit blinking. Although both pseudoconditioned reactions were unstable, this paradigm clearly deserves further research effort. Similarly, pseudoconditioning type SR should provide interesting data regarding relationships between various stimulus modalities and response systems.

The biological significance of sensitization has been discussed by Wells (1967) and Razran (1971) and is discussed by Dyal in Chapter 5 of this volume. The adaptive nature of the process is apparent on several grounds. First, it may be viewed both "in principle" and in specific behavior instances as a mechanism for countering the effects of habituation (Razran, 1971). Second, it permits an animal to respond adaptively to critical features of its environment, since food sensitizes forward-going behavior and noxious stimulation sensitizes stopping or withdrawal (Evans, 1966*a,b*; Young, 1958, 1960). Appetitive and aversive stimuli do not occur randomly—prey usually come in groups and persist for a time, as do predators—thus the potentiation of approach and avoidance behavior to a wide variety of stimuli enhances the probability of survival. Third, the question might well be asked: Why rely on a nonassociative sensitization mechanism when an associative mechanism would provide finer discriminations? Wells (1967) suggests that the answer is based on two points: (a) some animals simply do not possess the necessary sensory capacity to make the appropriate discriminations which would be required for associative learning, and (b) even if the discriminative and associative capacities were present, responses based on association simply take longer to acquire and the increased discriminatory precision would not outweigh the adaptive advantage of rapid (often one-trial) sensitization processes. Fourth, Razran provides a provocative generalization with regard to the phylogenetic significance of sensitization: "that *just as at the dawn of life habituation is the most prevalent type of learned modification, sensitization is the most prevalent in subsequent early evolution.* And with this goes the plausible tenet that in later evolution and the rise of true associative learning, learned sensitization either becomes very specialized and vestigial or is but a pseudo-sensitization involving undiscerned operation or bona fide associative or memorial-associative mechanisms' " (1971, pp. 65–66).

C. Punishment

The punishment category of behavior modification has been called *passive avoidance conditioning* by those who choose to emphasize the similarities between this paradigm and that involved in active avoidance conditioning (e.g., Mowrer, 1960; Solomon, 1964). They point out that in passive avoidance the animal learns to inhibit making a response, whereas in active avoidance he learns to make a response. Their contention is that both of these types of learning are mediated by common mechanisms based on classical conditioning of "fear" responses and thus "one cannot think adequately about punishment without considering what is known about the outcome of both procedures" (Solomon, 1964, p. 239). Razran takes an opposite view. He calls the paradigm *aversive inhibitory conditioning* and strongly asserts that "aversive inhibitory conditioning is not an aversive *classical* conditioning, and the concept of 'avoidance' based on prior classical action is inapplicable to its typical existence" (Razran, 1971, p. 127).

Consonant with our previous stance, we believe that not enough is known about learning based on noxious stimulation to provide an adequate taxonomy based on *mechanism*. We thus choose the more neutral term *punishment* and seek to establish a more general and polythetic set of parametric characteristics. The basic operational paradigm defining punishment is as follows: A stimulus S_1 is followed by a response R_1, and if that response is followed by a noxious and aversive stimulus S_2 which elicits a distinctive internal and overt response R_2, and if a decrement in S_1–R_1 is observed, then *response suppression due to punishment* is said to have been demonstrated. In this paradigm, stimulus S_2 is designated as a *punishing stimulus*, the response R_2 is the *punishing response*, and the response R_1 is called the *punished response*. The many conceptual problems associated with definition of *noxious* and *aversive* are nicely elaborated in the excellent review of punishment in higher vertebrates by Church (1963). Suffice it to say, for the present purposes, that a noxious stimulus is defined on the basis of its capability to establish and maintain escape behavior. There are a large number of parametric statements which can be made about punishment based on the reviews of Church (1963), Azrin and Holz (1966), Solomon (1964), and Razran (1971). These may be enumerated as follows, and taken together they constitute the beginning of a polythetic operational definition of punishment *based on vertebrate (usually mammalian) preparations:*

1. *The effectiveness of an aversive stimulus in suppression of responding is a monotonic increasing function of the absolute and relative intensity of the stimulus and its associated punishing response* (Church, 1963; Solomon, 1964; Azrin and Holz, 1966).

As the stimulus intensity is increased, the quality of the response shifts along a continuum from detection to temporary suppression, then partial suppression and total suppression.

2. *Response suppression is a function of the temporal and/or spatial proximity of the punished response to the punishing response such that suppression is greatest just prior to punishment* (Azrin and Holz, 1966; Church, 1963; Razran, 1971).
3. *Response suppression increases with the proportion of occasions on which the punished response is followed by the punishing S–R* (Church, 1963; Azrin and Holz, 1966).
4. *Response suppression is an inverse function of the strength of the punished response* (Church, 1963).
5. *Response suppression is a direct function of the degree of antagonism between the punished response and the punishing response* (Razran, 1971).
6. *The degree of response suppression depends on whether or not non-punished alternative responses are available* (Azrin and Holz, 1966).

The above generalizations must be regarded as specific to the punishment of learned responses based on appetitive food rewards. The degree to which they apply to punishment of consummatory responses, innate responses, or responses learned in an aversive situation (escape/avoidance) is unclear and often uninvestigated. It is also clear that the effects of these parameters interact with (a) familiarity of the subject with the punishing stimulus, (b) characteristics of the response topography, (c) age, sex, and species or strain of the subject, and (d) each other. With all of these qualifications, it is apparent that we are only beginning to elucidate the critical parameters of punishment for the "standard laboratory rat" and that in the case of most invertebrate preparations we have only the faintest glimmer of the possible effects of punishment. This is true despite the fact that the punishment paradigm would seem to be especially useful for investigating behavior modification throughout the phylogenetic scale. Indeed, it has already been used in a wide variety of invertebrate preparations, including paramecia (Soest, 1937), planarians (Hovey, 1929; Jacobson, 1963), earthworms (Smith and Dinkes, 1971), copepods (Applewhite, 1967), octopuses (Young, 1961), cuttlefish (Sanders and Young, 1940), and spiders (Drees, 1952).

The reader who is interested in the variety of theories advanced to account for the behavioral effects of the punishment operation may refer to Brush (1971), Campbell and Church (1969), Church (1963), Mowrer (1960), Razran (1971), Skinner (1938), Solomon (1964), Thorndike (1932), and Walters *et al.* (1972).

D. Classical Conditioning and Associated Phenomena

Associative Response Increment

In its simplest, most rudimentary form, Pavlov's law of excitation states that "a CS consistently paired with a US acquires excitatory properties" (Rescorla and Solomon, 1967). The implication is that a stimulus which prior to conditioning does not elicit a particular response comes to do so by being paired with a stimulus which does elicit the response. To elaborate further, consider two stimuli S_1 and S_2 which elicit R_1 and R_2, respectively. Assume that R_1 and R_2 are distinctive, easily detectable responses which have no overlap in their topography; that S_1 does not elicit R_2 prior to conditioning and that S_2 does not elicit R_1; and that S_1–R_1 is characterized by *relatively* rapid habituation, whereas S_2–R_2 is relatively unhabituable. Following consistent pairing of S_1 and S_2 in a critical temporal order, two changes in S–R relationships are typically observed: (a) the probability of S_1 eliciting R_1 has decreased, and (b) the probability of S_1 eliciting R_2 has increased. The increment in S_1–R_2 is said to demonstrate excitatory classical conditioning based on the *association* of S_1 with S_2. However, because the increment could reflect the operation of various types of sensitization, it is necessary to conduct a variety of control procedures to exclude these nonassociative effects. It will be recalled that response increments due to various types of sensitization result from the repeated presentation of a single stimulus (*cf.* Table II). In the present case, it is necessary to exclude the possibility that S_1–R_2 would have been observed if S_1 or S_2 had been presented by itself (unpaired) during training. Thus two control groups would be necessary: one which received the CS (S_1) only and one which was presented US (S_2) only. If the performance of the classical conditioning group were superior to that of both of the control groups, then a simple pseudoconditioning explanation would not be adequate. On the other hand, both of these control groups have the flaw that they not only eliminate the *pairing* between S_1 and S_2 but they also eliminate the other stimulus. What is needed is a control group in which both S_1 and S_2 are presented, but in an unpaired fashion. Several control procedures have followed this general strategy of "breaking up the forward pairing": (a) *backward conditioning*—the S_1-S_2 relationship is reversed so that S_2 is always followed by S_1 but S_1 is never followed by S_2; (b) *quasi-random control*—in this procedure, S_1 and S_2 are explicitly unpaired such that S_1 is *never* followed (within an effective interval) by S_2; (c) *differential conditioning*—S_1 is paired with S_2 in a forward conditioning manner but S_3 is never paired with S_2, which is to say that S_3 is always explicitly unpaired with S_2; (d) *long trace conditioning*—S_1 always precedes S_2 but with such a long interstimulus interval that classical conditioning is unlikely to occur. Rescorla (1967) has argued persuasively that these control procedures provide an overestimate of the "true" magnitude of excitatory conditioning because each of them involves a *negative contingency* between S_1 and S_2; that is, S_1 predicts the *absence* of S_2. Further, a negative contingency is shown to be the necessary and sufficient condition for the conditioning of response inhibition. Thus when the performance of the classical conditioning group is compared with any of these controls, the size of the difference is greater than would be obtained with a neutral or "zero" comparison condition.[5] Rescorla contends, quite rightly, that the preferred control of the S_1-S_2 contingency is a truly random (as opposed to quasi-random) control in which S_1 and S_2 occur an equal number of times in a training session but are programmed completely independently; thus any pairing forward or backward is strictly chance.[6]

As we saw in the section on sensitization, reflex potentiation and pseudo-conditioning are *nonassociative* phenomena which result in an increment in response probability between S_1 and R_1 or R_2. We have noted that these nonassociative effects must be excluded when an attempt is made to assert the existence of *associative* conditioning of S_1-R_2 based on the positive S_1-S_2 contingency (Pavlovian excitatory conditioning). Now it remains to examine two other types of *associative* conditioning which depend on the positive contingency of S_1-S_2. These S-R increments do not involve the learning of a *new* response but rather the associative strengthening of already existing S-R

[5]While this argument is technically correct, its force seems to be lessened when the object of the investigation is simply to detect the presence of excitatory conditioning. Further, the argument is of course irrelevant if one accepts a definition of learning such as that proposed by Eisenstein (1967), in which there is no necessity to establish a "zero" or "nonlearning" condition.

[6]Recent research has indicated that zero contingency programming of CS and US may provide an adequate neutral point (Ayres and Quinsey, 1970; Bull and Overmier, 1968; Rescorla, 1968), or it may result in excitatory conditioning (Kremer and Kamin, 1971; Quinsey, 1971), or it may result in either, depending on the nature of the first few events in the random series (Benedict and Ayres, 1972).

Table III. Classical Excitatory Conditioning and
Associated Phenomena[a]

Pretraining	Training	Effect	Phenomenon
$S_1 \nrightarrow R_2$	S_1–S_2	$\uparrow \Delta S_1$–R_2	Classical excitatory conditioning
S_1–R_1	S_1–S_2	$\uparrow \Delta S_1$–R_1	Alpha conditioning
S_2–R_2	S_1–S_2	$\uparrow \Delta S_2$–R_2	Beta conditioning
$S_2 \nrightarrow R_1$	S_1–S_2	$\uparrow \Delta S_2$–R_1	Delta conditioning

[a] Symbols: –, elicits; $\nrightarrow$, does not elicit, or elicits weakly; $\uparrow \Delta$, response increment between pretraining baseline and post-training test.

connections. These are *alpha conditioning* and *beta conditioning*. The basic procedures in each are represented in Table III.

Alpha conditioning may be seen to be an increment in S_1–R_1 probability as a result of the positively contingent pairing of S_1 and S_2. In order to interpret the obtained increment as an *associative* process, it must be shown that the S_1–R_1 increment is greater in the forward pairing case than in the single-stimulus, reflex potentiation procedure (*cf.* Table II). Furthermore, in order to measure precisely the size of the alpha conditioning effect, a truly random, strictly noncontingent control would need to be run just as in the case of classical excitatory conditioning.

Beta conditioning is an increment in the S_2–R_2 connection which occurs as a result of the positive contingency between S_1 and S_2. In order to be interpreted as an associative process, both single-stimulus (pseudoconditioning) and truly random contingency controls would need to be run.

It should also be noted that neither alpha nor beta conditioning is a widespread, generally observed effect. They may be specific to certain specialized preparations such as human eyelid conditioning (Cason, 1922; Grant and Adams, 1944; Grant and Norris, 1947) or the early phases of the conditioning process (Razran, 1971). Furthermore, although Razran maintains that these phenomena are *associative* rather than nonassociative, the evidence to support this interpretation is totally lacking, since the proper nonassociative control groups have never been run in an experiment in which the phenomena were observed.[7]

In addition to alpha and beta conditioning what we have called *delta conditioning* has been extensively reported in the Russian literature (Beritoff, 1965). The effect is manifested by an increased S_2–R_1 probability as a result of the positive contingency between S_1 and S_2. Although the effect has not been reported by western researchers, it is at least superficially similar to

[7]Razran has pointed out that Hullett and Homzie (1966), Kreps (1925), and Sergeyev (1962) have run controls and that "Literally hundreds of Soviet protocols of salivary conditioning compared with a number of studies of unpaired repetitions of unconditioned salivation attest to it." (Personal communication, August, 1972.)

Premack's (1962, 1965) notion of "reversibility of the reinforcement relation." Beritoff's doctrine of "two-way conditioning" is discussed more extensively by Razran (1971).

Decremental Phenomena in Classical Conditioning

Although Pavlov (1927) gave considerable attention to inhibitory conditioning, it is only within recent years that Western psychologists have shown an interest in the inhibitory phenomena associated with the classical conditioning paradigm (e.g., Lubow, 1965; Rescorla, 1967, 1969; Siegel, 1969, 1971; Siegel and Domjan, 1971). Rescorla (1969) has written an incisive and provocative review of Pavlovian conditioned inhibition in which he makes a strong case that the critical operation in *conditioned* inhibition is the occurrence of a *negative contingency* between S_1 and S_2; i.e., the CS is consistently followed by the *absence* of the US "despite the continued occurrence of US at other times" (Rescorla, 1969, p. 89). Rescorla discusses two test procedures (summation and retardation-of-excitatory-conditioning) whereby active inhibitory properties of a stimulus can be detected. He also points out that the inhibitory effects should be specific to the US which was used in the inhibitory conditioning and that control procedures to eliminate possible nonassociative inhibitory effects must be included in the experimental design.

The experimental procedures necessary to show that a stimulus is a conditioned inhibitor may be illustrated by considering the negative stimulus in differential conditioning. First, it is necessary that there be a conditioning period in which the conditioned inhibitor is related to the US in a negatively contingent manner. This may be seen to be the case for the negative stimulus in differential conditioning, since it is explicitly unpaired with the US. Second, it is necessary to make tests of the supposed inhibitory stimulus to determine if it in fact has active inhibitory properties. As noted, two tests may be applied.

In the summation test, the inhibitory stimulus (S^i) is presented with another stimulus (S^e) known to elicit a particular excitatory response (R^e). The eliciting stimulus (S^e) could be a previously conditioned excitatory stimulus (CS), a US, or a stimulus (S^x) which increases the general level of excitation. If the presentation of the compound stimulus $S^e S^i$ results in a decrement in responsivity, then the hypothesis that S^i is an inhibitory stimulus is supported. The second test is the "retardation-of-acquisition-of-an-excitatory-CR" procedure. The assumption is that excitatory and inhibitory processes are antithetical. Thus if the stimulus S^i is used as an excitatory CS in a new conditioning procedure, and if it does indeed possess active inhibitory properties, then these "negative valences" have to be overcome before "positive" excitatory effects can become manifest and a retardation of acquisition will occur.

Two control treatments are necessary in order to show that (a) the inhibition is *associative* or conditioned and (b) it is *specific* to the original US. To control for nonassociative factors, it is necessary to run the same control procedure which we noted as the preferred control in excitatory conditioning, namely, a truly random noncontingent presentation of the CS and the US. If this random control group performs significantly *better* in the suppression and retardation tests than does the inhibitory conditioning group, then the inhibition can be inferred to be associative rather than nonassociative. Since the inhibitory properties of the CS^i should be specific to the US utilized during inhibitory conditioning, a control group in which the CS^i is negatively contingent on a different US should be run. If the response decrement in suppression and retardation tests is greater for the experimental group than this control group, then specificity is demonstrated.

It should be apparent that the same procedures would be necessary to demonstrate the inhibitory properties which would accrue to a stimulus in a backward conditioning procedure, simple conditioned inhibition, or conditioned delay of response (*cf.* Rescorla, 1969).

Two major types of nonassociative inhibitory phenomena are *latent inhibition* and *external inhibition*. Latent inhibition is measured by preexposing the to-be-conditioned stimulus for a large number of trials prior to its use as a CS in excitatory conditioning, i.e., a retardation test. Whether the retardation effect should be thought of as resulting from habituation of orienting responses to the CS, or from counterconditioning of "orienting away," or from changes in some central process of "attention" is at the moment quite unclear.

External inhibition is demonstrated by the presentation of a novel stimulus and observing a decrement in the magnitude of a previously conditioned response. An external inhibitor may inhibit either excitatory processes or inhibitory processes (disinhibition). It is also known that the capacity of a novel stimulus to either inhibit or disinhibit tends to habituate with repeated presentation.

The "Neutral" Stimulus and the Problem of Overlap in Response Morphology

When Pavlov spoke of the desirability of using a neutral stimulus as a CS, he did not imply by the word *neutral* that the CS did not elicit a response from the organism. Rather, the CS was neutral in that it did not elicit the to-be-conditioned response, or a response similar to it, prior to the conditioning process. The existence of the phenomenon of alpha conditioning, in which the unconditioned response to the CS, the alpha response (R^a), increases in probability as a result of the CS–US pairing, presents a methodological problem if the two responses R^a and UR are not readily distinguishable. This was not a problem for Pavlov, since the orienting response to a bell (R^a) is quite easily distinguished from the salivary response to food powder. However, the problem becomes more acute in the case of conditioning of human GSRs in which both the US (shock) and the CS (tone) elicit a GSR (Stewart *et al.,* 1961; Kimmel, 1964). It occurs with even more force when we attempt to differentiate alpha conditioning from classical conditioning in simple systems in which the response repertoire is quite limited and the probability of considerable overlap between R^a and CR is high. There are several approaches to the solution of this problem, some of which are only partially satisfactory. In many instances, it may simply be a matter of being cognizant of the problem and searching for appropriate CS and US which *do* provide differentiable responses. For example, the original demonstrations of classical conditioning in planarians were criticized on the ground that the head-turning response which had been used as the CR was also an R^a to the light CS. Later work showed that the full contraction response was also conditionable (e.g., Jacobson *et al.,* 1967), and the battle over conditioning in planarians had to be fought on other grounds (*cf.* Corning and Riccio, 1970, for a complete review of the planarian controversies). It would thus appear that even a relatively simple response repertoire is occasionally rich enough to provide two separate handles if the experimenter is sufficiently manipulative.

Some investigators have tried to solve the problem by habituating R^a prior to the beginning of conditioning. This approach is not conceptually clear, since it hopelessly confounds the role of the US as a "reinforcer" with its possible effect as a dishabituating stimulus. A similar criticism is applicable to those experiments which have attempted to show the effect of "pairing" of CS and US on retardation of habituation of R^a (*cf.* attempts at classical conditioning of polychaetes in Chapter 5).

Assuming that in some preparations the problem of separating alpha conditioning from classical excitatory conditioning cannot be solved, it is nonetheless quite important to determine what behavior modifications are available in a particular species. The presence of *either* alpha conditioning or classical conditioning would demonstrate *associative* learning. It will be recalled that in order to assert the presence of alpha conditioning it must be shown that the increased probability of CS–R^a is not due simply to the repeated presentation of the CS alone, or the US alone, or both together in a noncontingent manner. In both alpha conditioning and classical conditioning, a response increment is dependent on a positive contingency between the CS and the US, and each demonstrates the ability of the organism to be modified by stimulus contingencies; i.e., they both represent *contingency conditioning.* They differ in that alpha conditioning represents an increment in response strength between a stimulus and a response which it already elicited prior to conditioning. Classical conditioning, on the other hand, represents a response increment (or decrement in the case of inhibitory conditioning) between a stimulus and a *new* response and is specific to the response elicited by the US.

In conclusion, we may admonish those who would study classical conditioning in a simple system as follows: (a) seek a preparation in which R^a and CR are clearly distinguishable; (b) explore the many phenomena associated with the classical conditioning paradigm, e.g., reflex potentiation, pseudoconditioning, alpha conditioning, and conditioned inhibition; (c) in order to evaluate these effects, include in the complete experimental design the following control groups: CS only, US only, CS and US presented in a truly random noncontingent manner with simultaneous monitoring of R^a, UR, and CR.

E. Instrumental Learning

Instrumental learning is an evolutionarily powerful form of response modification based on a response–reinforcement contingency. It is truly associative in that its presence depends on the temporal association between a response and some special consequences of that response (the reinforcement). At the strictly empirical level, a "reinforcer" may be defined as any event which will decrease the latency or increase the probability/magnitude of an antecedent response. The potential conceptual circularity which exists between "learning" and "reinforcement" in such a definition is obvious but fortunately does not constitute either a theoretical or an empirical impasse (*cf.* Hilgard and Bower, 1966, p. 481; Meehl, 1950; Premack, 1959, 1965). The theoretical mechanisms proposed to account for this reinforcement-contingent learning have relied on a variety of explanatory fictions, such as "satisfaction or discomfort" (Thorndike, 1932), "drive reduction" (Hull, 1943), "confirmation" (Tolman, 1932), and "drive induction, drive reduction" (Mowrer, 1960). Our view is that the "hypothetical mechanism" approach has not yet been particularly helpful in providing a coherent taxonomy of learning even if we restrict our consideration to the laboratory rat. It seems quite unlikely that we are at a stage in our analysis of invertebrate learning when learning theory will be of more than superficial aid in the generation of useful taxonomies of behavior modification. It is thus necessary to rely on a taxonomy based primarily on distinctive operations involving reinforcer types, response characteristics, stimulus characteristics, and control of trial spacing. We recognize that distinctions which seem phenotypically obvious may turn out to be based on the same mechanism, whereas operations which are lumped together may hide fundamental differences. Furthermore, we stress the importance of continual refinement of our taxonomy toward truly polythetic definitions which include an expanding set of functional relations between major parameters. A similar point of view has been expressed by D'Amato (1969), and an excellent treatment of many of the conceptual problems in learning taxonomy may be found in Melton (1964).

A Taxonomy of Instrumental Learning

The first distinction to be made is between those learning paradigms which require "choice" and those which do not. In the latter case, there is one response which is being measured, and reinforcement is contingent on that response occurring. No external, experimenter-controlled stimuli are differentially related to the probability of reinforcement. Such nondiscriminative instrumental learning is called *instrumental conditioning* (Spence, 1956; D'Amato, 1969).

When the learning paradigm provides for the differential reinforcement of two different responses and where external cues are correlated with the response–reinforcement contingency, then we use the generic term *discrimination learning*. Discrimination learning may in turn be divided into two types, which we shall designate as *differential conditioning* and *choice learning*. The differential conditioning paradigm has also been called successive discrimination learning. In this procedure, a *single* response is reinforced and external stimuli provide cues regarding which response–reinforcement contingency is in effect. Thus when S+ is present, making a specified response results in a reinforcement, whereas when S− is present the response is not rewarded. The animal thus learns to respond to S+ but to inhibit responding to S−. This simple "yes–no" choice is often referred to as a go/no-go situation. In differential conditioning, "to respond or not to respond" is the issue.

It should be noted that this taxonomy does not distinguish between what has been called discriminative instrumental conditioning and differential conditioning. Discriminative instrumental conditioning has been used to refer to the case in which S+ signals a positive contingency between the response and reinforcement, whereas the negative contingency is not signaled by a "different" stimulus but by *the absence of S+* (D'Amato, 1969). For example, if a light in the goal box signaled that food was available, the absence of the light would signal that food was not available. In differential conditioning, on the other hand, S+ would be a light and S− would be a tone, or S+ would be a black alley and S− would be a white alley. It seems to us that both procedures represent go/no-go paradigms in which S+ is an excitatory stimulus and S− is an inhibitory stimulus. It remains for future research to determine if an explicit S− (e.g., tone) results in basically different learning than an implicit S− (absence of S+). In any event, we choose to "lump" rather than "split" at this point.

Differential instrumental conditioning differs from choice learning in that the former involves either making a response or not making it (or making it at a reduced magnitude/probability) on the basis of discrimination between *stimuli*. The response to S+ differs *quantitatively* but not qualitatively from that to S−. Choice learning, on the other hand, not only requires stimulus discrimination but also a choice between qualitatively different responses such as turning left or turning right. Choice learning may involve either simultaneous or successive presentation of the discriminative stimuli, whereas in differential conditioning the discriminative stimuli are always presented successively.

Two other bases for classification need to be mentioned. First, instrumental conditioning, differential conditioning, and choice learning may be based on *appetitive reinforcement* or on *aversive reinforcement*. In appetitive reinforcement, the reinforcing event is typically the consumption of a drive-relevant incentive such as a food pellet by a hungry animal. In aversive reinforcement, the animal learns to escape or avoid a noxious stimulus. Not only are the operations quite different in these two procedures, but there is some evidence suggesting that the functions relating the two types of reinforcement to major learning and retention parameters may be different (*cf.* D'Amato, 1969, p. 82). However, such comparisons are more than suspect, since the few attempts made to equate the motivational–incentive conditions in the two paradigms have not been particularly successful. Once again we have little choice but to rely on monothetic phenotypic taxonomic categories. It should be noted that both appetitive and aversive reinforcement may be used in the same learning paradigm, as in a choice-learning procedure in which the correct choice is "rewarded" and the wrong choice is "punished."

The final basis for categorization depends on the way in which the response opportunities are controlled. Here we distinguish between *free response* or *operant procedures* and *controlled response* or *discrete trial procedures*. In the former, the interval which occurs between opportunities to respond (intertrial interval) is primarily controlled by the animal's own response rate. In the discrete trial procedure, on the other hand, the intertrial interval is programmed by the experimenter.

The taxonomic table which results from combining the above operational distinctions is presented in Table IV. We have indicated in each cell one defining or representative study with mammals as well as some of the experiments which have utilized invertebrates. We will now describe each of these instrumental learning procedures in somewhat more detail.

Cell 1. Instrumental Conditioning: Controlled–Appetitive. Cell 1 is the simplest and most frequently used paradigm for demonstrating instrumental conditioning in mammals. Two typical procedures used with rats involve a straight-alley runway and a Skinner box with retractable bar. The dependent measure is response latency. This procedure has not been used very often with invertebrates. In the instances when it has been used with annelids, the interpretation of the results has been complicated by possible nonassociative factors (Reynierse and Ratner, 1964; Reynierse *et al.*. 1968). On the other

Table IV. A Taxonomy of Instrumental Learning

	Instrumental conditioning		Discrimination learning			
			Differential conditioning		Choice learning	
	Appetitive	Aversive	Appetitive	Aversive	Appetitive	Aversive
Controlled response	Cell 1 Vertebrates: Graham and Gagne (1940) Invertebrates: Longo (1970) Reynierse and Ratner (1964)	Cell 2 Vertebrates: Bower et al. (1959) Invertebrates: Lacey (1971) Preyer (1886–1887) Ven (1922)	Cell 5 Vertebrates: Spence et al. (1959) Invertebrates: Young (1961)	Cell 6 Vertebrates: D'Amato and Fazzaro (1966) Solomon and Wynne (1953) Invertebrates: Ragland and Ragland (1965) Ray (1968)	Cell 9 Vertebrates: Bitterman and Wodinsky (1953) Grice (1949) Invertebrates: Alloway (1969, 1970) Best and Rubinstein (1962) Schöne (1961) Vowles (1964)	Cell 10 Vertebrates: Muenzinger and Fletcher (1936) Invertebrates: Schmidt (1964) Turner (1913) Zellner (1966)
Free response	Cell 3 Vertebrates: Skinner (1938) Invertebrates: Dews (1959)	Cell 4 Vertebrates: Winograd (1965) Invertebrates: Best (1965) Crawford and Skeen (1965) Lee (1963)	Cell 7 Vertebrates: Dinsmoor (1952) Invertebrates: None	Cell 8 Vertebrates: Hoffman et al. (1961) Invertebrates: None	Cell 11 Vertebrates: Blough (1966) Invertebrates: None	Cell 12 Vertebrates: Kelleher and Cook (1959) Invertebrates: None

hand, Longo (1970) has obtained evidence of instrumental conditioning of cockroaches in a straight alley using food as a reinforcer.

Cell 2. Instrumental Conditioning: Controlled–Aversive. The basic procedure in cell 2 involves measuring the speed with which an animal escapes from an aversive stimulus. Escape conditioning has seldom been tested with invertebrate preparations, although it would seem to be a perfectly feasible test of simple instrumental conditioning for almost all of the phyla. The phylum in which this phenomenon has been most extensively studied is Echinodermata. The escape reactions of a number of species of ophiurids and asteroids, when placed in situations which restrict movements of one or more rays, have been studied. Only two experiments have obtained an orderly decrease in escape time over trials (Preyer, 1886–1887; Ven, 1922). Escape conditioning has also been demonstrated in planaria (Lacey, 1971).

Cell 3. Instrumental Conditioning: Free–Appetitive. The paradigm of cell 3 is best represented by the standard operant conditioning procedure in which a bar-press response is conditioned in rats or a key-pick response in pigeons. The typical response measure is the rate of responding (i.e., number of responses per unit time). Appetitive operant conditioning has not been widely used as an instrumental conditioning procedure with invertebrates, although Dews (1959) has demonstrated that it would be a quite feasible technique with higher invertebrates. In any preparation in which drive-incentive parameters can be controlled and the animal has a structure which permits manipulation of objects, appetitive operant procedures have much to recommend them as techniques for investigating instrumental conditioning. It would appear that this procedure should be especially useful with certain arthropods and mollusks.

Cell 4. Instrumental Conditioning: Free–Aversive. In the paradigm of cell 4, both escape and avoidance conditioning can be demonstrated. Escape conditioning would involve irregular schedules of shock and measurement of latency of bar press which terminates the shock. Operant conditioning of this type was demonstrated in planarians by arranging a clear plexiglas field so that when the worm swam through a photobeam it turned off a bright, aversive light (Best, 1965; Lee, 1963; Crawford and Skeen, 1967). This procedure would seem to be a most promising one for use with a wide variety of invertebrate species.

Avoidance conditioning without an external warning stimulus is based on the experimenter's contriving a contingency between the desired response and postponement of aversive stimulation. In this "Sidman avoidance procedure," there are two critical parameters: the *shock–shock interval* and the *response–shock interval*. The experimenter might program the shock–shock interval to be 5 sec and the response–shock interval to be 20 sec. Under these constraints, the shock comes on every 5 sec until the animal makes a response, at which time the shock is postponed for 20 sec and every subsequent response starts the 20 sec delay again. This procedure typically generates quite stable rates of responding (Sidman, 1953, 1960) and is also adaptable to free responding in a shuttle box with animals which can locomote easily. It has been used to demonstrate avoidance conditioning in mammals and fish (Behrend and Bitterman, 1963) and would seem to be a useful technique with such highly mobile invertebrates as arthropods and mollusks.

Cell 5. Differential Instrumental Conditioning: Controlled–Appetitive. Differential instrumental conditioning in a straight runway could be demonstrated by using two discriminably different alleys (e.g., one black and one white) with a drive-relevant incentive available in one goal box but not in the other, the alleys being presented successively rather than requiring a choice. Significantly faster running to S+ than to S− would be taken to demonstrate differential conditioning. Similarly, in a discrete-bar Skinner box S+ could be "left light on" whereas S− could be "right light on."

Although used extensively with vertebrates, this procedure has received little attention in the analysis of behavior modification in any but the highest invertebrates (Young, 1961). It would appear that the paradigm would be quite suitable for use with leeches, crustaceans, and a variety of insects, and thus it merits further exploration.

Cell 6. Differential Conditioning: Controlled–Aversive. A single-bar Skinner box with a retractable lever can be programmed to demonstrate this type of differential conditioning in a manner analogous to that discussed with appetitive reinforcement, except that in this case S+ would signify shock and S− would indicate no shock. Differential conditioning would be indicated by significantly shorter response latencies to S+ than to S− (also by more refusals to respond to S−).

A second procedure which could be regarded as representing this paradigm is discriminative avoidance learning in which S+ is a warning stimulus and S− is simply the absence of that stimulus. This is in fact the case in typical "simple" avoidance conditioning in a shuttle box. In this case, a warning stimulus (S+) is presented, and the animal must shuttle to the other side of the box during a specified interval (e.g., 5 sec) in order to avoid being shocked. That the absence of S+ during the intertrial interval may be legitimately conceived as S− is demonstrated by the fact that intertrial responding is viewed as evidence that the animal has not yet properly discriminated the contingencies. Thus perfect responding would be evidenced by a large number of successful avoidances of shock via shuttling during S+ and few "false positive" runs during the S− intertrial interval.

The basic phenomenon of discriminative avoidance conditioning has been demonstrated with a variety of vertebrates using several different tasks, involving two-way shuttle boxes, discrete-bar Skinner boxes, and activity wheels or platforms. In the use of these last devices, the noxious stimulus can be avoided by engaging in a specified amount of activity during the avoidance interval. This procedure would seem to be especially promising for work with invertebrates "since it makes no special motor demands on the subject beyond the ability to move in a detectable manner" (Bitterman, 1966, p. 480).

Discriminative avoidance conditioning has been demonstrated in annelids (Kirk and Thompson, 1967; Ray, 1968) and planarians (Ragland and Ragland, 1965).

There is little doubt that this powerful paradigm should be more extensively exploited in future investigations of behavior modification in invertebrates.

Cell 7. Differential Conditioning: Free–Appetitive. In cell 7, the procedure is the same as for cell 5 except the animal is free to respond at any time. Typically, S+ (e.g., "left light on") would be presented for a specified period of time and then S− would be presented for an equal period of time so that S+ and S− periods would be represented an equal number of times throughout the training session. The response measure would be the rate of responding in the presence of the S+ and S−; differential conditioning would be reflected in a higher response rate to S+ than to S−. Appetitive operant differential conditioning has not been used in the study of invertebrate learning.

Cell 8. Differential Conditioning: Free–Aversive. The paradigm of cell 8 involves the use of S+ and S− in the same manner as discussed under cell 6, but with a free operant task rather than a discrete trial procedure. In the simplest case, S+ would be a warning stimulus always followed by shock unless the animal responded during the avoidance interval. S− would typically be the absence of S+. Differential conditioning would be reflected in a discrimination ratio in which response rate during S^D (S+) is compared to total responses. To our knowledge, this discriminated avoidance procedure has not been applied to an invertebrate preparation.

Cell 9. Choice Learning: Controlled–Appetitive. The single-choice T-maze is the apparatus which epitomizes the paradigm of cell 9. The animal is placed into a starting area and permitted to locomote through a length of straight alley (the stem) to a choice area. If he turns in one direction and enters the goal box, a positive reinforcer is encountered (e.g., food for a hungry rat); if he turns in the other direction and enters the goal box, no positive reinforcement is available. The trials may be run using either a correction procedure or a noncorrection procedure. In the former, the animal is permitted to retrace from the nonrewarded arm, and the trial is continued until he enters the correct goal box. In the noncorrection procedure, he is removed from the incorrect goal box and returned to the start box to begin a new trial. The rewarded arm could be on the left (or right), as in a spatial discrimination, or the correct choice could be positively correlated with a particular external cue such as brightness, as in a black–white discrimination. Several response measures are available in this single-choice task, including locomotion speeds in each maze segment, retraces, number of errors, and percent correct.

The single-choice paradigm has long been a favorite with comparative psychologists, who have used it to study discrimination learning at all levels of the phylogenetic scale. Positive reinforcers for invertebrate preparations include the opportunity to return to a preferred environment as well as typical food incentives. For the higher invertebrates, Vowles (1965, 1967) has shown that the wood ant *(Formica rufa)* can learn simple and difficult visual discriminations in a T-maze in which the reinforcer is the opportunity to return to the home cage. Similarly, Schöne (1961, 1965) has demonstrated both brightness and spatial discriminations in the spiny lobster *(Panulirus argus)* based on the opportunity to return to water. Grain beetles *(Tenebrio molitor)* have been taught spatial discriminations using cereal as a reinforcer (Alloway, 1969, 1970; Alloway and Routtenberg, 1967). Among the lower invertebrates, planarians have found happiness (and demonstrated discrimination learning) in single-choice mazes when restoration of water was used as the reinforcer (Best and Rubinstein, 1962).

Cell 10. Choice Learning: Controlled–Aversive. In the case of the single-choice maze, the procedure is identical to that discussed above except that choice of the incorrect arm results in aversive stimulation (punishment), whereas the correct arm is not rewarded but is "neutral." This procedure has been used extensively with rats, and several experiments on choice learning in annelids may be conceived as following this paradigm (e.g., Schmidt, 1955; Zellner, 1966). However, in most of these experiments it seems likely that the "neutral" goal box could also have had rewarding properties, especially when aversive stimulation is used to motivate the animal to move in the maze stem. This procedure thus shades into a combination of reward/punishment of the two alternatives.

Cell 11. Choice Learning: Free–Appetitive. A free operant procedure which would demonstrate choice learning would involve two stimuli and two response levers. When S_1 was presented, it would signify 100% reward on lever 1 and 0% reward on lever 2. Similarly, when S_2 was presented the response–reward contingencies would be 0% for lever 1 and 100% for lever 2. Instead of 100% (CRF) or 0% reward schedules, a wide variety of intermittent schedules might be used. For example, when S_1 was presented, a fixed-ratio (FR) schedule could be programmed on lever 1 and a variable-ratio (VR) schedule on lever 2. When S_2 was on, the schedules would be reversed. The response ratio would come to be controlled by the discriminative stimuli and would reflect the schedule of reinforcement.

Discrimination training in a free operant situation has not been attempted with any invertebrate preparation, although its success in the determination of sensory thresholds for mammalian preparations (Blough, 1966) would suggest that the use of a free operant procedure with cephalopods and crustaceans would merit exploration. (Imagine an eight-lever Skinner box for octopuses!)

Cell 12. Choice Learning: Free–Aversive. The basic considerations in cell 12 are the same as discrimination learning based on appetitive reinforcement. No invertebrate research has demonstrated discriminated avoidance conditioning in a free responding situation.

Complexifications and Combinations. The above 12 paradigms represent the most simple, basic procedures in instrumental learning. There are many combinations of these basic procedures which have been used extensively in both vertebrate and invertebrate research. For example, research on choice learning in annelids and cephalopods has typically made use of appetitive and aversive reinforcement combined (e.g., Datta, 1962; Wells, 1965; Young, 1961). The procedure has much to recommend it when the experimental concern is whether or not an animal can learn a particular discrimination.

Successive and simultaneous discriminations may be sequentially scheduled in both a single-choice maze and in free-responding operant chambers. In the maze, for example, a light in the stem could signal "turn toward the white arm," whereas a tone might signal "turn to the black arm." The light/tone would thus be presented on successive trials, whereas the black–white discrimination would be presented simultaneously each trial. Even more complex relationships can be programmed in an operant chamber, between multiple (successive) schedules and concurrent (simultaneous) schedules involving both appetitive and avoidance components. However, it seems likely that the behavioral technology applicable to invertebrates must be considerably more advanced before these more powerful techniques for the analysis of behavior are useful to the invertebrate researcher.

F. Other Major Learning Phenomena

The above discussion of basic behavioral modifications concentrates on the basic types of response acquisition. It has not treated such critical phenomena as extinction or retention, or more complex aspects of discrimination learning. These topics will be discussed briefly in this section.

Extinction

Extinction of an acquired response involves the elimination of the contingency on which the original learning was based. In classical excitatory conditioning, the traditional procedure has been to eliminate the US and observe the waning of the conditioned response. The newer conceptualization of classical conditioning as involving positive and negative contingencies (Rescorla, 1967, 1969) also has implications for the study of response extinction (Ayres and DeCosta, 1971). An extinction procedure which would eliminate *only* the critical associative factors (positive or negative contingencies) would involve a shift from the conditioning contingency to a completely random noncontingent presentation of the CS and US. Similarly, in the case of instrumental conditioning it would involve disrupting the response– reinforcement contingency by programming reinforcement in a noncontingent manner. The use of noncontingent extinction procedures is not widespread; the traditional extinction procedure of eliminating the reinforcer continues to be the most typical approach.

As far as research on invertebrates is concerned, extinction via elimination of the reinforcer has been demonstrated to occur in planarians (e.g., Corning and John, 1961) and in annelids (e.g., Wyers *et al.,* 1964). It is unfortunate that all studies of contingency learning (i.e., punishment, classical conditioning, and instrumental conditioning) in invertebrates have not followed the response acquisition phase with a period of extinction. It is hoped that future investigations will always include this procedure using both techniques of extinction (elimination of reinforcement and disruption of the critical contingency).

Retention

The basic operations involved in the measurement of retention (memory, forgetting, or learning[8]) involve an original learning phase, a retention interval, and a retention test. The retention test may be a *direct* test which involves performing the same task as originally learned or it may involve a more indirect test in which performance on a different, but related, task is measured. The direct measure we designate as a *simple retention test.* An example of the use of simple retention tests may be taken from a recent article by Cook (1971), who investigated habituation in a freshwater snail *(Limnaea stagnalis).* He habituated a large group of snails to 20 stimuli (light off) presented at 10 sec intervals and then tested at various retention intervals from 20 to 500 sec. Cook included an important control procedure not often used in such studies. The control groups were not given the original habituation but were tested following the same duration of time in the experimental environment as each retention group. The use of this control for time-dependent changes in responsivity independent of the original learning experiences resulted in a quite different conclusion regarding the course of retention of habituation than would have been inferred from the retention groups alone.

---

[8]Hall (1971) has argued cogently that these concepts all involve the same basic operations and thus cannot be appropriately distinguished. He opts for a single term, *memory,* to be substituted for these undifferentiable terms.

**Table V. Experimental Paradigms for Simple Retention and
Transfer Tests of Retention**

Simple retention

	Original learning	Retention test
Experimental group	Learn task A	Relearn task A
Control group	Retained in learning environment but does not learn task A	Learn task A

Proaction test

	Original learning	Retention test
Experimental group	Learn task A	Learn task B
Control group	Retained in learning environment but does not learn task A	Learn task B

Retroaction test

	Original learning	Interpolated experience	Retention test
Experimental group	Learn task A	Learn task B	Relearn task A
Control group	Learn task A	Exposed to learning situation but not task B	Relearn task A

The experimental paradigm for the simple retention test is represented in Table V.

The proactive transfer test of retention is a less direct, and thus presumably less satisfactory, test of retention *per se*. On the other hand, it permits examination of the question of how the retained learning interacts with new learning, perhaps a more significant issue in the long run. The paradigm for the proaction test is represented in Table V. The effect of retention of task A may be either to facilitate the learning of task B (positive transfer or proactive facilitation) or to interfere with the learning of task B (negative transfer or proactive interference). In either case, a necessary but not sufficient condition for the occurrence of transfer effects is that some degree of retention of task A has occurred. On the other hand, it is never possible to disentangle the effects of retention of A from those due to interaction of the two learning processes. Examples of the use of proaction paradigms in studying invertebrate retention are presented in Dyal's discussion of habituation in annelids (Chapter 5) and in Cook (1971). The use of this paradigm is common in choice learning experiments and in that context is called discrimination reversal; task A is a discrimination and task B is its reverse.

Retention tests may be further complicated by introducing a specific critical experience during the retention interval. This intervention has as its rationale the disruption of normal memory processes occurring during the retention interval. The disruption may be accomplished by introduction of chemical or electrical events or by introducing a new learning experience and testing for its effects on retention of the original learning. Regardless of the nature of the interfering event the experimental paradigm is that of *retroaction*. From Table V it may be seen that this paradigm involves an experimental group which learns task A and then, following a retention interval, the disruptor (e.g., task B) is introduced and the effect of introduction of task B on retention of A is measured by immediately retesting on task A. The control group is retested on task A and as such represents a simple retention measure uncomplicated by transfer. In this case, then, it is possible to get a separate estimate of retention *per se,* plus the interaction of amount retained with the new learning.

It is our contention that the analysis of retention functions is much more likely to provide systematic and evolutionarily meaningful comparisons among (and within) invertebrate phyla than are studies of simple response acquisition. We thus strongly encourage future investigators to routinely include retention tests in their studies of behavioral modification of any preparation.

Additional Aspects of Discrimination Learning

The recent excellent book by Sutherland and Mackintosh (1971), *Mechanisms of Animal Discrimination Learning,* attests to the fact that discrimination learning has generated extensive empirical and theoretical effort by animal behaviorists. As such, many issues and effects important for understanding discrimination learning cannot be reviewed in the present context, and the reader is referred to Sutherland and Mackintosh for a more extensive treatment.

Learning Sets. Harlow (1949) showed that rhesus monkeys "learned to learn" how to solve discrimination problems so that after several hundred such problems they were almost perfect in their choices on the second trial of each new problem. Similar learning set performance has been demonstrated in a variety of mammals (e.g., chimpanzees, Schusterman, 1964; monkeys and cats, Warren, 1966; and bluejays, Hunter and Kamil, 1971).

In the typical learning set problem used with monkeys, each of the stimuli is characterized by multiple critical dimensions. Considering that *Octopus* is able to form a variety of visual, tactile, and chemotactile discriminations, it would seem quite feasible to study learning set in this advanced invertebrate. The possibility that *Octopus* is capable of multidimensional learning sets is made more credible by the fact that a simple variant of this phenomenon (improvement in successive discrimination reversal) has been demonstrated in *Octopus* by Mackintosh and Mackintosh (1964).

Discrimination Reversal. As we noted in our discussion of retention, discrimination reversal learning involves both retention of the original learning and proactive transfer effects to the new task. The typical effect in a variety of specific tasks has been that substantially more errors are made on the first reversal problem than were made during original learning. This effect is of course a manifestation of proactive interference. If a series of successive discrimination reversals are given, it is found that these negative transfer effects are overcome and the animal eventually solves the reversal problems quite rapidly, sometimes within a single trial. These effects have been obtained with a variety of mammals, birds, and reptiles, including rats (Dufort *et al.,* 1954; North, 1950), sea lions (Schusterman, 1966), chimpanzees (Schusterman, 1962), opossums (Friedman and Marshall, 1965), horses and raccoons (Warren and Warren, 1962), pigeons (Bullock and Bitterman, 1962), chickens (Bacon *et al.,* 1962), several other avian species (Gossette, 1967), and turtles (Holmes and Bitterman, 1966; Kirk and Bitterman, 1963). However in Bitterman's laboratory, fish have proven to be notably recalcitrant about exhibiting the progressive improvement over reversals, even though they show the typical negative transfer effect on the first reversal (Behrend and Bitterman, 1967; Behrend *et al.,* 1965; and Bitterman *et al.,* 1958). On the other hand, "Canadian fish" show reliable improvement in reversals, albeit the rats are very much better than fish (Mackintosh, 1969; Mackintosh and Cauty, 1971; Setterington and Bishop, 1967).

In the case of invertebrates, the bulk of the research suggests that there is no improvement over a series of reversals which could not be more parsimoniously attributed to general adaptation to the experimental conditions rather than experience with reversals (learning to learn) *per se* (isopods, Thompson, 1957; annelids, Datta, 1962; cockroaches, Longo, 1964). However, there is *within-session* (as opposed to between-session) improvement over successive reversals (Datta, 1962; Datta *et al.,* 1960) and a reliable proactive interference from original learning to the first reversal (Datta, 1962). A significant exception to the failures to obtain improvement over successive reversals has been reported by Mackintosh and Mackintosh (1964), who obtained substantial improvement by *Octopus.* Numerous questions surrounding proactive interference effects from original training to reversals, their relation to retention, the possibility of between-session improvement in higher invertebrates, and the meaning of within-session improvements in lower invertebrates all await further research. It seems possible that the solution to these and other problems involved with retention–transfer–learning functions will provide illumination of the phylogeny of mediational processes in learning in a way in which the more traditional tests of symbolic processes have failed to do (see below).

Another phenomenon which has received considerable attention by investigators of discrimination learning is the *overlearning reversal effect (ORE)*; i.e., the more trials given in OL, the fewer trials required to learn the reversal (Reid, 1953). Early attempts to check the species generality of this overlearning reversal effect suggested that the effect might be fairly specific to rats, since facilitating effects of overtraining were not found with chickens (Warren *et al.,* 1960), paradise fish (Warren, 1960), children (Stevenson and Weir, 1959), or octopuses (Boycott and Young, 1955). However, more recent attempts by Mackintosh and Mackintosh (1963) have shown that discrimination reversal by octopuses is facilitated by overtraining if there are irrelevant cues present in the discrimination but no facilitation occurs without these irrelevant cues.

In addition to testing for the facilitative effects of overtraining on discrimination reversal, the effects of overtraining can be observed in nonreversal or extradimensional (ED) shifts, in which case the effect is to interfere with the learning of the transfer task. Mackintosh and Holgate (1965) have observed this overtraining effect on extradimensional shifts in *Octopus.* Unfortunately, the "phyletic expectation" is not clear, since several investigators have failed to find impairment of ED shifts in monkeys (Cross and Boyer, 1966), cats (Beck *et al.,* 1966), and rats (Waller, 1971).

Symbolic Processes

Such tasks as double alternation, delayed response, and detours have traditionally been used by comparative psychologists to test for evidence of symbolic activity in lower animals. In his review of comparative psychology of learning, Warren has concluded that these "classical tests for symbolic behavior in animals have not fulfilled their early promise; they have not provided data of significant value for the comparative psychology of learning" (1965, p. 109). Nonetheless, it seems likely that symbolic processes do exist in higher invertebrates, and in any event the possibility cannot be dismissed out of hand. The ability of *Octopus* to make delayed responses (Dilly, 1963) and to solve rudimentary detour problems (Wells, 1965) invites further exploration of these capacities in mollusks and in other higher invertebrates.

The question of "latent learning" plagued the S–R reinforcement psychologies of the 1940s. Latent learning may be defined as learning which occurs without detectable reinforcement. In retrospect, it may be seen that the whole question was improperly posed from the point of view of comparative psychology. The issue surely should not hinge on arbitrary definitions of "reinforcement" and specious suppositions regarding its mechanism. Rather, the issue is whether or not animals can learn to go "places" rather than only to make "responses." That is, is the behavior of an animal in the natural environment guided by the use of landmarks whose significance depends on previous experience? Despite problems of control which require careful attention to the possible role of innate orientations based on compass reactions, polarized light, etc., the evidence now seems overwhelming that such learning does exist and is critical for successful foraging behavior of many arthropods (e.g., van Iersel and van den Assem, 1965; Thorpe, 1956).

At this point, the evidence is too meager to permit us to discern any systematic phyletic trends in latent learning, beyond the fact that it occurs in higher invertebrates but its existence in lower invertebrates is yet to be demonstrated.

G. Critical Parameters in Behavior Modification

The present section can deal only in a most cursory manner with the many parameters shown to be important in behavior modification of vertebrates. The degree to which these variables function in a similar manner in invertebrate preparations is in most instances unknown, and further complicating the picture is the fact that wide differences in the critical parametric functions occur among vertebrates. The purpose of the present section is threefold: (a) to suggest the variables which may be important in modification of invertebrate behavior, (b) to note the general functions obtained with vertebrate preparations, and (c) to indicate any relevant research on invertebrates.

Stimulus Variables

Stimulus Intensity. We have already noted that habituation rate is some positive function of stimulus intensity and that this seems to hold for most vertebrate and invertebrate preparations tested.

Intensity of the CS has been shown to influence rate of conditioning in human eyelid conditioning (Grice and Hunter, 1964), avoidance conditioning in rats (Kessen, 1953), and classical conditioning of leg flexion in dogs (Barnes, 1956). Similarly, as intensity of the US increases, the amount of conditioning increases for the case of human eyeblink conditioning (Spence *et al.,* 1958), classical salivary conditioning in the dog (Ost and Lauer, 1965), classical fear conditioning in rats (Goldstein, 1960), instrumental avoidance conditioning in rats (Moyer and Korn, 1964), and punishment training in rats (Church, 1963). The effects of CS intensity or US intensity have not been investigated in any systematic manner in any invertebrate preparation, although one study has reported both CS intensity and US intensity effects in classical conditioning of the leech (Henderson and Strong, 1971).

CS–US Interval. Although the search for the optimal CS–US interval occupied considerable efforts of early researchers, that quest is now seen to be a fruitless endeavor. Whereas it was originally thought that the optimal interval was about 0.5 sec, later research has shown that the optimal interval varies with the response system being conditioned, the type of learning involved, and the species of the subject. "Clearly much work remains to be done before generalizations about the role of contiguity of stimulation in conditioning can be made with any confidence. The effects of differences in procedure, in measures, and in species must be sorted out carefully in subsequent experiments" (Behrend and Bitterman, 1962). The reader should refer to Brookshire (1970) for a nice tabular summary of the optimal CS–US intervals obtained with a variety of species. No phyletic trends are discernible.

As with other critical parameters, relatively little work has been done on the effects of CS–US interval conditioning in invertebrates, although Coppock and Bitterman (1955) found that the optimal CS–US interval for two polychaetes "differs markedly from that found in mammals."

Intertrial Interval. In the case of habituation, the interstimulus interval is conceptually equivalent to the CS–CS interval in classical conditioning or avoidance conditioning and to the intertrial interval in other types of controlled-response instrumental conditioning. We have previously seen that habituation rate increases as the interstimulus interval decreases and that this effect holds over a wide variety of vertebrate and invertebrate preparations. The effect of CS–CS interval in classical or avoidance conditioning of invertebrate preparations has not been investigated.

Appetitive reward conditioning generally tends to be more effective as intertrial interval is increased. However, the results of experiments with mammals are quite conflicting, indicating that the intertrial interval probably interacts with a wide variety of other parameters. The choice learning of polychaetes seems to deteriorate if the trials are massed more closely than 5 min (Evans, 1963).

Reinforcement Characteristics

Reinforcement Magnitude. Rate of response acquisition by rats tends to be a positive function of the magnitude of the reward used as the reinforcer. This general relationship holds for all of the types of appetitive instrumental learning represented in Table IV. It also seems to be characteristic of many mammalian and avian species (chickens, Grindley, 1929; chimpanzees, Cowles and Nissen, 1937; rhesus monkeys, Schrier and Harlow, 1956). The amount of reward interacts with amount of training in determining resistance to extinction such that with small or moderate training persistence is greatest for large rewards, whereas with a large amount of training small rewards result in more persistence. The above relations hold for rats but not for goldfish (Gonzalez *et al.,* 1967). Similarly, shifts in reward magnitude result in shifts in behavior "appropriate" to the new magnitude for mammals but not for fish (Lowes and Bitterman, 1967).

The effects of reward magnitude have not been studied in invertebrates; thus it is not known whether they are more analogous to fish or to mammals with regard to this variable. Manipulation of reward magnitude would appear to be feasible with a variety of insects, crustaceans, mollusks, and perhaps even leeches.

Delay of Reinforcement. Delay of reinforcement is a potent parameter in determining the reinforcing value of a reward. In the case of rats, the learning asymptote is an inverse function of the length of a constant delay of reward (Sgro *et al.,* 1967), and variable delays of reward result in greater persistence in extinction (Capaldi and Poyner, 1966).

To our knowledge, there have been no studies of delay of reinforcement among invertebrate preparations in instrumental conditioning paradigms. The likelihood of strong delay of reward effects in invertebrates in which appetitive instrumental conditioning is feasible should stimulate research on this parameter.

Reinforcement Schedules. It is typically found in mammalian preparations that a partial reward schedule (PRF) (reward given on less than 100% of the trials) results in greater persistence during extinction than does continuous reward (CRF). This partial reinforcement extinction effect (PRE) is an exceptionally robust phenomenon over a wide variety of instrumental learning tasks in the case of higher vertebrates (birds and mammals) and in some reptiles (e.g., turtles, Murello *et al.,* 1961). Again it appears that fish are different in that the PRE appears under only a special limited set of experimental parameters (Brookshire, 1970). The PRE has been subjected to intensive theoretical analysis (Amsel, 1967; Capaldi, 1967), but its generality has yet to be systematically investigated in invertebrate preparations. One of the few experiments on the PRE in invertebrates found an increased persistence resulting from partial reinforcement of a classically conditioned response in earthworms (Wyers *et al.,* 1964). Further, the partial-reinforcement group performed as well during the classical conditioning phase as did the continuously reinforced group. While these results are consistent with *instrumental* conditioning data from mammals, they are in distinct opposition to the usually expected effects for *classical* conditioning in mammals. Partial reinforcement results in quite inferior classical conditioning of humans (Razran, 1955; Grant *et al.,* 1950) and of dogs (Fitzgerald, 1963), and the obtained extinction differences are difficult to interpret because of substantial terminal acquisition differences. Whether other invertebrates show similar "anomalous" effects of partial reinforcement remains to be determined.

Capaldi (1967) has shown that the sequential order of reinforced and nonreinforced trials is an important determinant of resistance to extinction. Similarly, rats can learn a pattern of responding in accord with the single alternation of rewarded and nonrewarded trials. Whether or not sequential variables also influence the performance of invertebrates is at this point unknown and uninvestigated.

An obvious conclusion from this brief review of some of the major parameters of learning in mammals is that a whole domain of research remains to be explored, namely, determination of the degree to which these parametric functions hold for invertebrates. Indeed, it has been argued (Bitterman, 1965) that the primary strategy of a comparative psychology of learning should be the comparison of such parametric functions.

III. PERSPECTIVES ON THE COMPARATIVE ANALYSIS OF LEARNING

A. Snarks and Boojums

Beach (1950) has discussed the dangers of relying on the white rat to generate data and theory in psychology and goes on to describe the decline of comparative research in American psychology. With respect to invertebrate research, in the period between 1911 and 1948, only 9 % of the articles in the "comparative" publication of the APA *(Journal of Comparative and Physiological Psychology)* were devoted to invertebrates. Since 1950 there has been little change in the emphasis in the JCPP *vis à vis* invertebrates; the number of invertebrate papers remains at less than 5 %. In other journals, invertebrates fare better, representing at least 27 % of the papers in *Behaviour* and in *Animal Behaviour* (Kutscher, personal communication). It should be noted that there are something like 1.2 million reported species (a conservative estimate) in the animal kingdom. Of these, invertebrates comprise 1.1 million, demonstrating that there is still a relative lack of invertebrate interest among behaviorists. Furthermore, there is a concomitant lack of invertebrate data in theoretical developments.

This dearth of research on the behavior modification in invertebrates could be justified by arguing that because of psychology's basic interest in man, it makes little sense to invest heavily in animals that are so far removed. This argument does not find support in other disciplines, where invertebrate data have contributed heavily to our basic understanding of living systems. For example: (a) Ionic movements during nerve action potential conduction were determined by studies on the squid giant axon. (b) Photoreceptor transduction and image-sharpening mechanisms were first studied in *Limulus,* and the octopus eye has also provided basic photochemical information. (c)

Basic principles of modern genetics continue to be developed with *Drosophila*. (d) Stretch receptor principles and the types of neuroelectrical data transmitted to the central nervous system were obtained in the crayfish. (e) Synaptic pharmacology and the pharmacology and electrophysiology of heart cells have been studied in crayfish. (f) Presynaptic and postsynaptic events during learning are currently being determined in aplysians. (g) The chemistry of learning received considerable impetus and direction from research on the planarian.

B. Simple System Research

Because of the complexity of higher organization, it is frequently necessary to work with what is called a simple system. Neurological structure can then be limited, or the preparation may possess a feature particularly suited for the research goals. There are two ways to arrive at a reduced system: to surgically or chemically reduce the system (the vertebrate spinal preparation and the isolated prothoracic ganglion of the cockroach, for example) or to select an animal with a less complex system (the hydra, planarian, and protozoan) (Bullock and Quarton, 1966; Corning, 1968). There are pitfalls with these reductions, the major one being the qualitative and quantitative change in function with increased complexity. There is the likely possibility that the basic and simplified neural circuitry acts much differently in the intact system or in the higher-order system.

One way to assess whether an oversimplification or overgeneralization has been made is to compare simplified and complex versions of the same system with respect to development of some learning capacity. For example, in *Limulus,* and in other arthropods, there are some preparations well suited for this type of question in which ganglia are readily identifiable and surgically separable, permitting single ganglion *vs.* multiple ganglia comparisons to be made. Within the same system, input and output can be held the same for simplified or complex preparations. Using this type of "comparative" study, it has been shown, for example, that the acquisition of habituation is less dependent on complexity than is the retention of the response decrement (Lahue and Corning, 1971*a,b*). Another valid approach to a comparison of simple and complex systems is the comparison of functional relations advocated by Bitterman (1965).

There has been little systematic variation of factors to determine true capacity differences between species, there has been little attempt to ask what complexity adds to a system's capacity, and there has been little exploration of the capacities of thousands of species. The present review volume provides a primitive platform from which it is hoped systematic and ever expanding research programs will be launched. We are convinced that the "lower" animals can add greatly to understanding of the "higher" forms, including man, just as they have been used to enrich the physiological and medical sciences.

REFERENCES

Abraham, F., Palka, J., Peeke, H., and Willows, A. O. D., 1972. Model neural systems and strategies for the neurobiology of learning. *Behav. Biol., 7*, 1–24.

Alloway, T. M., 1969. Effects of low temperature upon acquisition and retention in the grain beetle, *Tenebrio molitor. J. Comp. Physiol. Psychol., 69*, 1–8.

Alloway, T. M., 1970. Methodological factors affecting the apparent effects of exposure to cold upon retention in the grain beetle, *Tenebrio molitor. J. Comp. Physiol. Psychol., 72*, 311–317.

Alloway, T. M., and Routtenberg, A., 1967. Reminiscence in the cold flour beetle *(Tenebrio molitor). Science, 158*, 1066–1067.

Amsel, A., 1967. Partial reinforcement effects on vigor and persistence. In Spence, K. W., and Spence, J. A. (eds.), *The Psychology of Learning and Motivation*, Vol. 1, Academic Press, New York, pp. 2–65.

Appel, S. H., Davis, W., and Scott, S., 1967. Brain polysomes: Response to environmental stimulation. *Science, 157*, 836–838.

Applewhite, P. H., 1967. Memory and the microinvertebrates. In Corning, W. C., and Ratner, S. C. (eds.), *Chemistry of Learning: Invertebrate Research*, Plenum Press, New York,

Ayres, J. J. B., and DeCosta, M. J., 1971. The truly random control as an extinction procedure. *Psychon. Sci., 24*, 31–33.

Ayres, J. J. B., and Quinsey, V. L., 1970. Between groups incentive effects in conditioned suppression. *Psychon. Sci. 21*, 294–296.

Azrin, N. H., and Holz, W. C., 1966. Punishment. In Honig, W. K. (ed.), *Operant Behavior: Areas of Research and Application*, Appleton-Century-Crofts, New York, pp. 380–397.

Bacon, H. R., Warren, J. M., and Schein, M. W., 1962. Non-spatial reversal learning in chickens. *Anim. Behav., 10*, 239–243.

Barnes, G. W., 1956. Conditioned stimulus intensity and temporal factors in spaced-trial classical conditioning. *J. Exptl. Psychol., 51*, 192–198.

Barondes, S. H., 1965. Relationship of biological regulatory mechanisms to learning and memory. *Nature, 205*, 18–21.

Beach, F. A., 1950. The snark was a boojum. *Am. Psychologist, 5*, 115–124.

Beck, C. H., Warren, J. M., and Sterner, R., 1966. Overtraining and reversal learning by cats and rhesus monkeys. *J. Comp. Physiol. Psychol., 62*, 332–335.

Behrend, E. R., and Bitterman, M. E., 1962. Avoidance conditioning in the goldfish: Exploratory study of the CS–US interval. *Am. J. Psychol., 75*, 18–34.

Behrend, E. R., and Bitterman, M. E., 1963. Sidman avoidance in the fish. *J. Exptl. Anal. Behav. 6*, 47–52.

Behrend, E. R., and Bitterman, M. E., 1967. Further experiments on habit reversal in the fish. *Psychon. Sci., 8*, 363–364.

Behrend, E. R., Domresick, V. B., and Bitterman, M. E., 1965. Habit reversal in the fish. *J. Comp. Physiol. Psychol., 60*, 407–411.

Benedict, J. O., and Ayres, J. J. B., 1972. Factors affecting conditioning in the truly random control procedure in the rat. *J. Comp. Physiol. Psychol., 78*, 323–330.

Beritoff, J. S., 1961, 1965. *Neural Mechanisms of the Behavior of Higher Vertebrates*, Izd. Akad. Nauk SSSR, Moscow (English translation by W. T. Liberson, Little, Brown, Boston, 1965).

Best, J. B., 1965. Behaviour of planaria in instrumental learning paradigms. *Anim. Behav., 13*, Suppl. 1, 69–75.

Best, J. B., and Rubinstein, I., 1962. Maze learning and associated behavior in planaria. *J. Comp. Physiol. Psychol., 55*, 560–566.

Bitterman, M. E., 1960. Toward a comparative psychology of learning. *Am. Psychologist, 15*, 709–712.

Bitterman, M. E., 1965. Phyletic differences in learning. *Am. Psychologist, 20*, 396–410.

Bitterman, M. E., 1966. Animal learning. In Sidowsky, J. P. (ed.), *Experimental Methods and Instrumentation in Psychology*, McGraw-Hill, New York.

Bitterman, M. E., and Wodinsky, J., 1953. Simultaneous and successive discrimination. *Psychol. Rev., 60*, 371–376.

Bitterman, M. E., Wodinsky, J., and Candland, D. K., 1958. Some comparative psychology. *Am. J. Psychol. 71*, 94–110.

Blough, D. S., 1966. The study of animal sensory processes by operant methods. In Honig, W. K. (ed.), *Operant Behavior: Areas of Research and Application*, Appleton-Century-Crofts, New York.

Bower, G. H., Fowler, H., and Trapold, M. A., 1959. Escape learning as a function of amount of shock reduction. *J. Exptl. Psychol., 58*, 482–484.

Boycott, B. B., and Young, J. Z., 1955. A memory system in *Octopus vulgaris* Lamarck. *Proc. Roy. Soc. Ser. B, 143*, 449–480.

Brattgärd, S. O., 1952. The importance of adequate stimulation for the chemical composition of retinal ganglion cells during early post-natal development. *Acta Radiol. 96*, Suppl., 1–80.

Brookshire, K. W., 1970. Comparative psychology of learning. In Marx, M. H. (ed.), *Learning: Interactions*, Macmillan, London, pp. 291–364.

Brush, F. R., 1971. *Aversive Conditioning and Learning*, Academic Press, New York.

Bull, J. A., III, and Overmier, J. B., 1968. Additive and subtractive properties of excitation and inhibition. *J. Comp. Physiol. Psychol., 68*, 511–514.

Bullock, D. H., and Bitterman, M. E., 1962. Habit reversal in the pigeon. *J. Comp. Physiol. Psychol., 55*, 958–962.

Bullock, T. H., 1948. Properties of a single synapse in the stellate ganglion of squid. *J. Neurophysiol., 11*, 343–364.

Bullock, T. H., and Quarton, G. C., 1966. Simple systems for the study of learning mechanisms. *Neurosci. Res. Program, 4*, 106–233.

Bush, R. R., and Mosteller, F., 1955. *Stochastic Models for Learning*, Wiley, New York.

Campbell, B. A., and Church, R. M., 1969. *Punishment and Aversive Behavior*, Appleton-Century-Crofts, New York.

Capaldi, E. J., 1967. A sequential hypothesis on instrumental learning. In Spence, K. W., and Spence, J. A. (eds.), *The Psychology of Learning and Motivation*, Vol. 1, Academic Press, New York, pp. 67–156.

Capaldi, E. J., and Poyner, H. B., 1966. After effects of delay of reward. *J. Exptl. Psychol., 71*, 80–88.

Cason, H., 1922. The conditional eyelid reaction. *J. Exptl. Psychol., 5*, 153–196.

Church, R. M., 1963. Effects of punishment of behavior. *Psychol. Rev., 70*, 369–402.

Clark, R. B., 1960a. Habituation of the polychaete *Nereis* to sudden stimuli. I. General properties of the habituation process. *Anim. Behav., 8*, 82–91.

Clark, R. B., 1960b. Habituation of the polychaete *Nereis* to sudden stimuli. II. The biological significance of habituation. *Anim. Behav., 8*, 92–103.

Cohen, M. R., and Nagel, E., 1934. *Logic and Scientific Method*, Harcourt, New York.
Cook, A., 1971. Habituation in a freshwater snail *(Limnaea stagnalis)*. *Anim. Behav., 19*, 463–474.
Coppock, W., and Bitterman, M. E., 1955. Learning in two marine annelids. *Am. Psychologist, 10*, 501, Abst.
Corning, W. C., 1968. The study of brain functions: Some approaches and techniques. In Corning, W. C., and Balaban, M. (eds.), *The Mind, Biological Approaches to Its Functions*, Wiley, New York, pp. 1–22.
Corning, W. C., and John, E. R., 1961. Effect of ribonuclease on retention of conditioned response in regenerated planarians. *Science, 134*, 1363.
Corning, W. C., and Lahue, R., 1971. Reflex "training" in frogs. *Psychon. Sci., 23*, 119–120.
Corning, W. C., Lahue, R., and Dyal, J. A., 1973. The invertebrate strategy in the behavioral sciences. *Am. Zoologist, 12*, 455–469.
Corning, W. C., and Riccio, D., 1970. The planarian controversy. In Byrne, W. L. (ed.), *Molecular Approaches to Learning and Memory*, Academic Press, New York, pp. 107–149.
Cowles, J. T., and Nissen, H. W., 1937. Reward expectancy in delayed responses of chimpanzees. *J. Comp. Psychol., 24*, 345–358.
Crawford, F. T., and Skeen, L. C., 1967. Operant conditioning in the planaria: A replication study. *Psychol. Rep., 20*, 1023.
Cronholm, J. N., Warren, J. M., and Hara, K., 1960. Distribution of training and habit reversal by cats. *J. Genet. Psychol., 96*, 105–113.
Cross, H. A., and Boyer, W. N., 1966. Influence of overlearning in single habit reversal in naive rhesus monkeys. *Psychon. Sci., 4*, 245–246.
D'Amato, M. R., 1969. Instrumental conditioning. In Marx, M. H. (ed.), *Learning: Processes*, Macmillan, London, pp. 35–118.
D'Amato, M. R., and Fazzaro, J., 1966. Discriminated lever-press avoidance learning as a function of type card intensity of shock. *J. Comp. Physiol. Psychol., 61*, 313–315.
Datta, L.-E., 1962. Some experiments on learning in the earthworm, *Lumbricus terrestris*. *Am. J. Psychol., 75*, 531–533.
Datta, L.-E. G., Milstern, S., and Bitterman, M. E., 1960. Habit reversal in the crab. *J. Comp. Physiol. Psychol., 53*, 275–278.
Dews, P. B., 1959. Some observations on an operant in the octopus. *J. Exptl. Anal. Behav., 2*, 57–63.
Dilly, P. N., 1963. Delayed responses in *Octopus*. *J. Exptl. Biol., 40*, 393–401.
Dinsmoor, J. A., 1952. The effect of hunger on discriminated responding. *J. Abnorm. Soc. Psychol., 47*, 67–72.
Drees, O., 1952. Untersuchungen über die angeborenen Verhaltenweisen bei Springspinnen. *Z. Tierpsychol., 9*, 169–207.
Dufort, R. H., Guttman, N., and Kimble, G. A., 1954. One-trial discrimination reversal in the white rat. *J. Comp. Physiol. Psychol., 47*, 248–249.
Eisenstein, E. M., 1967. The use of invertebrate systems for studies on the basis of learning and memory. In Quarton, G. C., Melnechuk, T., and Schmitt, F. O. (eds.), *The Neurosciences: A Study Program*, Rockefeller University Press, New York.
Evans, S. M., 1963. Behaviour of the polychaete *Nereis* in T-mazes. *Anim. Behav., 11*, 379–392.
Evans, S. M., 1966a. Non-associative avoidance learning in nereid polychaetes. *Anim. Behav., 14*, 102–106.
Evans, S. M., 1966b. Non-associative behavioural modifications in the polychaete *Nereis diversicolor*. *Anim. Behav., 14*, 107–119.
Fitzgerald, R. F., 1963. Effects of partial reinforcement with acid on the classically conditioned salivary response in dogs. *J. Comp. Physiol. Psychol., 56*, 1056–1060.

Franzisket, L., 1951. Gewohnheitsbildung und bedingte Reflexe bei Rückenmarksfroshchen. *Z. Vergl. Physiol., 33,* 142–178.

Franzisket, L., 1963. Characteristics of instinctive behavior and learning in reflex activity of the frog. *Anim. Behav., 11,* 318–324.

Friedman, H., and Marshall, D. A., 1965. Position reversal training in the Virginia opossum: Evidence for the acquisition of learning set. *Quart. J. Exptl. Psychol., 17,* 250–254.

Gaito, J., 1963. DNA and RNA as memory molecules. *Psychol. Rev., 70,* 471–480.

Goldstein, M. L., 1960. Acquired drive strength as a joint function of shock intensity and number of acquisition trials. *J. Exptl. Psychol., 60,* 349–358.

Gonzalez, R. C., Holmes, N. K., and Bitterman, M. E., 1967. Resistance to extinction in the goldfish as a function of frequency and amount of reward. *Am. J. Psychol., 80,* 269–275.

Gossette, R. L., 1967. Successive discrimination reversal (SDR) performance of four avian species on a brightness discrimination. *Psychon. Sci., 8,* 17–18.

Graham, C. H., and Gagne, R. M., 1940. The acquisition, extinction, and spontaneous recovery of a conditioned operant response. *J. Exptl. Psychol., 26,* 251–280.

Grant, D. A., and Adams, J. K., 1944. "Alpha" conditioning in the eyelid. *J. Exptl. Psychol., 34,* 136–142.

Grant, D. A., and Norris, E. B., 1947. Eyelid conditioning as influenced by sensitized beta-responses. *J. Exptl. Psychol., 37,* 423–433.

Grant, D. A., Riopelle, A. J., and Hake, H. W., 1950. Resistance to extinction and the pattern of reinforcement. I Alternation of reinforcement and the conditioned eyelid response. *J. Exptl. Psychol., 40,* 53–60.

Grice, G. R., 1949. Visual discrimination learning with simultaneous and successive presentation of stimuli. *J. Comp. Physiol. Psychol., 42,* 365–373.

Grice, R. R., and Hunter, J. J., 1964. Stimulus intensity effects depend upon the type of sexperimental design. *Psychol. Rev., 71,* 247–256.

Grindley, C. C., 1929. Experiments on the influence of amount of reward on learning in young chickens. *Brit. J. Psychol., 20,* 173–180.

Hall, J. F., 1971. Verbal learning, retention and memory. *Can. J. Psychol., 25,* 412–418.

Harlow, H. F., 1949. The formation of learning sets. *Psychol. Rev., 56,* 51–65.

Held, R., 1965. Plasticity in sensory-motor systems. *Sci. Amer., 213,* 84–94.

Henderson, T. M., and Strong, P. N., 1971. Classical conditioning in leeches. Paper presented at meetings of Psychonomic Society, St. Louis, Nov. 11–13.

Herrick, C. J., 1924. *Neurological Foundations of Animal Behavior,* Holt, New York.

Hilgard, E. R., and Bower, G. H., 1966. *Theories of Learning,* Appleton-Century-Crofts, New York.

Hoffman, H. S., Fleshler, M., and Chorny, H., 1961. Discriminated bar-press avoidance. *J. Exptl. Anal. Behav., 4,* 309–316.

Holmes, P. A., and Bitterman, M. E., 1966. Spatial and visual habit reversal in the turtle. *J. Comp. Physiol. Psychol., 62,* 328–331.

Horn, G., 1971. Habituation and memory. In Adam, G. (ed.), *Biology of Memory,* Plenum Press, New York.

Horn, G., and Hinde, R. A., (eds.), 1970. *Short-Term Changes in Neural Activity and Behavior,* Cambridge University Press, London.

Hovey, H. B., 1929. Associative hysteresis in marine flatworms. *Physiol. Zool., 2,* 322.

Hull, C. L., 1943. *Principles of Behavior,* Appleton-Century-Crofts, New York.

Hullett, J. W., and Homzie, M. J., 1966. Sensitization effect in the classical conditioning of *Dugesia dorotocephala. J. Comp. Physiol Psychol., 62,* 227–230.

Hunter, M. W., and Kamil, A. C., 1971. Object-discrimination learning set and hypothesis behavior in the northern bluejay *(Cynaocitta cristata). Psychon. Sci., 22,* 271–272.

Hyden, H., 1959. Biochemical changes in glial cells and nerve cells at varying activity. In *Proceedings of the IV International Congress of Biochemistry,* Pergamon, London.

Jacobson, A. L., 1963. Learning in flatworms and annelids. *Psychol. Bull., 60*, 74–94.
Jacobson, A. L., Fried, C., and Horowitz, S. D., 1967. Classical conditioning, pseudo-conditioning, and sensitization in the planarian. *J. Comp. Physiol. Psychol., 64*, 73–79.
Jensen, D., 1967. Polythetic operationism and the phylogeny of learning. In Corning, W. C., and Ratner, S. C. (eds.), *Chemistry of Learning: Invertebrate Research*, Plenum Press, New York.
Kelleher, R. T., and Cook, L., 1959. An analysis of the behavior of rats and of monkeys on concurrent fixed-ratio avoidance schedules. *J. Exptl. Anal. Behav., 2*, 203–211.
Kessen, W., 1953. Response strength and conditioned stimulus intensity. *J. Exptl. Psychol., 45*, 82–86.
Kimble, D. P., and Ray, R. S., 1965. Reflex habituation and potentiation in *Rana pipiens*. *Anim. Behav., 13*, 530, 533.
Kimble, G. A., 1961. *Hilgard and Marquis' Conditioning and Learning*, Appleton-Century-Crofts, New York.
Kimble, G. A., and Ost, J. W. P., 1961. A conditioned inhibitory process in eyelid conditioning. *J. Exptl. Psychol., 61*, 150–156.
Kimmel, H. D., 1964. Further analysis of GSR conditioning: A reply to Stewart, Stern, Winokur and Freedman. *Psychol. Rev., 71*, 160–166.
Kimmel, H. D., 1965. Instrumental inhibitory factors in classical conditioning. In Prokosy, W. F. (ed.), *Classical Conditioning*, Appleton-Century-Crofts, New York.
Kimmel, H. D., 1966. Inhibition of the unconditioned response in classical conditioning. *Psychol. Rev., 63*, 232–240.
Kimmel, H. D., and Pennypacker, H. S., 1962. Conditioned diminution of the unconditioned response as a function of the number of reinforcements. *J. Exptl. Psychol., 64*, 20–23.
Kirk, K. L., and Bitterman, M. E., 1963. Habit reversal in the turtle. *Quart. J. Exptl. Psychol., 15*, 52–57.
Kirk, W. E., and Thompson, R. W., 1967. Effects of light, shock, and goal box conditions on runway performances of the earthworm, *Lumbricus terrestris. Psychol. Rec., 17*, 49–54.
Konorski, J., 1948. *Conditioned Reflexes and Neuron Organization*, Cambridge University Press, London.
Krech, D., Rosenzweig, M. R., and Bennett, E. L., 1966. Environmental impoverishment, social isolation and changes in brain chemistry and anatomy. *Physiol. Behav., 1*, 99, 104.
Kremer, E. F., and Kamin, L. J., 1971. The truly random control procedure: Associative or nonassociative effects in the rat. *J. Comp. Physiol. Psychol., 74*, 203–210.
Kreps, E. M., 1925. The reactions of Ascidians to external stimuli. *Arkhiv Biologicheskikh Nauk, 25*, 197–226.
Lacey, D. J., 1971. Temporal effects of RNase and DNase in disrupting acquired escape behavior in regenerated planaria. *Psychon. Sci., 22*, 139–140.
Lahue, R., and Corning, W. C., 1971a. Habituation in *Limulus* abdominal ganglion. *Biol. Bull., 140*, 427–439.
Lahue, R., and Corning, W. C., 1971b. Plasticity in *Limulus* abdominal ganglia: An exercise in paleopsychology. CPA meetings, St. Johns.
Lee, R. M., 1963. Conditioning of a free operant response in planaria. *Science, 139*, 1048.
Livingston, R. B., 1966. Brain mechanisms in conditioning and learning. *NRP Bull. 4*, 235–347.
Longo, N., 1964. Probability learning and habit-reversal in the cockroach. *Am. J. Psychol. 77*, 29–41.

Longo, N., 1970. A runway for the cockroach. *Behav. Res. Meth. Instrument., 2,* 118–119.

Lowes, G., and Bitterman, M. E., 1967. Reward and learning in the goldfish. *Science, 157,* 455–457.

Lubow, R. E., 1965. Latent inhibition: Effects of frequency of nonreinforced preexposure of the CS. *J. Comp. Physiol. Psychol. 60,* 454–457.

Mackintosh, N. J., 1969. Comparative studies of reversal and probability learning: Rats, birds and fish. In Gilbert, R., and Sutherland, N. S. (eds.), *Animal Discrimination Learning,* Academic Press, London, pp. 137–162.

Mackintosh, N. J., and Cauty, A., 1971. Spatial reversal learning in rats, pigeons and goldfish. *Psychon. Sci. 22,* 281–282.

Mackintosh, N. J., and Holgate, V., 1965. Overtraining and the extinction of a discrimination in octopus. *J. Comp. Physiol. Psychol., 60,* 260–262.

Mackintosh, N. J., and Mackintosh, J., 1963. Reversal learning in *Octopus,* with and without irrelevant cues. *Quart. J. Exptl. Psychol., 15,* 236–242.

Mackintosh, N. J., and Mackintosh, J., 1964. Performance of *Octopus* over a series of reversals of a simultaneous discrimination. *Anim. Behav., 12,* 321–324.

McGeoch, J. A., and Irion, A. L., 1952. *The Psychology of Learning,* Longmans, Green, New York.

Meehl, P. E., 1950. On the circularity of the law of effect. *Psychol. Bull., 47,* 52–75.

Mekler, L. B., 1967. Mechanism of biological memory. *Nature, 215,* 481–484.

Melton, A. W., (ed.), 1964. *Categories of Human Learning,* Academic Press, New York.

Miller, N. E., 1967. Certain facts of learning relevant to the search for its physical basis. In Quarton, G. C., Melnechuk, T., and Schmitt, F. O. (eds.), *The Neurosciences: A Study Program,* Rockefeller University Press, New York.

Mowrer, O. H., 1960. *Learning Theory and Behavior,* Wiley, New York.

Moyer, K. E., and Korn, J. H., 1964, Effect of UCS intensity on the acquisition and extinction of an avoidance response. *J. Exptl. Psychol., 67,* 352–359.

Muenzinger, K. F., and Fletcher, F. M., 1936. Motivation in learning. VI. Escape from electric shock compared with hunger food tension in the visual discrimination habit. *J. Comp. Psychol., 22,* 79–91.

Murello, N. R., Dierks, J. K., and Capaldi, E. J., 1961. Performance of the turtle, *Pseudomys scripta troostii,* in a partial reinforcement situation. *J. Comp. Physiol. Psychol., 54,* 204–206.

North, A. J., 1950. Improvement in a successive discrimination reversal. *J. Comp. Physiol. Psychol., 54,* 204–206.

Ost, J. W. P., and Lauer, D. W., 1965. Some investigations of classical salivary conditioning in the dog. In Prokasy, W. E. (ed.), *Classical Conditioning,* Appleton-Century-Crofts, New York, pp. 192–207.

Pantin, C. F. A., 1935a. The nerve net of the Actinozoa. I. Facilitation. *J. Exptl. Biol., 12,* 119–138.

Pantin, C. F. A., 1935b. The nerve net of the Actinozoa. II. Facilitation and the "staircase." *J. Exptl. Biol., 12,* 389–396.

Pavlov, I. P., 1927. *Conditioned Reflexes,* Oxford University Press, London.

Pearl, J., 1963. Intertrial interval and acquisition of a lever press avoidance response. *J. Comp. Physiol. Psychol., 56,* 710–712.

Premack, D., 1959. Toward empirical behavior laws. I. Positive reinforcement. *Psychol. Rev., 66,* 219–233.

Premack, D., 1962. Reversibility of the reinforcement relation. *Science, 136,* 255–257.

Premack, D., 1965. Reinforcement theory. In Jones, M. R. (ed.), *Nebraska Symposium on Motivation,* University of Nebraska Press, Lincoln.

Pressman, Ya. M., and Tveritskaya, I. N., 1969. Quantitative characteristics of the summation (pseudoconditioned) reaction. *Zh. Vysshey Nervnoy Deyatel. I. P. Pavlov, 19,* 566–573.

Preyer, W., 1886–1887. Über die Bewegungen der Seesterne. *Mitt. Zool. Sta. Neapel., 7,* 27–127, 191–233.

Quinsey, V. L., 1971. Conditioned suppression with no CS–US contingency in the rat. *Can. J. Psychol., 25,* 69–82.

Ragland, R. S., and Ragland, J. B., 1965. Planaria: Interspecific transfer of a condition ability factor through cannibalism. *Psychon. Sci., 3,* 117–118.

Rasch, E., Swift, H., Riesen, A. W., and Chow, K. L., 1961. Altered structure and composition of retinal cells in dark-reared mammals. *Exptl. Cell Res., 25,* 348–363.

Ratner, S. C., and Stein, D. G., 1965. Responses of worms to light as a function of intertrial interval and ganglion removal. *J. Comp. Physiol. Psychol., 59,* 301–304.

Ray, A. J., 1968. Instrumental light avoidance by the earthworm. *Commun. Behav. Biol., 1,* 205–208.

Razran, G., 1955. Partial reinforcement of salivary CR's in adult human subjects: Preliminary study. *Psychol. Rep., 1,* 409–416.

Razran, G., 1971. *Mind in Evolution: An East–West Synthesis of Learned Behavior and Cognition,* Houghton Mifflin, Boston.

Reid, L. S., 1953. The development of non-continuity behavior through continuity learning. *J. Exptl. Psychol., 46,* 107–112.

Rescorla, R. A., 1967. Pavlovian conditioning and its proper control procedures. *Psychol. Rev., 74,* 71–80.

Rescorla, R. A., 1968. Probability of shock in the presence and absence of CS in fear conditioning. *J. Comp. Physiol. Psychol., 66,* 1–5.

Rescorla, R. A., 1969. Pavlovian conditioned inhibition. *Psychol. Bull., 72,* 77–94.

Rescorla, R. A., and Solomon, R. L., 1967. Two-process learning theory: Relationships between Pavlovian conditioning and instrumental learning. *Psychol. Rev., 74,* 151–182.

Reynierse, J. H., and Ratner, S. C., 1964. Acquisition and extinction in the earthworm, *Lumbricus terrestris. Psychol. Rec., 14,* 383–387.

Reynierse, J. H., Halliday, R. A., and Nelson, M. R., 1968. Non-associative factors inhibiting earthworm straight alley performance. *J. Comp. Physiol. Psychol., 65,* 160–162.

Roberts, M. B. V., 1966. Facilitation of the rapid response of the earthworm, *Lumbricus terrestris* L. *J. Exptl. Biol., 45,* 141–150.

Sanders, F. K., and Young, J. Z., 1940. Learning and other functions of the higher nervous center of *Sepia. J. Neurophysiol., 3,* 501–526.

Schapiro, S., and Vukovich, K. R., 1970. Early experience effects upon cortical dendrites: A proposed model for development. *Science, 167,* 292–294.

Schmidt, H., Jr., 1955. Behavior of two species of worms in the same maze. *Science, 121,* 341–342.

Schöne, H., 1961. Learning in the spiny lobster, *Panulirus argus. Biol. Bull., 121,* 354–365.

Schöne, H., 1965. Release and orientation of behavior and the role of learning as demonstrated in crustacea. *Anim. Behav., 13,* Suppl. 1, 135–143.

Schrier, A. M., and Harlow, H. F., 1956. Effect of amount of incentive on discrimination learning in monkeys. *J. Comp. Physiol. Psychol., 49,* 117–125.

Schusterman, R. J., 1962. Transfer effects of successive discrimination-reversal training in chimpanzees. *Science, 137,* 422–423.

Schusterman, R. J., 1964. Successive discrimination-reversal training and multiple-discrimination training in one-trial learning by chimpanzees. *J. Comp. Physiol. Psychol., 58,* 153–156.

Schusterman, R. J., 1966. Serial discrimination: Reversal learning with and without errors by the California sea lion. *J. Exptl. Anim. Behav., 9,* 593–600.

Sergeyev, B. F., 1962. Formation of temporary connections in lancelets. *Zhurnal Vysshey Nervnoy Deyatel'nosti imeni I. P. Pavlova, 12,* 757–761.

Setterington, R. G., and Bishop, H. E., 1967. Habit reversal improvement in the fish. *Psychon. Sci., 7,* 41–42.

Sgro, J. A., Dyal, J. A., and Anastasio, E., 1967. Effects of constant delay of reinforcement on acquisition asymptote and resistance to extinction. *J. Exptl. Psychol., 73,* 634–636.

Sidman, M., 1953. Avoidance conditioning with brief shock and no exteroceptive warning signal. *Science, 118.* 157–158.

Sidman, M., 1960. *Tactics of Scientific Research,* Basic Books, New York.

Siegel, S., 1969. Effect of CS habituation on eyelid conditioning. *J. Comp. Physiol. Psychol., 68,* 245–248.

Siegel, S., 1971. Latent inhibition and eyelid conditioning. In Black, A. H., and Prokasy, W. F. (eds.), *Classical Conditioning II,* Appleton-Century-Crofts, New York, pp. 231–247.

Siegel, S., and Domjan, M., 1971. Backward conditioning as an inhibitory procedure. *Learning and Motivation. 2,* 1–11.

Skinner, B. F., 1938. *The Behavior of Organisms.* Appleton-Century-Crofts, New York.

Smith, C. E., 1962. Is memory a matter of enzyme induction? *Science, 138,* 889–890.

Smith, G. E., and Dinkes, I., 1971. Passive avoidance learning in the earthworm. Paper delivered at Eastern Psychological Association meetings, New York City, Apr. 15–17.

Soest, H., 1937. Dressuversuche mit ciliaten und rhabdocoelen Turbellarien. *Z. Vergl. Physiol., 24,* 720–748.

Sokal, R. R., 1966. Numerical taxonomy. *Sci. Am., 215,* 106.

Solomon, R. L., 1964. Punishment. *Am. Psychologist, 19,* 239–253.

Solomon, R. L., and Wynne, L. C., 1953. Traumatic avoidance learning: Acquisition in normal dogs. *Psychol. Monogr., 67,* 1–119.

Spence, K. W., 1956. *Behavior Theory and Conditioning,* Yale University Press, New Haven.

Spence, K. W., Haggard, D. E., and Ross, L. E., 1958. UCS intensity and the associative (habit) strength of the eyelid CR. *J. Exptl. Psychol., 55,* 404–411.

Spence, K. W., Goodrich, K. P., and Ross, L. E., 1959. Performance in differential conditioning and discrimination learning as a function of hunger and relative response frequency. *J. Exptl. Psychol., 58,* 8–16.

Stevenson, H. W., and Weir, M. W., 1959. Response shift as a function of overtraining and delay. *J. Comp. Physiol. Psychol., 52,* 327–329.

Stewart, M. A., Stern, J. A., Winokur, G., and Fredman, S., 1961. An analysis of GSR conditioning. *Psychol. Rev., 68,* 60–67.

Sutherland, N. S., and Mackintosh, N. J., 1971. *Mechanisms of Animal Discrimination Learning,* Academic Press, New York.

Thompson, R., 1957. Successive reversal of a position habit in an invertebrate. *Science, 126,* 163–164.

Thompson, R. E., and Spencer, W. A., 1966. Habituation; A model phenomenon for the study of neuronal substrates of behavior. *Psychol. Rev., 873,* 16–43.

Thorndike, E. L., 1932. Reward and punishment in animal learning. *Comp. Psychol. Monogr., 8,* Whole No. 39.

Thorpe, W. H., 1956. *Learning and Instinct in Animals,* Methuen, London.

Tolman, E. C., 1932. *Purposive Behavior in Animals and Men,* Appleton-Century-Crofts, New York.

van Iersel, J. J. A., and van den Assem, J., 1965. Aspects of orientation in the chigger wasp *Bembix rostrata. Anim. Behav., 13,* Suppl., 145–161.

Ven, C. D., 1922. Sur la formation d'habitudes chez les asteries. *Arch. Neerl. Physiol.*, *6*, 163–178.

Vowles, D. M., 1965. Maze learning and visual discrimination in the wood ant *(F. rufa)*. *Brit. J. Psychol.*, *56*, 15–31.

Vowles, D. M., 1967. Interocular transfer, brain lesions, and maze learning in the wood ant, *Formica rufa*. In Corning, W. C., and Ratner, S. C. (eds.), *Chemistry of Learning: Invertebrate Research*, Plenum Press, New York.

Waller, G., 1971. The effect of overtraining on two extradimensional shifts in rats. *Psychon. Sci.*, *23*, 123–124.

Walters, R. H., Cheyne, A. J., and Banks, R. K., 1972. *Punishment: Selected Readings*, Penguin, Harmondsworth, Middlesex.

Warren, J. M., 1960. Reversal learning by paradise fish. *(Macropodus opercularis)*. *J. Comp. Physiol. Psychol.*, *53*, 376–378.

Warren, J. M., 1965. The comparative psychology of learning. *Ann. Rev., Psychol.*, *16*, 95–118.

Warren, J. M., 1966. Reversal learning and the formation of learning sets by rats and rhesus monkeys. *J. Comp. Physiol. Psychol.*, *61*, 421–428.

Warren, J. M., and Warren, H. B., 1962. Reversal learning by horse and raccoon, *J. Genet. Psychol.*, *100*, 215–220.

Warren, J. M., Brookshire, K. H., Ball, G. G., and Reynolds, D. V., 1960. Reversal learning by White Leghorn chicks. *J. Comp. Physiol. Psychol.*, *53*, 371–375.

Wells, M. J., 1965. Learning and movement in octopuses. In Thorpe, W. H., and Davenport, D. (eds.), Learning and Associated Phenomena in Invertebrates, *Anim. Behav., 13*, Suppl. 1, 115–127.

Wells, M. J., 1967. Sensitization and evolution of associative learning. In Salanki, J. (ed.), *Neurobiology of Invertebrates*, Academic Kiadó, Budapest, pp. 391–411.

Wickens, D. D., and Wickens, C. D., 1940. Study of conditioning in neonates. *J. Exptl. Psychol., 26*, 94–102.

Wickens, D. D., and Wickens, C. D., 1942. Some factors related to pseudoconditioning. *J. Exptl. Psychol., 31*, 518–526.

Wiersma, C. A. G., 1949. Synaptic facilitation in the crayfish. *J. Neurophysiol., 12*, 267–275.

Winograd, E., 1965. Escape behavior under different fixed ratios and shock intensities. *J. Exptl. Anal. Behav., 8*, 117–124.

Wyers, E. J., Peeke, H. V. S., and Herz, M. J., 1964. Partial reinforcement and resistance to extinction in the earthworm. *J. Comp. Physiol. Psychol., 57*, 113–116.

Young, J. Z., 1958. Responses of overtrained octopuses to various figures and the effect of removal of the ventral lobe. *Proc. Roy. Soc. Ser. B, 149*, 463–483.

Young, J. Z., 1960. Unit processes in the formation of representations in the memory of *Octopus*. *Proc. Roy. Soc. Ser. B, 153*, 1–17.

Young, J. Z., 1961. Learning and discrimination in the octopus. *Biol. Rev., 36*, 32–96.

Young, J. Z., 1962. Repeated reversal of training in *Octopus. Quart. J. Exptl. Psychol., 14*, 206–222.

Zellner, P. K., 1966. Effects of removal and regeneration of the suprapharyngeal ganglion on learning, retention, extinction and negative movements in the earthworm, *Lumbricus terrestris* L. *Physiol. Behav., 1*, 151–159.

Chapter 2

PROTOZOA[1]

W. C. CORNING

Division of Biopsychology
Department of Psychology
University of Waterloo
Waterloo, Ontario, Canada

AND

R. VON BURG

Division of Biology
Ramapo State College
Mahwah, New Jersey, U.S.A.

I. INTRODUCTION

Although all animals experience an ontogenetic stage when they are *unicellular* (the fertilized egg), the protozoans remain unicellular throughout life and have most successfully used this condition in the diverse environments in which they are found. While protozoans are usually thought of as being single-celled, they are not a single version of a metazoan cell nor are they generally as simple. It is probably more accurate to designate them as *acellular* organisms, as animals not divided into cells (Barnes, 1966; Hyman, 1940). Some of the organelles in this phylum (cilia, for example) perform functions taken over by highly specialized cells in metazoans. The term *unicellular* is

[1]Preparation of this chapter was facilitated by Grant A 0351 from the National Research Council of Canada.

also misleading because it tends to obscure the fact that some protozoans are colonial—they exist in multicellular aggregates. Accordingly, the following definition seems appropriate:

> *"The Protozoa are acellular animals without tissues or organs, existing singly or in colonies of a few to many individuals; such colonies differ from a metazoan in that their components are all alike except when engaged in reproductive activities."* (Hyman, 1940, p. 45)

The characteristic of "acellularity" is the only one applicable to all protozoans, as the diversity and range within the phylum are wide, perhaps due to their multiple origins.

The name *protozoan* (Greek *protos,* first; *zoon,* animal) and their microscopic size imply a "simple" animal, but it is erroneous to assume an uncomplicated structure and function. As Sonneborn (1950) notes in the case of *Paramecium,* it is "far more complex in morphology and physiology than any cell of the body of man." The "simplicity" occurs only in cell *number.* Additionally, the acellular condition should not imply a primitive state of evolution; there are both primitive and advanced forms.

The small size of members of this phylum would appear to preclude their procurement and study by behavioral scientists whose training and theoretical frameworks depend mainly on supply-house white rats and captive college sophomores. However, with the aid of optical magnification, many interesting species can be observed and photographed. Large species also exist—for example, *Spirostomum ambiguum* (Fig. 1) can attain a length of 4 mm, and in the order Foraminifera species of over 8 mm in diameter have occurred.

From a medical and physiological standpoint, research on protozoans has been and continues to be of prime importance. An understanding of their nutritional requirements, life cycles, and genetics is critical because of the parasitic species that can infect animals and man. Amoebic dysentery, sleeping sickness, and malaria are estimated to parasitize more than a third of the world's population. Protozoans are important in the diet of organisms that may eventually find their way into man's food. They can also be useful models for the study of basic biological processes, e.g., as bioassays to assess anticancer drugs (Aaronson, 1963).

It is estimated that there are as many as 40,000–100,000 species comprising the phylum Protozoa, but no more than a handful have ever been studied by behavioral scientists or incorporated in research projects attempting to solve some of the problems of interest to psychologists. It is apparent that a "protopsychology" is needed to complement the active and increasingly voluminous research in "protozoology." Certainly, if one is interested in cellular events that underlie functional plasticity, the protozoan is an obvious candidate for experimentation; the problem of finding which of the billions of cells in the vertebrate brain are involved in learning is dispensed with. Gelber

(1962*a*) has provided one of the better rationales in the introduction of a study concerned with paramecium learning:

> This paper presents a new approach to behavioral problems which might be called molecular biopsychology . . . . Simply stated it is hypothesized that the memory engram must be coded in macromolecules . . . one of the psychologist's areas of interest might be described as the study of acquired characteristics. As the geneticist studies the inherited characteristics of the organism the psychologist studies the modification of this inherited matrix by interaction with the environment. Possibly the biochemical and cellular processes which encode new responses are continuous throughout the phyla (as genetic codes are) and therefore would be reasonably similar for a protozoan and a mammal. (p. 166)

II. GENERAL CHARACTERISTICS

Microscopic size, wide variation in form, and diversity of habitat may best characterize animals comprising phylum Protozoa. Examples of various species are included in Fig. 1. Some workers have suggested that the acellular group be treated as a subkingdom rather than a phylum because of the lack of common characteristics above a class (Barnes, 1966). The acellular group are generally more "animal-like" than "plantlike," but this and their "single-celled" nature appear to be the only unifying descriptions. They are *poly-phyletic* (i.e., a phylum whose members derive from many different protistan lines), and it is this variegated ancestral origin that makes any attempt at a detailed, unifying portrayal impossible.

A. Habitat

Protozoans are found where there is water; in freshwater streams and ponds, in saltwater bodies, in moist earth, and within the moist confines of host organisms. They are *free-living* as well as *parasitic*. With respect to the latter mode of adaptation, there are at least 25 parasitic varieties in man; some occupy the intestine, others the blood. The adaptations of some species to their microcosm can be elegant. A *Trichonympha* species exists as a wood-digesting symbiont in the digestive tracts of termites. The *Plasmodium* exists in blood corpuscles and causes malaria; the characteristic fever is produced with each wave of reproductive fission activity. In short, protozoans are found in all ecosystems and exhibit forms specifically adapted to the particular environs. Generally speaking, the form is indicative of where the animal survives—spherical forms are located in fresh- and saltwater media, animals that move with a definite anterior–posterior orientation are elongated, those that remain attached for periods of time are radial, and the "crawlers" are flattened dorsoventrally.

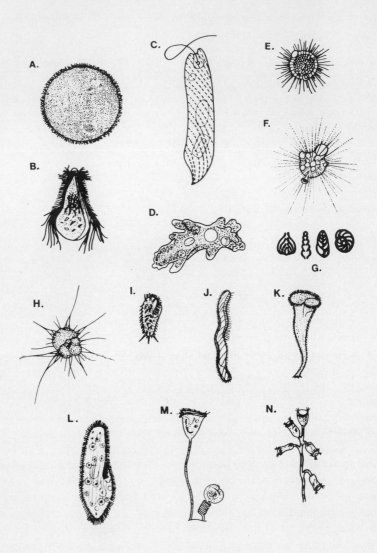

Fig. 1. Some examples of various members of phylum Protozoa, demonstrating the diversity and elegance of form. A, Flagellata, *Volvox globator*. (After Brown, 1950.) B, Flagellata, *Trichonympha collaris*. (After Brown, 1950.) C, Flagellata, *Euglena spirogyra*. D, Sarcodina, *Amoeba proteus*. (After Buchsbaum, 1948.) E, Sarcodina, *Actinosphaerium eichornii*. (After Brown, 1950.) F, Sarcodina, *Actinophrys sol*. (After Brown, 1950.) G, Foraminiferan shells. (After Hyman, 1940.) H, Sarcodina, *Globigerina*. (After Hyman, 1940.) I, Ciliata, *Stylonychia mytilus*. (After Hegner, 1933.) J, Ciliata, *Spirostomum ambiguum*. (After Hegner, 1933.) K, Ciliata, *Stentor coeruleus*. (After Hegner, 1933.) L, Ciliata, *Paramecium aurelia*. (After Hegner, 1933.) M, Ciliata, *Vorticella*. N, Ciliata, *Carchesium polypinium*. (After Hegner, 1933.)

B. Cell Surface and Associated Organelles

The *pellicle* or membranous surface of the animal can be simple as in the amoeba or quite elaborate as in order Radiolaria, where the shells and membranelles form exotic arrays (for examples, see Fig. 1). In the Foraminifera, the principal component of the shell is $CaCO_3$. It is the accumulation of these shells that account for the white cliffs of Dover and also for about one third of the sediments deposited on the ocean bottom. Geologists use these deposits to date various layers. The location of oil may be aided by the fact that certain species of fossil foraminiferans are found in oil-containing strata. In addition, some species of Radiolaria are characteristic of warm water and others of cold. Geological oceanographers can use this to estimate the temperatures and currents of ancient bodies of water.

The *cilia* and *flagella* associated with the surface are used primarily for movement and food capture. Recent research indicates that they both have a similar ultrastructure, the only difference being that cilia are smaller and more numerous. They are remarkably similar in organisms from protozoans to mammals. The beat of the cilium originates in the *basal body* located near the base of each cilium which by some unknown mechanism affects the contractile elements within the organelles themselves. In some ciliates, rodlike or oval structures at right angles to the body surface are found. These *trichocysts* serve as protection, provide a means of attachment, and also capture food. One type of trichocyst has a thornlike tip which discharges explosively. Another type, a *tyxicyst,* contains a fluid that can paralyze prey. The advantages of cilia and flagella are speed and maneuverability. While the amoeba moves at rates of 0.2–3.0 μ/sec, flagellates can manage 15–300 μ/sec, and the ciliates attain rates of 400–2000 μ/sec. The protozoans have usually been classified according to their locomotor apparatus, but this may no longer be acceptable in modern systematics. Further details on particular locomotor mechanisms will be presented below.

C. Cell Inclusions

Protozoans are generally *uninucleate,* but many, especially the ciliates, are *multinucleate.* In the multinucleated species, two kinds of nuclei occur: *macronuclei* and *micronuclei.* The macronuclei control vegetative functions and development, while the micronuclei are involved in sexual functions and in the production and control of the macronuclei. Animals will survive after removal of the micronuclei.

Vacuoles are abundant, particularly in freshwater species, and serve an important role in water regulation, since freshwater organisms are hyperosmotic with respect to their environment and the influx of water

threatens rupture. The contractile vacuoles remove this water by forcing their
contents back to the external medium. Marine and parasitic forms typically
do not contain vacuoles, as they live in an environment where the solute
concentration is the same as their cytoplasm.

Other cellular organelles that may be found in the protozoan cell include
the following (Anderson, 1967; Manwell, 1961): *mitochondria*, a "power
plant" concerned with respiratory activities; *kinetoplast*, which lies near the
base of the flagellum, its function is not understood; *chloroplast*, a chloro-
phyll-containing organelle that is involved in photosynthesis and contains
DNA; *golgi complex*, membranous or "canalicular" system that may have a
secretory function; *endoplasmic reticulum*, both "rough" (RNA associated
with it) and "smooth" (no RNA) varieties of cytoplasmic membranous
structures are found in protozoans; *stigma*, a light-sensitive organelle; and
food vacuoles which occur when food is ingested.

In addition, an incompletely understood but highly interesting complex
of *fibrils* may be found in many forms. Many have referred to these as
"neurofibrils," obviously implying a conductile and perhaps an integrative
function. There is some evidence that they may be involved in electrical
conduction, but this is by no means well established (Bullock and Horridge,
1965). Even less understood is the sensitivity of protozoans to temperature,
*p*H, and a wide variety of physical agents when differentiated sensory receptors are not found (Manwell, 1961).

D. Respiration

The small size of protozoans enables them to rely on *diffusion* as a means
of exchanging molecules across the membrane. Calculations have shown that
in an organism of fairly high metabolic rate, an adequate supply of oxygen
by diffusion is possible if the body shape does not keep organelles more than
0.5 mm from the oxygen source. In larger organisms, special transport mechanisms are required unless the metabolic rate is low. The larger varieties of
protozoans have avoided this problem by being elongated rather than spherical. For excretion purposes, the process of diffusion is aided by contractile
vacuoles.

E. Digestion and Nutrition

Most protozoans are *holotrophic* (i.e., they digest other organisms),
but all forms of nutrition are observed *(heterotrophism)*. Some flagellates
are *saprotrophic;* they rely on dissolved substances such as decayed material
or material derived from other animals. Feeding mechanisms vary. Amoeboid
species capture and ingest food by first encircling it with pseudopodia. Once

the particle is surrounded, the tips of the pseudopodia fuse, a membrane forms to produce a food vacuole, and the vacuole is moved into the endoplasm. After digestion by enzymes, nutrients are absorbed from the vacuole. More advanced feeding mechanisms are found in the free-living ciliates, which possess a mouth and a gullet at the end of which food vacuoles form. In *Paramecium,* food is directed into the *cytostome* (mouth) by the beat of the cilia; undigested food is ejected at a specific point on the body surface known as the *cytopyge.*

F. Reproduction

The usual method of reproduction is the asexual mechanism of *binary fission,* where the organism is divided into two parts. Most division is transverse, although longitudinal splitting is observed in some flagellates. Sexual reproduction provides not only for an increase in the number of individuals but also for a change in the genetic composition. One type of sexual reproduction observed in *Sporozoa* and in some members of *Mastigophora* and *Sarcodina* involves the *fusion* of two individuals followed by meiosis and the production of offspring with altered genetic characteristics. *Conjugation* involves the lateral attachment and exchange of nuclear material, a subsequent separation, and then division. Further details of reproduction will be provided below.

G. Protective Devices

During unfavorable conditions when there is a lack of moisture or food, many species can encyst, or form a protective coating. During encystment, they lose their locomotor apparatus and decrease in size. In the case of one *Trichonympha* species, a wood-digesting flagellate in the cockroach gut, it must encyst when the roach molts and sheds its digestive tract along with its exoskeleton. The roach must have the symbiont to survive, and it eats the cysts after molting.

Perpetuation of organisms is also assisted by their ability to adapt to changing conditions. For example, Hyman (1940) notes that while most protozoans will die at temperatures of 30–40°C, a temperature as high as 70°C can be tolerated if it is increased gradually. This ability to adapt to environmentally induced variations can be passed on to the progeny in some species.

III. PHYLOGENY AND TAXONOMY

Protozoans are part of the category Protista; Protista, Monera, Metazoa, and Metaphyta comprise the living world. Monera represents a category of organisms distinguished by a lack of nuclear membranes, a lack of a clearly segmented nucleus, and the possession of photosynthetic capacities. Within Monera, there are two major phyla: Schizophyta (bacteria) and Cyarrophyta (blue-green algae). Metazoa encompasses the multicellular animals, while Metaphyta includes the plants.

A. Classification of Protozoa[2]

Phylum Protozoa: the acellular animals.
  Subphylum Plasmodroma: locomotion with pseudopodia and flagella.
    Class Flagellata (Mastigophora): the flagellates. The locomotor organelles (flagella) are temporary or permanent. These are considered to be the most primitive members of Protozoa.
      Subclass Phytomastigophora; chloroplasts present.
        Example members:
        Volvox (order Volvocales).
        Euglena (order Euglenoidina).
        Paranema (order Euglenoidina).
      Subclass Zoomastigophora: no chloroplasts.
        Example member: *Trichonympha* (order Polymastigina).
    Class Sarcodina (rhizopods): members have fluid pseudopodia and pseudopodia with internal structure (axopodia).
      Subclass Rhizopodia: no axopodia.
        Example members:
        Amoeba (order Amoebozoa).
        Globigerina (order Foraminifera).
      Subclass Actinopodea: axopodia present.
    Class Sporozoa: no clear locomotor apparatus. Members are parasitic and form spores. Many reproduce by multiple fission.
      Subclass Telosporidia:
        Example member: *Plasmodium* (order Haemosporidia).
  Subphylum Ciliophora: locomotor organelles are cilia.
    Class Ciliata:
      Example members:
      Stylonychia, Spirostomum, Stentor, Euplotes (order Spirotricha).
      Tetrahymena, Paramecium (order Holotricha).

[2]Not all subclasses and orders are included. Examples of members and their orders are provided for species discussed in the text. It should also be noted that alternative classifications have placed the classes at a phyletic level and have combined Ciliata and Suctoria.

Vorticella, Carchesium (order Peritricha).

Class Suctoria: cilia while young, sessile.

B. Protistan Characteristics and Evolution of Protozoans

Comparisons with moneran cells generally indicate an increase in specialization of cell organelles among the protistans. The nucleus and nucleolus are clearly set off, a complex network of intracellular membranous network appears, and numerous cellular inclusions such as vacuoles are present. With the development of organelles to permit efficient movement, the probability of successful adaptation is improved. Concomitant with the development of specialized cell inclusions, variations in nutrition have emerged. The protistan groups have specialized, developing either photosynthetic or heterotrophic means of nutrition.

The protistan category contains the algae, fungi, and slime molds, as well as the protozoans. Of these groups, the algae appear to be the most primitive, retaining the plant–animal characteristics of ancient protistans. Of the protozoans, the flagellates are considered to be the most primitive. Their retention of the plantlike capacity to carry on photosynthesis with chloroplasts is the major evidence. The photosynthetic capacity makes a clear demarcation between protozoans and algae difficult, since order Euglenoidina includes photosynthetic as well as heterotrophic and mixed species. In fact, the euglenoids are frequently included in the plant kingdom. The flagellates also resemble plants and sponges in their life stages, and they can form colonies. Given this evidence, a tentative assumption is that Metazoa and Metaphyta derived from flagellate-like protistans. The heterogeneity in this class would provide a basis for the diversity observed in Metaphyta and Metazoa.

Evolution of protozoan groups would begin with a form of ancient flagellate. Pseudopodia formation appeared later, giving rise to species having both flagella and pseudopods and then to species possessing only pseudopodia for movement and food capture. The flagellates and sarcodinians are probably closely related; in Phytomastigophora, some members are amoeboid but have a flagellum. The relationship of sporozoans to the flagellates and sarcodinians is indicated by the fact that immature sporozoans can possess flagella or pseudopodia. Ciliates can also be traced to flagellates; intermediate forms possess many flagella and more than one nucleus. Of the major classes, the Ciliata appear to be a more distant group.

The ciliates would seem to be the most highly evolved protozoans, developing in size and complexity to attain the ultimate in acellular existence. The limitation in size is dictated by respiratory and circulatory mechanisms, and the limitation in functional differentiation is dictated by the lack of highly specialized organs. In the multicellular organism, cells give up their "jack-of-all-trades" role and are capable of performing one function well, leaving other functions to other cells.

C. Metazoan Origins?

The usual assumption is that the single-cell existence preceded the multicellular organism in evolution. It is difficult to base an accurate reconstruction on present-day species, but what scant evidence there is suggests that the flagellates gave rise to the sponges and Metaphyta and that the main line of Metazoa derived from the ciliates. The flagellates are similar to poriferan cells, and the presence of photosynthetic capacities links them to plants. The flagellates are also sufficiently unspecialized to permit later diversification—many have a unicellular state, others exist as a colonial flagellate, and others as a mixed organism containing both amoeboid and flagellate cells (the latter would be similar to the blastula of the sponge). *Naegleria,* for example, could be an ancestor with its capacities to behave as a single flagellate or an amoeboid cell, or to form colonies. Its protean nature has led to some debate over whether it is a protozoan or a member of Myxophyta, the slime molds.

The move toward colonialism and differentiation of cellular function is still very much apparent in protistans. In the phylum Myxophyta, the slime mold *Dictyostelium discoideum* is an excellent example of a combined unicellular–multicellular protistan organism. Members exist independently until the food supply is exhausted. They aggregate, forming a slug with metabolically differentiated anterior and posterior ends. The slug then forms a stalk with a ball of spore cells on top. This aggregation and differentiation of function is sometimes cited as a means by which multicellular existence came to be. Functional differentiation in a protozoan colonial form is observed in *Volvox.* Although a sphere formed by approximately 2000 flagellates, the organism moves with one portion as an "anterior," while the remaining half engages in reproductive activities. There is another way in which the multicellular organism could have evolved from the acellular animals. If the single-celled protozoan possessed several nuclei (such as the ciliates), the appearance of membranes between nuclei would produce the multicellular organism.

IV. PROTOZOAN SPECIALIZATIONS RELEVANT TO BEHAVIORAL RESEARCH

A. General Phylogenetic Advances and Limitations
Advances:
1. The existence of specialized receptor and effector organelles with the

achievement of permanent differentiation, especially in ciliates. These developments provide some flexibility and refinement in movement and adaptation. For example, the flagella of *Paranema* pull the animal forward but may also communicate a "stationary object" signal when the animal bumps into something.

2. The presence of a permanently differentiated protoplasm or "physiological gradient" (Herrick, 1924). The advancing anterior end is more sensitive and already differentiated. Early evidence concerning the presence of gradients in the less stable (amoeboid) as well as the permanently differentiated forms was derived from the application of toxic substances; in *Amoeba,* the newest pseudopodia are affected first, while in *Vorticella,* the peristome will disintegrate before the stalk.

3. Possible integrative mechanisms. The roles of "neurofibrils" and "myofibrils" are not clearly understood in all cases, but it is possible that they provide for integration and for more refined responses.

4. The appearance of colonial forms and the differentiation of structure and function.

Disadvantages of acellular existence:

1. The necessity for a more total commitment to a function. While there is organelle differentiation in the acellular animal, the involvement in any one function may utilize all the capacities of the organism.

2. The confinement to a single cell restricting size and differentiation. One consequence of being small would be the limitation in the size of prey; predators usually need to be larger (Telfer and Kennedy, 1965).

3. The lack of differentiation restricting spatiotemporal patterning. Summation and association between differing stimuli and the translation of these processes into precise effector patterns cannot develop highly as long as the organism must act more or less as a totality.

B. Reproduction in Paramecia

The micronuclei of paramecia give rise to the macronuclei and are responsible for controlling reproduction. The role of the macronuclei is vegetative; species will survive the loss of micronuclei but not of the macronuclei. Nuclear changes in the ciliates are complex and may take several forms. For example, the macronucleus may undergo disintegration and resynthesis in the vegetative stage, with no changes occurring in the micronuclei. This internal reorganization may be a form of rejuvenation, but the evidence defining the functions of various nuclear changes is unclear.

Classes of Reproduction

Asexual. Fission is common among ciliates. It may be a transverse or longitudinal division, and it is a *vegetative* form of reproduction, perhaps serving a rejuvenation function. Both macro- and micronuclei divide, being amitotic and mitotic divisions, respectively. The frequency of division is generally dependent on nutrition and can be prevented by starving the animals.

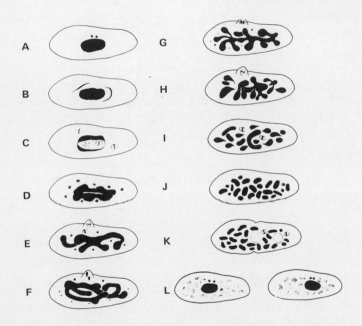

Fig. 2. General summary of nuclear changes during autogamy in *Paramecium aurelia.* (After Diller, 1956; reprinted by permission.) A, Typical animal with two micronuclei and a macronucleus. B, "Crescent" micronuclei; this is a characteristic prophase of the first prezygotic division. C, Second prezygotic division; four micronuclei, two of them represented as lying in the concavity of the cup-shaped macronucleus, D. Eight micronuclear products of the second division; macronucleus preparing for skein formation. E, Variable numbers of nuclei continuing to divide a third time; in the case illustrated, two are functional, one of which lies in a bulge near the mouth ("paroral cone"). F, Potential gamete nuclei arising after the third division; degenerating nuclei from the second and third divisions may be present. G, Two gamete nuclei fusing in the paroral cone; synkaryon formation. H, First division of the synkaryon. I, Second division of the synkaryon; macronucleus fragmenting. J, Four products of the second division; two transform into the new macronuclei and two remain micronuclei. K, Macronuclear anlagen well developed; the first cell division under way and micronuclei dividing. L, Macronuclear anlagen segregated to the two daughter cells; old macronuclear fragments disintegrating.

Autogamy. Self-fertilization or autogamy is a form of *sexual* reproduction involving a single animal. Homozygous offspring result, with the heterozygote producing only one type; i.e., *Aa* yields *AA* or *aa* as progeny. The nuclear changes are summarized in Fig. 2. Autogamy begins with three divisions of the micronuclei and the production of gametic nuclei. The gametic nuclei fuse and the fused nuclei subsequently divide, the various portions differentiating into micro- and macronuclei. *Paramecium aurelia* frequently undergoes autogamy, and it appears to be a rejuvenation process. It occurs when the proper mating types are lacking, when there is crowding, or when there is a lack of food.

Conjugation. A second form of sexual reproduction in ciliates is conjugation, which involves the exchange of nuclear materials between individu-

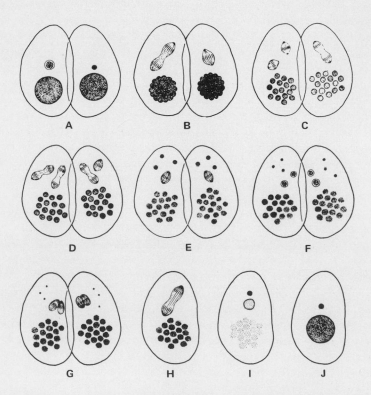

Fig. 3. Conjugation in ciliates. (After Diller, 1956; reprinted by permission.) A and B, First pregamic division. C and D, Second pregamic division. E, Disintegration of three of the products of the second pregamic division and third division of the fourth nucleus. F, Gametic nuclei. G, Synkaryon formation. H, First postzygotic (metagamic) division. I, Young macronuclear anlage and new micronucleus, old macronuclear fragments being absorbed. J, Reconstituted ex-conjugant, before first fission.

als. Haploid micronuclei are exchanged, the animals separate, and the result is two genetically identical heterozygotes. The major stages of conjugation are represented in Fig. 3. Conjugation provides a new nuclear apparatus, which is passed on to descendant cells. It is presumably a means of genetic variation and recombination, with the transfer of new genes to the progeny.

C. Behavior and Ciliary Mechanisms

Paramecia appear to be quite adaptive with respect to their movements. Their ability to reverse direction and quickly back away from barriers and noxious substances and their great variability in movement are due to the cilia. In behavioral research where avoidance or escape behavior and approach behavior are being conditioned, the mechanisms subserving ciliary action and changes in direction are of interest.

Jensen (1959) has proposed the existence of pacemakers to account for directionality and ciliary control in *P. aurelia*. Direction of movement is a function of the combined influences of the anterior, buccal, and posterior pacemakers. The activation of the posterior pacemaker produces movement in the forward direction since the cilia bend backward, the buccal pacemaker is responsible for aboral turning, and the anterior pacemaker is responsible for backward locomotion as the cilia move anteriorly. Localized microbeam (ultraviolet) radiation of a particular pacemaker region was found to produce specific alterations in movements. For example, during irradiation of the anterior portion, the animals swam backward, presumably because the irradiation increased the degree of the anterior pacemaker's activity.

Evidence for mechanisms somewhat less hypothetical than a "pacemaker" to account for differential orientations and changes in ciliary beat in *P. caudatum* was obtained by Naitoh (1966). Nickel ions placed in the animal's medium were found to inhibit ciliary beat. During this inhibition, Naitoh was still able to reverse cilia orientation with an increase in external potassium ionic concentration or by electrical current. Microelectrode recordings demonstrated that normal and nickel-treated animals possessed a similar inside negativity level. However, when the cilia orientation was reversed, a membrane depolarization was observed. Further differentiation of ionic events and the animal's orientation were obtained in subsequent investigations (Naitoh and Eckert, 1969*a,b*). It is known that the direction of ciliary orientation reverses when *P. caudatum* strikes an object, while stimulation from behind accelerates the animal's movement in whatever direction it is headed for. This differential responsivity is dependent on particular membrane changes at the anterior and posterior ends. Mechanical stimulation of the anterior end was found to produce membrane depolarization, stimulation of the posterior end produced hyperpolarization, and stimulation in the middle of the animal yielded no effects. This suggested a variation in permeability at the anterior and posterior ends with respect to ionic influx. Tests of the ionic hypothesis indicated that anterior stimulation increased Ca^{2+} influx, while posterior stimulation increased posterior K^+ permeability. These events are summarized as follows (Naitoh and Eckert, 1969a):

Anterior stimulation:
1. Membrane deforms.
2. Ca^{2+} permeability increases.
3. Ca^{2+} moves down electrochemical gradient.
4. Potential approaches E_{Ca}; depolarization.
5. Depolarization spreads electrotonically.
6. Cilia reverses beat direction.
7. Locomotion is reversed; avoiding.

Posterior stimulation:
1. Membrane deforms.
2. K^+ permeability increases.
3. K^+ moves down electrochemical gradient.
4. Potential approaches E_K; hyperpolarization.
5. Hyperpolarization spreads electrotonically.
6. Beat frequency increases.
7. Forward locomotion accelerates.

Whether there is an intracellular system that coordinates ciliary orientation and activity in protozoans has been a subject of debate. As discussed earlier, it has been suggested that perhaps the fibril system acts as an intracellular "nervous system," providing a reticulum of pathways for integration and coordination. Early research was quick to assign a coordinative function to the "neuromotor" fibrils, especially when there was a reported loss of ciliary coordination when the fibril system was disrupted. Recent evidence suggests that ciliary coordination is not attributable to the fibril system in *Euplotes* (Naitoh and Eckert, 1969b). Animals were immobilized with $NiCl_2$, and the potential changes in response to applied currents were recorded. In intact animals, the potential changes were the same at both ends of the animal, and a deep transverse cut halfway across (interrupting the neuromotor fibrils) failed to interfere with the identical nature of the anterior and posterior recordings. Mechanical stimulation at the anterior or posterior end elicited the same results as reported in *P. caudatum,* depolarization with stimulation at the anterior and hyperpolarization with stimulation at the posterior. The cilia reversed orientation with anterior stimulation. Ciliary coordination and orientation were unaffected by the fibril sectioning.

D. Amoeboid Movement

The typical textbook picture of a cell suggests a rather static state, but this is far from reality. Indeed, the observations of Weiss (1961, 1969) suggest a continual movement and renewal of processes. Nowhere is this more dramatic than in *Amoeba,* where peripheral and internal structures show considerable flexibility in the process of pseudopod formation. Actually, pseudopod formation is characteristic of many cells in metazoans as well, e.g., the amoeboid white blood cells.

The sol–gel theory of movement initially proposed by Mast (1926) remains one of the tenable hypotheses. This theory suggests that the liquid endoplasm streams forward to the tip of the pseudopod, where there is a temporary liquefaction of the plasmagel. The weakening of the tip combined with a gelation at the posterior end of the animal forces a pseudopod extension. The theory essentially proposes the interconvertibility of peripheral and internal structures. The gel or protein network at the rear of the animal contracts, pushing the less fibrous endoplasm anteriorly. At the tip, the sol becomes more fibrous.

Allen (1962) postulates an active contraction in the anterior portion of the pseudopod, a pulling action by the frontal end. The fibrous gel moves into the anterior end and then contracts. This contraction pulls more sol along, which is then converted into gel. Breaking a part of the cell membrane (thereby reducing internal pressure) does not interfere with movement, providing some evidence against a posterior contraction and squeezing mechanism.

V. LEARNING STUDIES

A. Habituation

Selected Earlier Studies[3]

"Becoming Accustomed to Stimuli"—the Research of Jennings. Jennings (1901, 1906) has provided one of the earliest accounts of modifiability in ciliates. His work on the behavioral variability and adaptability displayed by *Stentor* when subjected to a steady stream of carmine is a classic. Interest in behavior which might be encompassed under the rubric "learning" was stimulated by an observation of a colleague, W. E. Castle, who noted that

[3]There are several dozen early reports on the "adaptability" of protozoans to varying pH, chemicals, etc. Most of these studies were not controlled attempts to specifically demonstrate habituation and are not reviewed here. A thorough overview and summary of these studies can be found in Warden *et al.* (1940).

when *Stentor* is repetitively struck by tubifex worms, it eventually ceases to contract. To provide an experimental model of this, Jennings repeatedly touched various ciliates with a glass rod or hair. The usual reaction was to contract for a minute and then reextend. The number of strokes necessary to elicit a contraction became the index of response decrement—the larger the number of strokes, the less responsive the subject. Jennings reported "typical" findings in a number of ciliates. For *Epistylis flavicans,* the sequences on two animals were as follows:

1. 1-1-1-1-1-1-1-2-33-25-7-13-36-20-14-13-13-33-9-30-3-31-226. (Thus on the first seven touches, the subject contracted after the first stroke; on the thirteenth attempt at eliciting a contraction, 36 were needed to produce a contraction, etc.)
2. 1-22-10-3-3-1-1-22-59-125, followed by "continuous blows" for 60 sec until the next contraction occurred, etc.

A typical sequence for *Stentor roselii* (the leaf on which the subject was attached was jarred):

1-1-1-1-1-1-1-1-1-1-1-3-1-5-1-1-3-1-3-3-48-40-2-250-36-36-154

Similar results were obtained with *Vorticella* and *Carchesium.* An interesting observation with the latter colonial form concerned the gradual localization of contraction. When a single *S* was stimulated, the entire colony contracted at first. Continuation of stimulation produced more localized contractions until finally only the branch containing the touched organism contracted.

Jennings was quite aware of possible alternative explanations of the response failure. To avoid sensory and motor fatigue, he tested the effects of a weak jet of water directed on the ciliates. This current caused the animal to contract at first, but if the current was continued the animals re-extended. If the stimulation was stopped and initiated again after a few seconds, the animals were typically unreactive. Jennings concluded that the response diminution could not be due to fatigue, since with a much stronger stimulus it was possible to keep a ciliate contracting for an hour. Although his discussion struck a conservative tone (relative to contemporaries interested in espousing "consciousness" and "intelligence" in lower forms), Jennings points out that the response decrement is action to obtain an end—an action aimed at abolishing the stimulus. The similarity between the ciliate's response decrement and that observed in higher organisms was noted.

Mechanical Stimulation of Vorticella. Having noted that *Vorticella* in natural conditions became habituated to certain stimulation, Danisch (1921) investigated the possibility of habituation to a vibratory stimulus of constant and strong intensity. By striking the platform containing the culture with specific weights, Danisch was able to specify stimulus strength at 500 ergs. When the stimulus was delivered at 10 sec intervals, it was found that most animals ceased contracting after the ninth stimulation. Increasing the stimulus intensity (1000 ergs) caused the contractions to persist longer, although after 25 stimulations it was reported that "several" animals ceased to respond. After a rest period of 25 sec, 18 of 20 subjects reacted to the initiation of stimulation again. More intense stimulation (1500–2000 ergs) resulted in an unabated contraction to each stimulation. For example, at 2000 ergs with a 20–25 sec intertrial interval (to allow the animal time to reextend), animals responded to 420 stimulations.

Danisch made it clear that fatigue could not be a factor in the response decrement observed at the weaker stimulus intensities. The cessation of response after nine trials at 500 ergs and the persistence of response for up to 420 trials at 2000 ergs supported his argument that something other than fatigue was involved. Furthermore, the recovery after 25 sec demonstrated that injury could be ruled out as an explanation.

Mechanical Stimulation of Spirostomum ambiguum. Probably the best and clearest demonstration of habituation was performed by Wawrzyńczyk (1937*a*) with *Spirostomum ambiguum*. The methodology and application of habituation criteria were quite adequate when compared to concurrent investigations and, in fact, would stand up well with present-day publication standards.

Subjects were placed in round, flat vessels on a slide mount and were mechanically stimulated by the stroke of a 4 g weight dropped from a height of 6 cm. The frequency of stimulation differed: Wawrzyńczyk applied regular (6 strokes/min) and irregular (6–12 strokes/min) stimulation and obtained response decrements for both. For the regular stimulation, an average of 194 min passed and an average total of 307.9 contractions were elicited before the subjects ceased responding, and for the irregular stimulation 156 min passed and 322.5 contractions occurred before response failure. The data for the regularly and irregularly stimulated group are summarized in Fig. 4. An estimate of retention was obtained by testing subjects after response cessation. The retention data are presented in Table I and indicate a retention span of about 50 min. Fatigue was eliminated as a factor in the response decrement by testing habituated subjects' contractability to a *slower* stimulus frequency. After habituating a group to 12/min, Wawrzyńczyk changed the stimulus frequency to 6/min and obtained dishabituation or a full return of contractability. Confounding factors such as chemical changes in the medium were avoided by changing the solution. The habituated subjects were also adjudged to be physiologically normal because the number of food vacuoles was the same as in unstimulated subjects.

Another interesting finding in these studies was the apparent setting of a

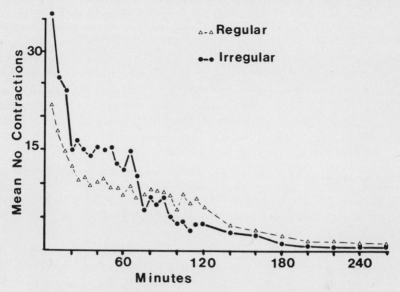

Fig. 4. The decrease of contractability with regular (6 strokes/min) and irregular (6–12 strokes/min) mechanical stimulation. (After Wawrzyńczyk, 1937*a*.)

contracting rhythm in *Spirostomum*. Animals were subjected to a $\frac{1}{10}$ min stroke. When the frequency of delivery was slowed down, it was found that subjects continued to respond at the previous rate— $\frac{1}{10}$ min, early evidence of temporal conditioning.

Recent Research

Behavioral Modification of Stentor coeruleus *(by C. M. Harden).*[4] In the experiments to be reported, groups of approximately 100 animals occupied 50 by 50 mm Stender dishes containing, unless otherwise specified, 20 ml of spring water. After preparation, all dishes were left undisturbed for 2 days until most animals had attached to the specially roughened glass bottom. The laboratory, maintained at 22–24°C, was illuminated continually throughout the experiments.

Ten minutes before testing, each dish individually was placed gently into a holder attached to a loudspeaker cone. An electronically controlled square pulse to the speaker displaced the dish laterally, and a strain gauge

[4]This paper represents a well-controlled recent demonstration of habituation in *Stentor*. Since it is unpublished, we are printing the bulk of Harden's original manuscript in this section (by permission).

Table I. Duration and Number of Contractions to Response Cessation in 25
Subjects and the Time for Response Recovery (Wawrzyńczyk, 1937a)

Subject No.	Time until response cessation	No. contractions before response cessation	Recovery time
1	245	182	40
2	150	243	40
3	255	147	45
4	160	321	55
5	215	197	45
6	195	421	30
7	235	209	45
8	220	201	40
9	210	428	35
10	185	459	35
11	205	392	40
12	205	276	40
13	250	441	40
14	155	337	45
15	180	323	30
16	190	409	40
17	180	254	45
18	155	328	80
19	255	312	50
20	200	300	45
21	145	409	40
22	165	165	45
23	155	201	55
24	170	389	85
25	170	354	85
x	194	307.9	47

monitored dish displacement of approximately constant velocity for 20
msec through an excursion of 0.26 mm. Two seconds after each such stimulus
the dish was photographed by a 35 mm camera mounted on a microscope
stand. After a series of 30 stimuli at 1 min intervals, the dish remained undisturbed for 2 min, when one additional photograph was made. Contraction
responses were recorded from the developed film starting with this last
frame so the experimenter could identify those animals (approximately
90%) which had remained attached to the bottom of the dish throughout the
procedure. Of these animals, the number contracted was counted in each
frame.

The contraction response was then investigated under the following
conditions:

1. Because *Stentor* contracts occasionally while undisturbed, five dishes

(group A) were run to obtain a series of 30 photographs, each taken at 1 min intervals, without stimulation. These photographs provided data for a base level of contraction under the conditions of the experiment.

2. To investigate Jennings' (1901, 1906) observation that the frequency of contractions decreases with repeated stimulation, nine dishes (group B) were photographed after each of 30 successive stimuli at 1 min intervals. One minute after the last pulse, the dish was exposed to the flash of a strobe light. A different dish in group B was run at 9:30 AM on successive days.

3. Following the run of each group B dish, a single dish (group C) was run in the same manner. Thus, this dish was run at 10:30 AM on 9 successive days to determine the effects of stimulation at 24 hr intervals.

4. Beginning at 11:30 AM daily, one dish (group D) was subjected to four runs per day with a 1 hr undisturbed period between each successive run. Photographs were taken after each stimulus only during the fourth run. This procedure yielded data on the effects of repeated prior stimulation after a brief interval without disturbance.

5. In six dishes of previously untested animals (group E), 10 ml of the 15 ml of water in each dish was removed by a pipette immediately after the fifteenth stimulus. The water was replaced after the sixteenth stimulus with 10 ml of water which had just been removed from a dish of undisturbed animals (group F), which, in turn, received the water removed from the group E dish. Each of the six group F dishes was run immediately following completion of a group E run. As a control, six dishes (group G) had 10 ml of water removed after the fifteenth stimulus and replaced after the sixteenth stimulus. Throughout this procedure, dishes in all three groups were photographed after each stimulus.

The results obtained with groups A, B, C, and D are shown in Fig. 5. If undisturbed, about 5% of the animals were contracted at any time (group A). However, the first stimulus produced contraction in about 50% of the animals. With repeated stimulation (group B) at 1 min intervals, this proportion decreased, approaching the resting level of group A by the thirtieth stimulus. The contraction of 39% of the animals in response to the strobe flash stimulation at the end of the run (group B) indicates that the decrease in response to mechanical stimulation cannot be explained simply in terms of motor fatigue. There was little difference in the responses of animals tested for the first time (group B) and those run at 24 hr intervals (group C). However, the 1 hr interval (group D) response was initially lower, and the decrease

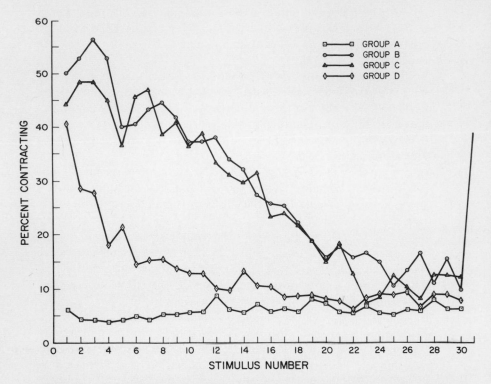

Fig. 5. Contraction response of *Stentor* (Harden, unpublished). Group A, baseline activity with no stimulus, $n = 290$. Group B, naive animals; the thirty-first stimulus was a strobe flash, $n = 250$. Group C, animals tested each 24 hr for 9 consecutive days, average n/day = 33. Group D, animals tested at four hourly intervals for 9 consecutive days with only fourth test recorded, average n/day = 36.

to the baseline (group A) faster, than for animals run the first time (group B) or those run 24 hr previously (group C). This suggests that the behavioral change associated with repeated stimulation persists for at least 1 hr.

The result of changing the water after stimulus 15 is shown in Fig. 6. If the response decrease with repeated stimulation is related to some substance in the water, introduction of water as in group E from an undisturbed dish of *Stentor* should increase the proportion of animals responding to the stimulus. Conversely, introduction of water as in group F from a dish of stimulated, relatively nonresponsive animals should tend to reduce the number of post-stimulus contractions. A marked increase in percentage of animals contracted was observed in group E immediately following the water change and in group G following water disturbance without change. However, there was a

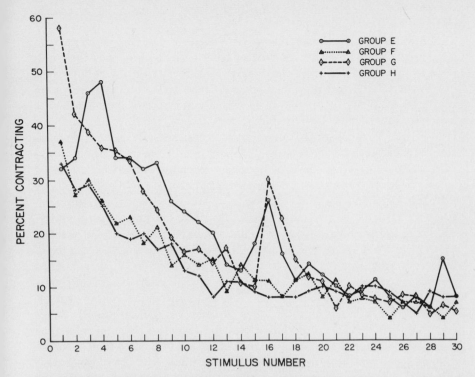

Fig. 6. Environmental effects on behavior (Harden, unpublished). Group E, water replaced after sixteenth stimulus with water from an undisturbed dish of animals, $n = 170$. Group F, 30 min before testing received water removed after fifteenth stimulus from group E dishes, $n = 180$. Group G, water removed after fifteenth stimulus and returned to dish after sixteenth stimulus, $n = 320$. Group H, water removed and immediately replaced 30 min before testing, $n = 360$.

rapid return to previous response levels in all dishes in both groups, indicating that the transient increase was associated with mechanical disturbance of changing the water rather than a substance in the water.

The behavior of animals in dishes (group F) which, before the run, had water added from previously stimulated dishes (group E) is shown in Fig. 6. If the water from previously stimulated group E animals contained some substance affecting responsiveness, one would expect the initial percent of responses of animals in group F to be lower than normal (i.e., lower than groups B, C, D, and E). Because the initial responsiveness of group F did seem a little lower than normal, five dishes (group H) were run after the water was withdrawn and immediately replaced. The response of this group was the same as that of group G. These results indicate that response decrease with repeated stimulation is not associated with accumulation of, or change in, some substance in the water.

Stentor *Habituation and Electrophysiological Correlates.* Procedures similar to those of Harden with *Stentor* have been recently used by Wood (1970*a,b*) in a carefully conducted study attempting to draw comparisons between the habituation phenomenon in *Stentor* and the data obtained in metazoans. Mechanical stimulation of animals at a rate of once per minute for an hour resulted in a drop of average response probability from over 90% to less than 25%. A comparison of different magnitudes of mechanical stimulation indicated that, like metazoans, the more intense the stimulus, the weaker the habituation effect. Tests of optimum interstimulus intervals showed 1/min to be best for habituation. Intervals less than this could not be reliably used because the animals could take up to 45 sec to recover from a mechanical shock. Tests of recovery time showed that after an hour response probability rose to over 60%. However, when habituated animals were retested, savings were obtained up to 6 hr after the initial habituation. This degree of retention is rather remarkable when compared to the retention levels obtained by Harden, Applewhite, and others who have also used mechanical shock. What is even more remarkable is the degree of specificity in the response decrement. The application of a more intense mechanical stimulus after habituation resulted in a dishabituation—the response probability jumped to almost 100% for that trial. This dishabituation was not retained, for on the subsequent trial, with the stimulus level that had been used to habituate the animal, the response level was again low. A different type of stimulus (electrical shock) also failed to produce any persisting dishabituation. Measurements of threshold changes after mechanical stimulation demonstrated that there was no interaction between the two stimulus modalities, i.e., the response decrement to mechanical stimulation was specific.

Subthreshold stimulation of an animal for a period of time has been shown to lead to faster habituation when a suprathreshold version of the same stimulus is subsequently used. Animals presented below-threshold stimuli for 30 trials were found to habituate with a steeper negative gradient than animals with no "prehabituation" experience. Their initial responsivity to the suprathreshold stimulus was the same as the controls.

Initial research on the relationship of recorded potentials and contractile behavior indicated that an action spike may trigger contraction (Wood, 1970*b*). With animals held extended in methylcellulose, it was possible to position a microelectrode for recording within *Stentor* and an electrode for stimulation external to the animal. To obtain a record of contractions, the animal was mounted so that a spot of light fell on the tail and also illuminated a photomultiplier tube. The contraction of an animal was represented as a voltage deflection recorded on the oscilloscope. Electrical and mechanical stimulation were found to induce negative-going, all-or-nothing spikes of 23 and 27 mv, respectively, with animals contracting 7–11 msec after the shock onset. With mechanical stimulation, a negative prepotential was observed (Fig. 7), and this potential could be dissociated from the spike; electrical stimulation only produced the spike, while mechanical stimulation would occasionally produce prepotentials without subsequent spikes. Additionally, the prepotential magnitudes were correlative with stimulus intensity. Wood suggests that these potentials represent the electrical concomitants of mechanical stimulus reception, while the spike is indicative of effector (contractile) activity. The following model is offered:

$$\text{Stimulus} \longrightarrow \text{prepotential} \longrightarrow \text{spike} \longrightarrow \text{contraction}$$

With respect to the habituation or response-decrement demonstration, Wood suggests that this form of behavior modification is a function of the receptor. Accordingly, habituation ought to be accompanied by a cessation or diminution in the magnitude of the prepotential. Localized mechanical stimulation was applied at or near the membranellar band, the most sensitive area. Stimulation that was effective in eliciting a contraction was applied at the rate of 1/min, and the decrement in responsivity that was observed corresponded with previous data (Wood, 1970a). Analysis of the electrophysiological correlates of this decrement in animals placed in methylcellulose was then carried out (Wood, 1970c). The prepotential was shown to drop in amplitude and increase in latency during habituation, while other potentials remained essentially the same (see Fig. 8). Wood points up similarities between these findings and those obtained with metazoans. Although the homologies are not by any means proven, it is tempting to compare the prepotential to the EPSP (excitatory postsynaptic potential), as Wood does, and to consider the

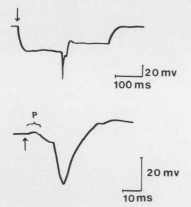

20 mv
100 ms

P

20 mv
10 ms

Fig. 7. Intracellularly recorded responses obtained to cathodal stimulation (top) and mechanical stimulation (bottom). *P* indicates prepotential observed with mechanical stimulation. (After Wood, 1970c.)

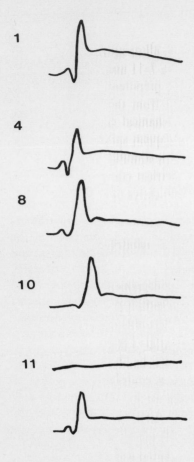

Fig. 8. Example of potentials observed during mechanical stimulation at 1/min. Numbers indicate the stimulus number. Bottom response is that obtained 3 min after cessation of stimulus series. (After Wood, 1970c.)

changes during habituation as due to EPSP diminution observed in some multicelled animals (Wood, 1970c).

Research on Spirostomum. The ability of the ciliate *Spirostomum* to habituate to mechanical stimulation has been well established (Applewhite and Morowitz, 1966, 1967; Kinastowski, 1963a,b; Osborn et al., in press; Osborn and Eisenstein, unpublished), and details of the contractile behavior are being examined (Hamilton et al., 1971). Kinastowski, using a water drop with varying force to vibrate the culture drop, observed a relationship between intensity and the number of stimulations before a response decrement was observed. At the stimulation rate of 10/min and a force of 1600 erg the animals were at a low rate of contraction by 120–150 stimulations. Higher-intensity stimulation (3200–4800 erg) did not permit the same low level of responsivity. With lower-intensity stimulation (400–800 erg), the animals were fairly unresponsive by 70–80 stimulations. These data are represented as percent responses in Table II.

Varying the number of stimuli delivered per minute was also found to be critical. At 1/min, there was no habituation, while at 5, 10, or 15/min there was considerable response decrement. Standard stimulus parameters for subsequent studies were 10/min at 1600 erg.

Tests for retention or savings were carried out at different rest periods, ranging from 1 to 30 min. Kinastowski found retention up to 10–15 min, a result recently replicated by Eisenstein's group (Osborn *et al.*, unpublished). This level of retention is superior to that observed by Applewhite (1968*b*), whose subjects began to return to initial rates after 1–2 min. Subjects were also tested after repeated periods of stimulation with either 30 min or 10 min rest intervals between periods. With 30 min intervals, there was little carryover from session to session, while with 10 min intervals, faster habituation was observed in the sessions that followed the initial one (Table III).

An interest in exploring the learning capacities and physiological substrates of what were termed "microinvertebrates" has also uncovered some interesting data in *Spirostomum* (Applewhite and Morowitz, 1966, 1967; Applewhite *et al.*, 1969*a*). Microinvertebrates were defined as those species less than 1 mm in length. These microscopic animals are preferred to those typically used because of the small number of cells and the reduction in neuronal interactions. It is reasoned that if electron microscopy and au-

Table II. Average Number of Contractions (Percent) at Various Intensities in *Spirostomum* (Kinastowski, 1963*a*)

Time (min)	Strength of stimulus (erg)				
	400	800	1600	3200	4800
1	52	58	80	90	94
2	30	40	64	80	86
3	26	26	52	72	76
4	12	26	38	62	80
5	14	18	38	52	60
6	8	12	28	48	66
7	6	12	18	40	48
8	2	8	14	42	52
9			20	32	40
10			10	32	40
11			8	34	26
12			6	30	34
13			10	32	22
14			4	24	26
15			2	20	18
	150	200	392	690	768

Table III. Average Number of Contractions in Successive Periods of
Stimulation with a 10 min Interval Between Periods
(Kinastowski, 1963*b*)

Time (min)	Successive periods of stimulation					
	I	II	III	IV	V	
1	78	44	32	23	28	208
2	54	18	22	20	20	144
3	42	26	14	16	12	112
4	34	14	14	10	6	78
5	20	12	12	8	10	62
6	18	8	4	4	4	38
7	10	8	2	2	2	24
	256	144	96	88	82	666
Median number of contractions	$x_1 = 36.6$ $x_5 = 11.7$	$x_2 = 20.6$	$x_3 = 13.7$	$x_4 = 12.6$		

toradiography are to be used to detect meaningful structural and metabolic
changes correlative with learning, it is necessary to reduce the system size
in order for thorough analyses to be made in a reasonable period of time. The
assumption is that an eventual generalization can be made from protozoans
to metazoans. Whether it is a safe assumption remains to be determined, but
as exemplified by the findings of Wood in the above section there are some
interesting similarities, even at the electrophysiological level.

The techniques for demonstrating habituation in *Spirostomum* were
similar to those used by others. Subjects were placed in a culture drop on a
depression slide which was positioned on a ring stand. A solenoid, activated
by a stimulator, mechanically displaced the stand, causing *Spirostomum* to
contract. The criterion for habituation was three successive trials without
contraction. By positioning a camera over the slide, it was possible to photo-
graphically record the number of contracting animals before and after the
mechanical shock. The results of the initial experiments are summarized in
Table IV, and the data support Kinastowski's findings that there is a response
diminution with repeated shock. Comparative analyses with metazoans
demonstrate similar time courses (Figs. 9 and 10). Injury and fatigue were
eliminated as alternative explanations by the observations that after habituation to a specific stimulus a more intense stimulus or a different stimulus
(electric shock) will elicit a contraction (ruling out fatigue) and that the organisms can be habituated again after their retention of the response has
waned (ruling out injury). Another method of testing whether receptor or
contractile fatigue is involved in the response decrement is to habituate
Spirostomum to a stimulus of gradually increasing intensity (Applewhite

et al., 1969*a*). It had been shown by others that this procedure is just as effi-
cacious in producing habituation as when a constant stimulus is presented the
animal (Davis and Wagner, 1969). It was reasoned that if *Spirostomum* can
be habituated to a gradually increasing stimulus, fewer contractions would
be involved, thus providing further evidence that fatigue was not a factor.
The results demonstrated that the level of habituation was the same for
constant and gradually increasing stimuli. A summary of these studies and
some of the physiological factors involved in *Spirostomum* habituation is
provided in Table V.

The mechanisms responsible for retention of the response decrement
were found to be nonlocal and temperature dependent (Applewhite, 1968*a,b,c;*
Gardner and Applewhite, 1970*a,b*). *Spirostomum* individuals, habituated at
5/sec until they did not respond to three successive stimuli, were transected
into equal portions; 15 sec later, they were tested for retention and compared
with naive, transected controls and with habituated, intact animals. The results of this study, summarized in Table V, indicate that both the anterior and
posterior halves retained the response decrement and that the retention of
these halves was equal. Investigations of temperature effects demonstrated
that the persistence of the response decrement could be enhanced by low
(5°C) temperature (Table V). Furthermore, higher temperature (37°C) was
found to have more of an effect on retention than acquisition. While animals
at 5 and 12°C and 22 and 32°C all demonstrated equal *rates* of initial habituation, dishabituation was observed at the higher temperature (32°C) during
retention tests 90 sec later. This suggested a passive diffusion mechanism
(perhaps ionic) as being responsible for the *acquisition* of the response
decrement and an active *metabolic* process for its recovery. Preliminary
assessments of the role of ions on habituation indicated that, while Na, K,
and Ca had no effect, $MgCl_2$ prevented habituation (Applewhite and Davis,
1969).

Research on possible cellular substrates of the habituation capacity has
yielded a number of interesting findings. When *Spirostomum* is centrifuged at
2000 $\times g$, the macronuclei and 40% of the cytoplasm are forced into one end;
the subject can then be sectioned, leaving an organism without macronuclei

Table IV. *Spirostomum* **Habituation to Mechanical Shock**
(Applewhite and Morowitz, 1966)

	No. stimuli to habituate	SD	*p*	No. stimuli to habituate	SD	*p*
Initially	3.6	1.6		10.0	3.6	
+15 sec	1.6	0.8	<0.01	3.1	2.0	<0.01
+30 sec	2.6	0.8		5.1	2.1	
+60 sec	3.2	1.1		7.4	2.2	<0.01

Table V. Habituation and Associated Phenomena in *Spirostomum:*
A Summary of the Research of Applewhite

A. Gradual and constant-intensity habituation (Applewhite *et al.*, 1969*a*)

	Percentage of organisms contracting		
	Stimulus	Gradual	Constant
	1	0	69
	10	9	—
	20	16	53
	30	20	—
	40	22	31
	50	22	—
	59	18	18
Maximal intensity for both groups	60	23	22
	65	20	18
	+45 sec	29	37
Total percentage of contractions[a]		1000	2500

B. Retention of habituation after transection (Applewhite, 1968*a*)

	Weak stimuli	Strong stimuli
Whole *S*s	9.0 ± 6.9	—
Naive anterior sections	7.0 ± 4.1	—
Naive posterior sections	9.6 ± 6.1	—
Habituated anterior sections	3.1 ± 2.9	—
Habituated posterior sections	3.6 ± 3.1	—

C. Effects of cold on habituation (Applewhite, 1968*c*)

		Mean percentage of organisms contracting		
	Stimuli	Experimental group (cold)	Habituated control	Quiet control
Acquisition	1, 2	64	61	—
	15, 16	45	47	—
	29, 30	36	36	—
Retention	1, 2	26	44	52

D. Habituation after loss of macronuclei and cytoplasm (Applewhite *et al.*, 1969*b*)

	No. stimuli to produce habituation		
	Initially	30 sec	5 min
Experimental (centrifuged)	6.5	3.5	7.0
Control	5.8	3.0	6.4

Table V. (Continued)

	E. RNA during habituation (Applewhite and Gardner, 1968)			
Stimulus	Percent Ss contracting	Counts/min habituated Ss	Counts/min in habituated controls	p
1	65 ± 9	$7,920 \pm 920$	$8,160 \pm 938$	
20	42 ± 8	$9,140 \pm 950$	$8,053 \pm 970$	0.001
50	22 ± 6	$12,676 \pm 1378$	$12,634 \pm 1300$	
Retention test 4 min later	69 ± 8	$26,036 \pm 2040$	$27,730 \pm 1981$	

	F. Protein turnover changes during habituation (Applewhite *et al.*, 1969*a*)		
Stimulus	Counts/min in habituated Ss	Counts/min in control Ss	p
20	11,220	11,840	
50	24,340	20,374	0.01

*a*Area under curve when percentage of organisms contracting is plotted against stimulus number.

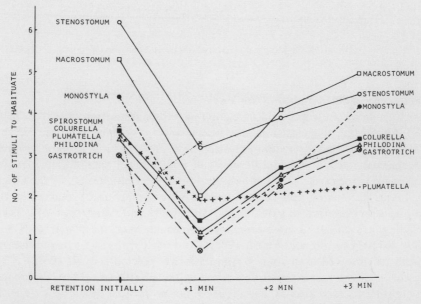

Fig. 9. Habituation to weak mechanical shock in microinvertebrates (Applewhite and Morowitz, 1967).

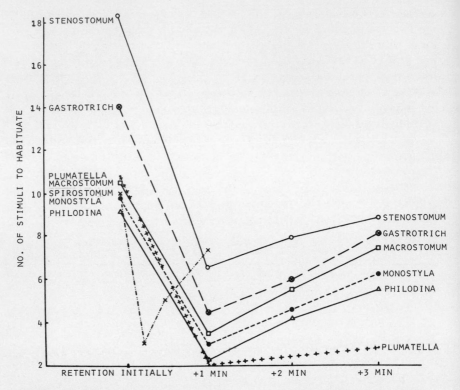

Fig. 10. Habituation to strong mechanical shock in microinvertebrates (Applewhite and Morowitz, 1967).

and some of its cytoplasm (Applewhite *et al.*, 1969*b*). Subjects prepared in such a manner could still habituate and retain this habituation for 30 sec (Table V). Applewhite suggests an infraciliature locus for cellular events subserving habituation. Measurements were also made of RNA and protein turnover during habituation. While differences in uridine-5-H^3 uptake in RNA were observed in the earlier phase of habituation (three stimulations), by the fiftieth stimulus habituated and control groups were similar and remained so for 4 min after cessation of stimulation (Applewhite and Gardner, 1968). These findings are similar to those observed with a metazoan, the planarian (Corning, 1971*a*; Corning and Freed, 1968). Conversely, protein turnover changes were not observed until after 50 stimulations (Applewhite *et al.*, 1969*a*). After removal of the macronuclei and some of the cytoplasm, RNA differences were still observed after 20 stimulations, but no protein differences were observed after 50 (Applewhite and Gardner, 1970). This indicated that protein

synthesis was not a necessary concomitant of habituation. Evidence support-
ing this was obtained in two subsequent studies. Inhibition of protein synthesis with chloramphenicol (95 % inhibition) or puromycin (92 % inhibition)
had little effect on habituation, whereas RNA inhibition with 5-fluorouracil
had a slight effect. Additional evidence against a change in protein synthesis
as being a necessary correlate of habituation was obtained in studies compar-
ing animals habituated with constant stimuli and those habituated with
stimuli of gradually increasing intensity. Although the two groups both
reached the same level of habituation, the "gradual" group showed no
differences in protein synthesis when compared with unstimulated controls
(Applewhite, 1970,a).

These various studies provide the basis for Applewhite's theory of the
cellular mechanisms underlying habituation (Applewhite and Gardner,
1971). The outline of the theory is as follows:

1. Mg^{2+} diffusion occurs during acquisition. It leaves a compartment
near the contractile elements and activates ATPase. ATPase inhibits
contraction by its hydrolytic action on ATP binding actin and
myosin.
2. Mg^{2+} active transport back into the compartment occurs during
recovery from habituation.
3. Ca^{2+} release accounts for dishabituation.

In summary, the basis for the theory is derived from the observations
that the acquisition phase is *temperature independent* (suggesting passive
diffusion of a molecular species) and *does not require protein or RNA synthesis*. Mg^{2+} is a critical candidate; it is involved in ATPase activation, its
presence decreases the rate of habituation, and Ca^{2+} is found to prevent
Mg^{2+} effects. Furthermore, an actinomyosin-like substance has been isolated
in *Spirostomum*.

During *recovery* or loss of response decrement, *Mg^{2+} is actively trans-
ported back*. Recovery is *temperature dependent*. Active transport can be
slowed at low temperatures, and in *Spirostomum* retention is improved at
5°C. For a different stimulus to be effective in the habituated animal (dishabituation), it is hypothesized that Ca^{2+} release is produced by the dishabituation stimulus allowing the return of contractability.[5]

[5]Osborn *et al.* (in press) have recently suggested that electrical and mechanical stimuli
release Ca^{2+} from different sites in *Spirostomum*. Although both stimulus modes were
demonstrated to release Ca^{2+}, mechanical stimulation led to a response decrement (habituation) whereas electrical stimulation did not. It is proposed that mechanical stimulation
is mediated by a sensory transducer in which Ca^{2+} is involved.

Table VI. Reversal Learning in Paramecia (Day and Bentley, 1911)

A. Average number of trials to reversal for the first 15 reversals

Reversal	1	2	3	4	5	6	7	8	9	10	11	12	13	14	15
Average	22.6	15.5	6.6	9.6	4.9	5.5	6.0	5.5	3.3	2.9	3.4	4.0	4.9	3.4	2.9
SE	7.6	11.3	4.6	7.7	2.9	4.6	4.8	4.0	1.9	2.1	2.3	2.3	3.6	2.3	1.9

B. Average time (sec) between reversals

Reversal	1	2	3	4	5	6	7	8	9	10	11	12	13	14	15
Average	88.6	79.3	43.1	63.6	30.6	40.4	51.4	37.8	33.9	36.8	30.5	35.8	41.9	26.3	26.5
SE	25.6	50.4	27.1	39.0	17.1	31.0	44.1	20.2	19.6	27.9	22.6	24.3	33.6	20.6	18.3

B. Associative Conditioning

Attempts to establish "associations" between stimuli in unicellular organisms have been confronted with two major difficulties. First, the CS and/or US may "condition" the environment so that the response alterations are due to tropistic relationships rather than any associative properties. Second, the reinforcing stimulation may sufficiently alter the animal's physiology so that its response to the previously neutral CS is changed independently of any CS–US pairing. The latter, usually referred to as sensitization, is considered by some as the more likely explanation of studies reporting associative learning. However, it could be argued that when dealing at the unicellular level, something like sensitization may actually be the gross physiological event that underlies system plasticity. What we observe as "learning" in the multicellular system is actually the product of a sensitization of specific cells, and the study of the nature and duration of it in protozoans can provide insights into unit activity in metazoans. Nonetheless, the question of whether protozoans are capable of establishing an association between two stimuli so that the appearance of one stimulus leads to an anticipatory response has not been settled yet, and controversy characterizes the history of this research.

Trial and Error Behavior

In his attempt to assess the range of learning capacities in paramecia, Smith (1908) observed the reactions of subjects to hitting the meniscus of a small-diameter tube ("of a bore smaller than the length of the *Paramecium* and larger than his width"). At first, the animals continually hit the meniscus, jerking backward each time. Later, the subjects succeeded in developing a "U" turn, thus reversing their orientation in the tube. These changes were considered by Smith to be "trial and error" behavior and not a tropism, since the responses varied over time and under optimum conditions there was a reduction of turning time from 5 min to 1–2 sec. In Smith's words, the organism is "avoiding any great difficulty and turning to a more easily accomplished movement." Another task examined the responses of paramecia to hot and cold areas in a tube. The organisms were placed in a tube, one half of which was heated and the other half cooled. Initially, animals might enter the heated area several times before eventually settling in the cooled half. The heated and cooled halves were then reversed several times, and Smith observed that the behavior of subjects became more regular and directed—they would move more quickly into the cold half.

In a more carefully conducted experiment, Day and Bentley (1911) obtained similar results and reported data from each subject representing the number of meniscus contacts before reversal and the time in seconds between reversals. The averages, ranges, and standard errors for these findings are summarized in Table VI and demonstrate an increased capability of reversal behavior in the tube. The possibility of an accrual of substances in the medium that would account for the changes was also considered. Subjects were transferred from the capillary tube to an open watch glass of culture medium for 10–20 min. After this period, they were returned to the tube and observed to maintain the ability to reverse.

Buytendijk (1919) objected to these findings mainly on the basis of experiments that showed the reversing or turning in the tube could be facilitated by adding chemicals to the medium. Although his studies also demonstrated an increased ability to reverse, he concluded that repeated tactile stimulation and the accumulation of chemicals could account for the improved flexibility of a paramecium. French (1940) avoided these criticisms by placing animals in a larger tube so that contact with the sides of the tube was reduced, and he also changed the culture fluid after each trial. The tube was placed into the drop containing the animal, and the animal was sucked into the tube. A small

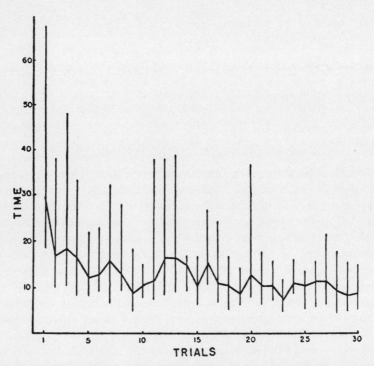

Fig. 11. Learning curve for the group of paramecia. The median time for escape from a glass tube is given in seconds. The ends of the vertical lines mark the upper and lower quartiles of the group for each of the 30 trials (French, 1940).

gap between the plate containing the drop and the end of the tube provided an escape route. Each subject was given 30 trials with 15 sec between each trial.

The median escape time is plotted in Fig. 11 and shows a modest decrement over trials. French noted that there was considerable variability between subjects and that only ten of the 20 subjects demonstrated consistent improvement. His qualitative observations confirmed that in these subjects there was directed behavior toward the bottom of the tube in the later trials.

Classical Conditioning in a Stalked Ciliate

By pairing weak light (CS) with the electric shock, Soest (1937) obtained suggestive evidence of stimulus association in both *Stentor coeruleus* and *Stentor polymorphous*. In *S. coeruleus*, the first CR appeared after five to 12 trials, although after approximately 15 trials the response modification (reaction to CS) appears to have degenerated. In *S. polymorphous*, many more trials were required to obtain a CR with the range of trials on which the first CR was elicited varying from 88 to 150. With the latter animal, it was also possible to obtain dark–shock conditioning. These data are summarized in Table VII. In these studies, as in so many others of this era, there was little attention paid to the effect of the US on light sensitivity or on the medium in which the animal was trained. This lack of adequate control data restricts interpretation and the results can only be termed "suggestive."

Classical Conditioning and "Transfer of Training" in Carchesium

The classical conditioning experiments of Plavilstchikov (1928) have received little attention in the current search for a rosetta stone in psychology. Aside from presenting data that suggests classical conditioning in a colonial ciliate *(Carchesium lachmanni)*, Plavilstchikov conducted a transplantation experiment where a portion of a conditioned colony was grafted to a "naive" host colony and observed what was termed an "induction" and "transfer" of training. With the existing interest and controversy over memory transfer in planarians and vertebrates (for reviews, see Byrne, 1970; Corning and Ratner, 1967; Gurowitz, 1969; McConnell, 1966; Ungar, 1970), it is surprising that a replication of this earlier, most provocative work has not been undertaken.

The attempts at obtaining a transfer with *Carchesium* were stimulated by previous successes with insects and earthworms. For the colonial ciliate, the basic procedure was to pair a light of a specific wavelength (red or blue), representing the CS, with a tactile stimulus provided by a rod with glass fibers on it (US). The general apparatus and viewing arrangement are re-

Table VII. Summary of Classical Conditioning Findings of Soest (1937) with *Stentor coeruleus* and *Stentor polymorphous*

A. Training *S. coeruleus* to contract to light[a]

			Animal				
1	2	3	4	5	6	7	8
8	11	8	6	12	6	5	6
+	+	+	+	+	+	+	+
+	1	+	4	+	+	1	+
1	+	2	+	2	1	+	1
+	+	+	+		+	4	+
+	3		5		+	+	5
+	+				4	+	
2	+					2	
+	+						
6	3						

B. Training *S. polymorphous* to contract to light[a]

					Animal						
1	2	3	4	5	6	7	8	9	10	11	12
108	101	88	109	121	150	108	103	90	138	123	139
+	+	+	+	+	+	+	+	+	+	+	+
+	3	10	1	5		6		20	10	+	2
10	+	+	+	+		+		+	+	2	+
+	2	4		+		1			+	+	10
	+	+		2		+				15	+
				+		8				+	2
						+					+

C. Training *S. polymorphous* to contract to cessation of illumination

					Animal						
1	2	3	4	5	6	7	8	9	10	11	12
44	91	102	51	106	49	61	50	110	100	106	95
+	+	+	+	+	+	+	+	+	+	+	+
+	7	3	1	3	+	12		8		2	7
3	+	+	+	+	2	+		+		+	+
+	3	4	+	2	+			70		8	
+	+	+	4	+	1			+		+	
2	12	75	+	+	+					1	
+	+	+	10	4	8					+	
	74	2	+	+	+					60	
	+	+	3	9						+	
	2		+	+						3	
	+		1	1						+	
				+							

[a]The numbers indicate the number of contractions elicited by shock. A plus sign indicates a contraction caused by the light.

presented in Fig. 12. CS onset was initiated by the movement of the red or blue glass across the aperture of the otherwise lightproof box positioned beneath the microscope. The change in light preceded the tactile stimulation by 1–2 sec, with the resultant excitation by the US lasting about a minute. Each colony received 20 trials/day with an intertrial interval of 15–20 min. Reactions of the colony were quantified with respect to the degree of contraction: a full or maximal contraction was recorded as 1.0, while partial contractions were ranked from 0.0 to 0.8. A "clear" response was 0.4 or higher. Once the colony began showing signs of conditioning (contracting to the light), Plavilstchikov presented trials with the CS alone interspersed with reinforced trials. After 100–200 trials, the colonies began to react to the light change. These data are summarized in Table VIII and possibly demonstrate the conditionability in most of the 82 colonies. It is reported that without the use of the US the response weakens and in several days disappears (extinction).

Transplantations were accomplished by removing a section from a host

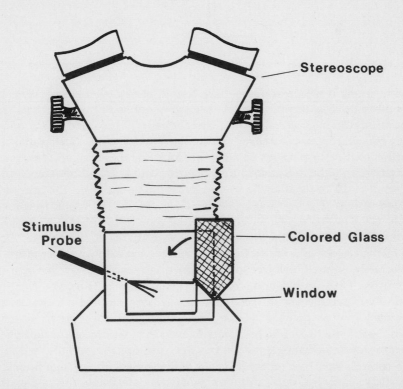

Fig. 12. General view of conditioning apparatus and viewing arrangement. The colored glass moves across the window to provide a change in wavelength (Plavilstchikov, 1928).

Host

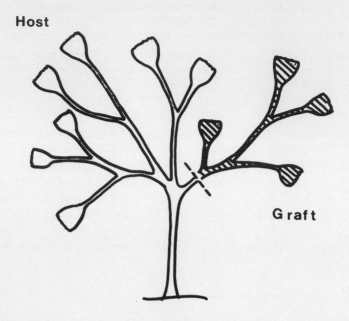

Graft

Fig. 13. *Carchesium* colony showing the graft–host arrangement (Plavilstchikov, 1928).

and replacing it with a similar section from a conditioned colony (Fig. 13). Not all transplants were successful—several were rejected, while others died. Several days after a successful transplant, the colony was tested and, astonishingly enough, began contracting during the first few trials. This initial contraction of the entire colony could be produced by the mechanical spread of excitation from the conditioned transplanted section. Plavilstchikov presented seven to 22 trials after the transplant and then removed the previously trained section. The host was tested again and found to continue to contract to the CS. Thus there was a transmission of some effect from the transplant to the host which altered the subsequent excitability of the "naive" host. To rule out injury as a reason for the increase in excitability, experiments on deliberately injured colonies demonstrated no acceleration in light responsivity. The transplantation data are summarized in Table IX. There are some obvious criticisms of these experiments. For one, the n is low for the transplantation experiment. Given Plavilstchikov's own criterion for a "clear" response (0.4 or higher), only five of the eight successful transplants demonstrated convincing signs of a transfer. Additionally, stimulus sensitization controls were not reported. These limitations blunt the significance of this study somewhat, but it still remains a potentially important finding that should be more thoroughly validated.

Table VIII. Summary of Classical Conditioning Experiments of
Plavilstchikov

Experiment No.	CS	Trial on which first CR appeared	Trial on which a maximum or strongest response occurred[a]
1	red	142	168
2	"	125	141
3	"	170	189 (0.5)
4	"	139	161
5	"	141	166
6	blue	151	187
7	"	162	198 (0.6)
8	"	174	193 (0.4)
9	"	159	196 (0.6)
10	"	171	192
11	"	149	173
12	red	138	162
13	"	112	149 (0.8)
14	"	126	151
15	"	163	192 (0.2)
16	"	119	138
17	blue	143	178
18	"	160	182 (0.4)
19	"	127	151
20	"	170	194 (0.3–0.4)
21	"	139	187
22	"	151	196
23	red	98	117
24	"	131	163 (0.8)
25	"	118	142
26	"	143	182 (0.6)
27	"	137	186 (0.4)
28	"	124	159
29	"	127	151
30	"	108	144
31	"	119	172 (0.6)
32	"	101	132
33	"	79	116 (0.8)
34	"	96	127
35	"	148	171
36	"	163	194 (0.4–0.6)
37	"	184	240 (0.1)
38	"	132	156
39	"	103	126
40	"	114	133 (0.8)
41	"	143	177 (0.2–0.3)
42	"	112	128
43	"	88	121 (0.8)
44	"	107	141 (0.6)

Table VIII. (Continued)

Experiment No.	CS	Trial on which first CR appeared	Trial on which a maximum or strongest response occurred[a]
45	red	132	164
46	″	112	died
47	″	92	124
48	″	163	200 (0.2)
49	″	142	179
50	″	128	165 (0.8)
51	″	116	141
52	″	114	152 (0.8)
53	″	128	169
54	″	86	121
55	″	117	141
56	″	133	169 (0.6)
57	″	148	200 (0.2)
58	″	139	182
59	″	141	176
60	″	152	188 (0.8)
61	″	114	151 (0.8)
62	″	170	200 (0.2)
63	″	101	139
64	″	143	184 (0.8)
65	″	137	192
66	blue	171	200 (0.1)
67	″	167	198 (0.8)
68	″	132	184 (0.6–0.7)
69	″	125	153
70	″	151	186
71	″	160	192 (0.6)
72	″	183	200 (0.1)
73	″	142	176
74	red	139	181
75	″	138	170
76	″	122	200 (0.2)
77	″	143	171 (0.8)
78	″	138	169
79	″	154	176 (0.8)
80	″	157	174
81	″	128	173
82	″	170	200 (0.1)

[a] Numbers in parentheses indicate strongest response obtained.

The conclusions of Plavilstchikov were a portent of later thinking and terminology:

Je considère surtout comme importante cette seconde série d'expériences avec le Carchesium, *car elle indique l'existence d'une particulière "induction," d'une*

Table IX. Summary of Transplantation Experiments

Subject No.	Trials on which first and maximum responses occurred during initial training (degree of contraction indicated in parentheses)	No. trials given naive colony with trained transplant	Trial on which the first CR appeared in host colony after removal of transplant (degree of contraction indicated in parentheses)
30	108/144 (1.0)	14	2 (0.6)
43	88/121 (0.8)	12	1 (1.0)
47	92/124 (1.0)	18	2 (0.8)
49	142/179 (1.0)	8	1 (0.6)
54	86/121 /(1.0)	7	1 (0.5)
64	143/184 (0.8)	22	1 (0.1)
67	167/198 (0.8)	19	3 (0.1)
78	138/169 (1.0)	12	1 (0.2)

"transmission" intracellulaire de la capacité de la réponse conditionnelle. Les réactions de la colonie "enseignée" provoquent des réactions correspondantes chez la colonie non "enseignée," une certaine "transmission" y a lieu. Pour le moment il n'est pas aise de déterminer qu'est-ce qui provoque cette "induction," quel est son mécanisme. (p. 23).

In 1928, Plavilstchikov was unable to offer possible mechanisms to explain the apparent transfer of conditioning; he probably would be surprised at the continuing ignorance in 1973.

Conditioned Approach Behavior in Paramecia?—The Gelber–Jensen Exchange

General Summary. Gelber (1952) detailed procedures and data which at first seemed to provide convincing evidence that the ciliate *Paramecium aurelia* was capable of a conditioned approach response. Basically, the conditioning technique consisted of presenting a culture with a wire coated with food reinforcement for a set number of trials and measuring the number of animals that approached the wire. In naive cultures, a bare wire elicits few attachments, but when the wire was coated with food (the bacterium *Aerobacter aerogenes*) and exposed to the culture for 40 trials it was found that there was a progressive increase in the number of animals attracted to it. In a final test, a bare wire elicited a significantly greater number of attachments when compared to attachments observed in naive cultures and when compared to a control culture which had been exposed to the wire without food.

The interpretation of these findings was disputed by Jensen (1957a), who felt that "learning" had not been demonstrated conclusively and that other interpretations were equally likely. Jensen's major points were that the repeated dipping of a wire coated with bacteria will produce a bacteria-rich area, that *Paramecium* will migrate to and remain within this area, and that the presence of bacteria will increase attachment responses (positive thigmo-tropism) in *Paramecium*. Thus the creation of a food-rich zone accounted for the increased number of animals at the point where the wire was dipped rather than an "association" between wire and food.

If Jensen's interpretations are correct, then the homogenization of the culture fluid prior to the final test with the sterile wire should provide a critical test. Katz and Deterline (1958) carried out this test and found that the approach response to the wire in "trained" cultures disappeared. This ex-periment would appear to have settled the issue, but subsequent work of Gelber has kept viable the possibility of approach conditioning in paramecia.

Outline of Gelber's (1952) Initial Experiment

1. Subjects and culturing conditions:

> it was desirable to avoid the reorganization of nuclear material that occurs in con-jugation and autogamy . . . . By using experimental cultures which are of a single-mating type (i.e., will not conjugate with one another), conjugation can be eliminated. Autogamy, however, will occur in fully fed cultures after a certain number of fissions. Under starvation, which inhibits fission, autogamy also occurs after sufficient time. A special study of the particular strain of *P. aurelia* used for this experiment showed that autogamy occurred in fully fed cultures every 16 to 22 fissions. Moreover, macronuclear rests (fragments of the old macronucleus) re-main in the cytoplasm during the 3 fissions following autogamy. By the fourth fission, these rests are usually completely resorbed into the cytoplasm, which may then be assumed to be in a relatively stable condition until the next autogamy . . . . Thus, uniformity can be approached by using experimental cultures which have started from the propagation of individuals past the third postautogamous fission and by using these cultures before they have completed the sixteenth fission. (Gelber, 1952, p. 59)

Each culture was derived from a single animal. A pair of cultures, experimental and control, were grown in parallel.

2. Apparatus: The cultures were grown and tested in depression slides; during the testing, the slides were placed on the stage of a microscope. Substage lighting was provided by a 25 w lamp. The wire was plati-num and positioned so that when it was lowered the tip was vertical and touching bottom in the center of the well.

3. Procedures:

   a. Test trials: A test trial was a 3 min exposure to a bare wire; the number of animals *attached* to the wire was counted.
   b. Training trials: A training trial consisted of a 15 sec exposure to the wire coated with the bacteria. The intertrial interval was 25 sec; the food reinforcement was put on the wire every third trial.

   During training trials, the number of animals entering a 3 mm dia-meter circle around the wire was counted.

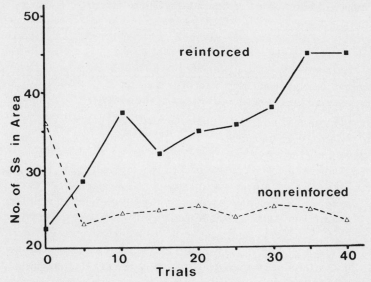

Fig. 14. Number of animals in 3 mm area around wire on every fifth trial during training period. Zero trial shows mean activity of animals before training (Gelber, 1952).

c. Treatments:

Groups	Initial	Training	Final
I (6 cultures)[6]	1 test trial	40 training trials	1 test trial
II (6 cultures)	1 test trial	no trials	1 test trial
III (6 cultures)	1 test trial	40 training trials with no food on wire	1 test trial
IV (6 cultures)	1 test trial	no trials	1 test trial

4. Results: In Fig. 14, a comparison of groups I and III during training trials is summarized. It can be readily seen that food reinforcement increased the number of subjects entering the 3 mm diameter area around the wire during training. Counts during test trials also demonstrated a significant increase in wire attachments in the trained group (group I) and a decided difference in attachments between group I and groups II–IV. These findings are summarized in Table X.

Gelber was cautious in her conclusions, stating that the reinforcement plus the wire produced a difference in behavior not obtained with the wire alone. However, the use of the word *training* and an introductory discussion of learning in paramecia leave little doubt as to the implications of this study. Jensen (1957a,b) took issue with these implications, and, using data from his own experiments and others, notably Jennings (1906), he offered

[6]Each culture contained approximately 128 *P. aurelia*.

Table X. Mean Number of Attachments Before and After Treatments in All Groups (Gelber, 1952)[a]

	Group I	Group II	Group III	Group IV
Before	2.00	1.33	0.33	1.67
After	10.75	4.00	3.58	3.92

[a]This is a simplified version of Gelber's original table.

alternative explanations for Gelber's results. The exchange, which appeared in *Science,* is summarized in the following section.

The Gelber–Jensen Exchange

1. Jensen (1957*a*):
   a. Gelber's reinforcement procedure created bacteria-rich areas which would attract paramecia. To test this, Gelber's reinforcement procedure was repeated in a pool of distilled water; a wire coated with bacteria was repeatedly lowered into the center of the pool. Samples of fluid were removed from the center and periphery of the pool, and it was found that there was a higher concentration of bacteria in samples drawn from the center than in those drawn from the periphery.
   b. Paramecia will collect in a bacteria-rich area. A 1 ml sample of culture fluid was divided in half. To one half, a drop of fluid containing reinforcement was added and to the other half a drop of distilled water. The pools were joined by a fluid bridge. Prior to joining, there were 40 paramecia in the drop which received reinforcement fluid and 35 in the drop receiving distilled water; after joining, there were 62 paramecia in the drop containing reinforcement and only 13 in the other (see Fig. 15 for a summary of this experiment and Gelber's replication attempt).
   c. The presence of bacteria will increase positive thigmotropic responses. The measure of attachment was the number of subjects that were on the bottom and immobile.[7] A culture was divided into six equal pools; two drops of reinforcement fluid were placed in each of three pools and two drops of distilled water in each of the others. Twelve minutes later, it was observed that there was a significant increase in the number of immobile animals in the cultures receiving reinforcement fluid. This remained higher for the 1 hr testing period.
2. Gelber's (1957*b*) reply:
   a. Wire attachment will occur when food reinforcement is inserted elsewhere in the culture drop. To obviate the possibility of a

[7]This particular experiment does not appear to be a very direct means of measuring "attachment." "Immobility" would seem to be a more accurate description of what was actually measured.

bacteria-rich zone being the only reason for the congregation of paramecia in the center of the pool, a training procedure was carried out with the wire being lowered into the center of the pool and with the *reinforcement being placed at the edge*. A control culture received food but no wire. After 30 min, a bare wire was introduced into both cultures, and the culture that was exposed to food plus the wire demonstrated a greater number of wire attachments than the culture receiving the food alone.

b. Frequency of exposure of wire–food is more critical than amount of food:

two wires were used, one 3 times the diameter of the other. On the larger wire, 3 times as many wipes of bacteria were applied as on the smaller, but the smaller wire was immersed in the paramecium culture 3 times as often, with shorter time intervals between immersions. Total duration of training period, amount of food, and area of wire exposed were equated for both groups, but the time of exposure to wire was 3 times as long in one group as in the other. (p. 1340)

Gelber found that the group receiving the greater frequency of exposure was significantly higher in attachment counts than the group receiving the larger wire but shorter exposure.

c. In all experiments, the culture fluid is stirred by rotating the slide before the final tests.

d. Immobile behavior is elicited by bare wire as well as by wire with food. The "lying on the bottom" behavior is typical; what is not typical is the attachment to a clean wire which is observed during the testing sessions in the trained cultures.

e. Paramecia do not necessarily approach a bacteria-rich area. Gelber attempted to replicate Jensen's experiments [see above under Jensen (1957a)] concerned with the approach of paramecia to bacteria-rich pools and obtained the opposite results. For a summary of these experiments, see Fig. 15.

f. Additional comments: The experiments of Jensen had more bacteria, and an increased concentration of bacteria can affect behavior. The addition of distilled water also has an effect; it will delay fission. Finally, the location of bacteria in a live culture where animals are moving around cannot be compared with a pool of distilled water into which bacteria are put.

3. Jensen's rebuttal (1957b):

a. Animals attached to wire may create an acid-rich zone which can trap them. With reference to Gelber's experiment (see section 2 above) in which the position of the wire and the location of the reinforcement were dissociated, Jensen agrees that this experiment

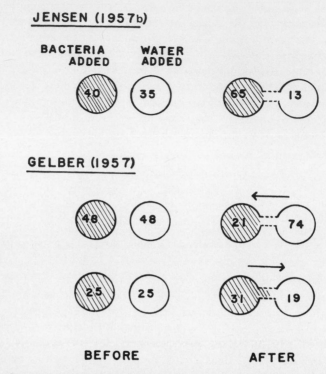

Fig. 15. Preference of paramecia for bacteria-rich areas. Jensen (1957b) found that *Paramecium aurelia* will collect in the bacteria-rich (shaded) drop when they are joined; Gelber (1957b) was unable to replicate these results and, in fact, obtained the opposite results when the "bridge" between the drops was drawn from the no-bacteria to the bacteria-rich culture.

rules out explanation in terms of bacterial concentrations. However, paramecia that become attached to the needle will produce by their metabolism an acid-rich zone, and this is known to restrict paramecia to the zone.[8]

b. Rotation of the slide is a difficult way to mix the fluid. The part of the pool at the bottom and center was not easily mixed; it was this area where Gelber's wire was positioned.

c. A replication of Gelber's "frequency of exposure" experiment (see 2b above) introduced an unequal concentration of bacteria. Jensen found that 70% more bacteria were introduced by the smaller wire that was introduced more frequently.

d. Gelber has no proof that her procedures introduced less bacteria in the area around the wire than Jensen's techniques.

[8]Why the animals would attach to the needle in the first place is not explained.

A Critical Test of Gelber's Experiment. A replication of Gelber's initial study was attempted by Katz and Deterline (1958). The basic experimental design was similar, but there was an additional experimental manipulation: Prior to a final test of retention with a bare wire, the cultures were thoroughly stirred to remove any gradients due to food or to any metabolic products of the paramecia. The procedures and groups are presented below.

Experimental and control groups. The 20 samples were distributed into four groups of 5 each.

Group I corresponds to Gelber's "reinforced" group . . . . The wire was lowered into the sample for a series of 40 trials. Each "trial" consisted in immersing the tip of the wire for 15 sec, and then raising it for 25 sec . . . . Before the first "reinforcement" trial, the sterilized platinum wire tip was lowered into the culture medium for 3 min., and an "initial" count was taken of the number of paramecia adhering to the wire, and the number present within a radius of 3 mm about the wire. In the course of the "reinforcement" procedures, counts were taken during the last of each five "trials." . . . At the end of "Trial" 40 (during which a count was made), the wire was raised, flamed, and washed before being lowered into the well for 3 additional minutes and a "final" count. After the "final" count, the wire tip was removed from the well for flaming, while the well fluid was stirred by a stream of air blown onto the surface through a micro-pipette. The wire tip was then lowered, and after another 3 min. the "post-stir" count was taken.

Group II is comparable with Gelber's "nonreinforced, trained" cultures. The same procedure was followed as in Group I, except that no bacteria were introduced at any time. The number and spacing of wire immersions and of counts were the same.

Group III parallels Gelber's "control" cultures. The "initial," "final," and "post-stir" counts were collected as in Group I and II, and a count was taken 26 min. after the "initial" count (a period of time equal to that required for the 40-trial series).

Group IV was treated like Group III except that a small drop of the bacterial mixture was lowered by means of the wire at the end of the Trial 40 count. Three minutes later the "final" count was taken, with stirring and a "post-stir" count following in sequence. (p. 244)

Gelber's acquisition curves were replicated by Katz and Deterline; in fact, there was a better experimental–control group differentiation than that obtained by Gelber. The reinforcement procedures produced a steady increase in the number of animals approaching the wire. What damages Gelber's position are the data obtained from the counts made after stirring. When the "trained" culture was stirred and then retested, the number of approaches dropped markedly and was no different from the approach frequency observed in controls. Furthermore, in group IV it was observed that the lowering of bacteria into the culture, at trial 40 only, increased the number of approaches during the final count even though this group had never experienced "training" trials of wire–food associations.

While the Katz and Deterline data cast doubts on the learning interpretations of Gelber, they were not entirely conclusive, because of some pro-

Table XI. Reference Summary for Gelber–Jensen Controversy

	A. Gelber's original research
Gelber (1952)	*P. aurelia* shows an increased attraction to a sterile wire that had previously been associated with food.
Gelber (1954)	Young cultures are able to show more wire-approach behavior with 15 trials than older cultures.
Gelber (1956a)	Behavior modification is reduced after conjugation but improved after fissioning.
Gelber (1956b)	Animals trained to the wire-approach task in light show more wire approach when tested in the light than when tested in the dark.
Gelber and Rasch (1956)	Percentage of autogamy in a culture has no effect on wire-approach training up to 21 fissions after previous autogamy.
Gelber (1957a)	Food alone does not increase wire approaching.
Gelber (1958a)	Retention after 3 hr with spaced training but not with massed trial training.
Gelber (1958b)	The trained wire-approach behavior can be extinguished.
Gelber (1962a,b)	Discusses cellular mechanisms underlying behavior.
	B. Debate
Jensen (1955)	Unpublished M.A. thesis providing data for alternative interpretations.
Jensen (1957a)	See text for summary.
Gelber (1957b)	See text for summary.
Jensen (1957b)	See text for summary.
Kellogg (1958)	Argues for the ubiquity of learning at all phylogenetic levels.
	C. Follow-up research: negative
Katz and Deterline (1958)	Food alone will increase a wire-approaching tendency.
	D. Summaries and critical reviews
Gelber (1965)	Poskocil (1966)
Jensen (1965)	Corning (1971a)
McConnell (1966)	Eisenstein (1967)

cedural deviations; for example, the "stirring" procedure itself could have been sufficiently disruptive (McConnell, 1966; Poskocil, 1966). Additionally, Katz and Deterline only counted animals within a 3 mm radius rather than including wire-attachment frequencies. In a summary of her research, Gelber (1965) reports that mixing of her cultures had been carried out and that when trained cultures were washed and placed in a clean medium the animals still demonstrated retention. The various studies relevant to this debate are listed in Table XI.

Simple Avoidance Learning

Smith (1908) attempted to delineate the "limits of educability" in paramecia with an interest in "that modification supposedly due to a re-

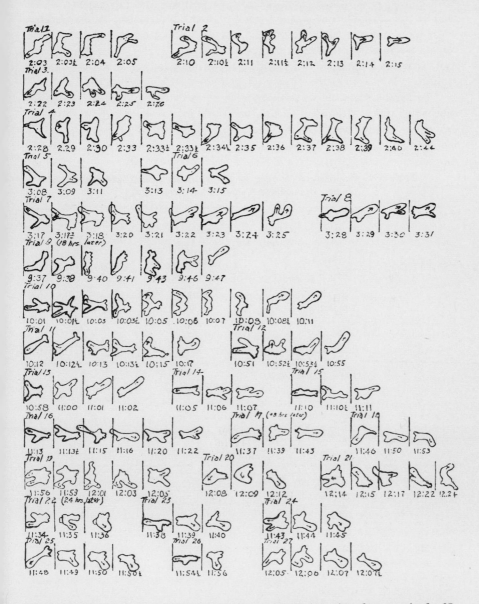

Fig. 16. A series of diagrammatic sketches illustrating the behavior of an amoeba for 27 trials. The straight lines of equal length represent a band of light. The outlines of the amoeba describe its position in relation to the light at the time given below in each individual sketch. The arrows show the direction of protoplasmic streaming (Mast and Pusch, 1924).

arrangement of structure more suitable to perform the movement, which
may be called adaptation through practice, and which is usually characterized by more rapid or exact movement" (p. 505). Smith obtained what can be
termed a simple avoidance in paramecia but was unable to demonstrate a
more complex associative conditioning, as will be discussed in the subsequent
section.

To demonstrate simple avoidance, paramecia were placed in a capillary
tube which was large enough so that they could turn around without touching
the sides. Heating and cooling tubes were arranged so that either end of
the capillary tube could be held at a high or low temperature. At first, the

**Table XII. Response of Amoeba to Contact with a High Illuminated
Region in the Field (Mast and Pusch, 1924)**

Designation of trials	Number of pseudopodia that came in contact with light in each trial					Total contacts for all individuals in each trial	Total contacts for all individuals in groups of three consecutive trials
	B	D	E	F	G		
1	4	0	0	0	1	5	
2	3	0	3	0	0	6	
3	3	3	1	0	0	7	18
4	0	0	7	0	0	7	
5	3	1	0	0	0	4	
6	0	1	0	0	0	1	13
7	1	0	2	0	1	4	
8	2	0	0	2	4	8	
9	0	0	2	1	2	5	17
10	0	0	4	0	1	5	
11	3	0	2	2	2	9	
12	0	0	0	0	0	0	14
13	0	0	0	1	2	3	
14	1	0	0	0	4	5	
15	0	0	1	2	1	4	12
16	1	0	1	0	1	3	
17	0	0	0	0	5	5	
18	0	0	0	0	0	0	8
19	0	0	1	0	0	1	
20	0	0	0	0	1	1	
21	2	0	2	0	0	4	6
22	0	0	0	1	0	1	
23	1	0	0	0	3	4	
24	0	0	0	0	3	3	8
25	4	0	0	0	1	5	
26	1	0	0	0	0	1	
27	0	0	0	0	0	0	6

animals that had contacted the heated end would back away but continue to return to that end. Eventually, they remained at the cool end. At this point, the temperature gradient was reversed. Continued reversals led to movements which were slower and more regulated, with subjects orienting more readily toward the cold end. According to Smith, the random movements initially observed gave way to more determined ones.

A similar modification was also reported for *Amoeba proteus* (Mast and Pusch, 1924). When an advancing pseudopod pushes into an area of intense light, it stops advancing. Other pseudopodia extend and stop, until finally one is extended in another direction, a response modification that is possibly dependent on the preceding experience with light. Each amoeba was observed until it moved away from the light, and the number of contacts with the light border was recorded. Animals received 27 trials with a 3 min interval between trials. The total behavior of one subject is presented in Fig. 16, and a summary of the five subjects is included in Table XII. These data indicate that there is a decreasing number of contacts with the edge of the lighted area, a modification that Mast and Pusch considered similar to learning in higher animals.

Associative Avoidance Learning

If one had to select a single study that was most instrumental in stimulating interest in protozoan learning, it would have to be the report published by Fritz Bramstedt in 1935 purporting to demonstrate avoidance conditioning in *P. caudatum* and *Stylonychia mytilus*. An earlier attempt by Smith (1908), although a failure, provided a basic paradigm utilized by Bramstedt and by a number of others. Smith placed a number of animals in a trough which was located in a water bath with two compartments. One half of the bath was heated, while the other was kept cool (see Fig. 17A). The heated half of the tube was kept dark. Since paramecia react negatively to heat, the experiment was an attempt to obtain an association between darkness and heat. After 40 hr of exposure to these conditions in the tube, subjects still maintained neutrality with respect to darkness. Smith also tried to obtain a gravity–heat association by using the arrangement shown in Fig. 17B. Subjects would swim into the vertical portion of the capillary tubing, receive the heat stimulus, and then swim downward. According to Smith, after 3 days the animals never learned to avoid the vertical portion. Bramstedt criticized Smith's attempt mainly on the basis of his mass conditioning[9] procedure and the use of a large trough. This procedure, according to Bramstedt, was unreliable because by trying to observe several subjects at once individu-

[9]*Mass conditioning* as used in this chapter refers to the attempt to train several subjects at once in a common container.

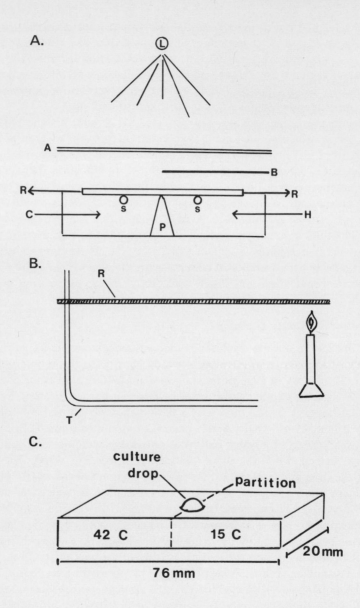

Fig. 17. Apparatus used by Smith and Bramstedt to test avoidance conditioning. A, Diagram of Smith's (1908) apparatus (A, heat screen; B, light screen; C, cold-water supply; H, hot-water supply; L, electric light; P, partition; R, overflow return pipes; S, supports for trough; T, trough). B, Apparatus used to test gravity–heat associative conditioning in paramecia (Smith, 1908) (T, tube in which subjects were placed; R, brass rod providing heat to tube). C, Schematic representation of Bramstedt's experimental arrangement.

al behavioral alterations were missed. Furthermore, Bramstedt claimed that the vessel in which the subject must acquire a response should be small enough so that it can be "comprehended" as a whole.

Bramstedt used the arrangement portrayed in Fig. 17C. The subject was put in a water drop placed on the dividing line of a two-compartment chamber in which one half was heated to 42°C while the other half was held at 15°C. In addition, the warm half was exposed to a light while the cold portion was kept dark. At first, a subject would spend an equal amount of time in both halves, but by the end of the first hour it would be on the dark side for most of the time. A representative result is summarized in Table XIIIA. Following the 90 min exposure to the temperature gradient, the temperature of the two halves was equalized (both halves at 15°C), and the subjects' preferences were then measured. The light avoidance was found to persist. The results with 11 subjects are summarized in Table XIIIB.

Table XIII. Light Preferences in a Paramecium During and After Training (Bramstedt, 1935)

A. During training[a] : light–warm/dark–cold	
Observation No. (each observation period is 3 min)	Time in light (sec)
1	95
2	25
3	21.5
4	16
5	4.2
6	0
7	0
8	0
9	5.1
10	2.1

B. After training: light–cold/dark–cold		
Subject No.	Duration of observation (sec)	Time in light (sec)
1	1440	0
2	660	11
3	1380	65
4	900	60
5	1500	26
6	1380	17
7	660	25
8	360	60
9	660	24
10	840	38
11	780	0

[a] Total training time 60 min (approximately).

As controls, Bramstedt was concerned about light–dark preferences in naive animals, the possibility of position learning confounding the data, and the possibility of heat-induced photosensitivity. The data summarized in Table XIVA clearly indicate that *P. caudatum* had no initial light–dark preference. When animals were exposed to a lighted drop that had one side heated and the other cooled, they tended to remain on the previously cool side when the temperature was equalized in both halves (Table XIVB). Bramstedt rules this out as an explanation for his conditioning results, since the position preference is smaller than the preference obtained after light–heat/dark–cold conditioning. To test the possibility of heat-induced photosensitization, Bramstedt measured the light–dark preferences of subjects that had

Table XIV. Control Studies of Bramstedt (1935)

A. Light–dark preferences in naive paramecia at 15°C

Subject No.	Time in light (sec)	Time in dark (sec)
1	132	100
2	176	94
3	90	186
4	96	118
5	97	146
6	117	74
	708	718

B. Position preferences in drop after exposure to a temperature gradient

Subject No.	Time in previously warm half (sec)	Time in constantly cold half (sec)
1	167	433
2	600	600
3	273	387
4	213	417
5	184	481
	1437	2218

C. Light–dark preferences after exposure to heated drop

Subject No.	Time (min) of exposure to heated drop	Observation period	
		Time on dark side (sec)	Time on light side (sec)
1	30	458	310
2	60	379	341
3	60	439	434
4	70	533	641
5	50	660	600
		2469	2326

been kept in a heated container for 30–70 min and found no particular shift in preference (Table XIVC).

An apparent associative conditioning between heat (US) and vibration (CS) was also reported. When a heated tube was lowered into a petri dish, subjects demonstrated an avoidance reaction. Repetitive lowering (1/min) resulted in some habituation of the avoiding response. Two hours later, a cold tube was found to elicit an avoidance reaction which was extinguished after eight to ten immersions.

Experiments with *Stylonychia mytilus* again relied on a light–dark discrimination but used a rough surface as a negative reinforcement. *Stylonychia* avoids rough surfaces, and when placed on a slide glass where one half is rough and the other smooth the animal is found to prefer the smooth portion. For example, in a 593 sec observation period one animal spent only 83 sec on the rough side. To condition the animals, Bramstedt kept the smooth portion dark and the rough portion lighted. After being exposed to this treatment, animals were tested for light–dark preferences on a smooth surface and were found to prefer the darkened half, presumably because they associated light with a negative stimulation.

Soest (1937) reported success in establishing associations in a number of different species. In *P. caudatum*, he initially demonstrated that naive subjects displayed no light–dark preferences and then proceeded to replicate Bramstedt's (1935) basic findings using light (CS) and electric shock (US). As in the previous studies, subjects were placed in a drop that had light and dark halves. Each time they entered the lighted portion, they were shocked, and, over trials, their avoidance of the lighted half increased (Table XV). Soest failed to obtain a dark avoidance with electric shock as the US. When cold was used as a negative reinforcement, he failed to observe light as well as dark avoidance. *Stylonychia mytilus* could be conditioned when a rough surface was used as the CS. When subjects crawled onto a roughened area, they were shocked and an increased avoidance of the area was obtained. The use of roughness is questionable in *Stylonychia*, since it had been shown that the organism will normally avoid rough surfaces (Bramstedt, 1935). Soest was unable to develop an avoidance of a smooth surface. Successful light–shock avoidance conditioning was also reported for *Spirostomum, Stentor coeruleus*, and *Stentor polymorphous*. Dark avoidance conditioning was only reported for *Stentor polymorphous*. Soest's data are summarized in Table XVI.

Grabowski (1939) attempted a careful replication of the "learning" demonstrations of Bramstedt. Initial control work indicated that at cold temperatures paramecia displayed little preference toward the lighted or darkened halves of a well. At room temperature (19.8°C), a slight negative phototaxis was observed. Repeating Bramstedt's basic stimulus conditions (one side of the well was lighted and heated; the other side was dark and

Table XV. Light-Avoidance Training in _P. caudatum_ (Soest, 1937)

	A. Example of light-avoidance training using electric shock			
Time in training	Cumulative number of contacts with light –dark boundary	Number of crossings of the boundary without avoiding reaction	Number of avoiding reactions at boundary	Avoiding reactions at boundary (%)
6	10	10	0	0
14	20	10	0	0
20	30	10	0	0
25	40	10	0	0
32	50	10	0	0
38	60	10	0	0
43	70	10	0	0
48	80	9	1	10
54	90	9	1	10
62	100	10	0	0
71	110	7	3	30
76	120	4	6	60
Shock terminated				
85	130	1	9	90
93	140	3	7	70
99	150	8	2	20
108	160	10	0	0

B. Time spent in light and dark immediately after training in five subjects

	Time (sec)	
Animal	Light	Dark
1	95	685
2	122	617
3	72	644
4	113	587
5	58	521
	460	3054

Table XVI. Reported Success (+) and Failure (−) of Soest (1937) with Associative Conditioning in Various Protozoans[a]

	Light–shock	Dark–shock	Roughness–vibration	Roughness–shock	Smooth–vibration	Light–vibration
Paramecium caudatum	+	−	0	0	0	0
Stylonychia mytilus	0	0	+	+	0	0
Spirostomum	+	0	0	0	−	+
Stentor coeruleus	+	0	0	0	0	0
Stentor polymorphous	+	+	0	0	0	0

[a]A zero indicates that no attempt was made.

cold), Grabowski observed similar behavior—during test sessions the organisms spent less time in the lighted half even though the temperature throughout the well was equilibrated and cool. However, Grabowski noted that the animals appeared to be reacting to a different kind of border rather than the light–dark border. The subjects were displaying a reaction *after* they passed the light–dark division, and Grabowski concluded that the avoidance-producing stimulus could not be light. The time spent in each half during testing is included in Table XVII along with a schematic indication of when the animals were showing what Grabowski calls a "fright" reaction.

It was suggested that the paramecia were not reacting to light but to a chemical gradient that had been set up by the stimulus conditions and which persisted for a time during testing, e.g., an oxygen deficiency or carbon dioxide build-up. After 55 min, the subjects were still spending less time in the lighted half and further testing did not change this, as the animals tended to "stick" at a certain point. Grabowski argued that if the animals were "forgetting" the light border during the test, the change in direction would be more casual and not a shifting line. The point at which the paramecia showed a fright reaction was precise and shifted slowly to the left (lighted half of the well).

A critical test of the chemical-gradient hypothesis consisted of placing naive subjects into a well that contained previously trained animals. These subjects behaved similarly to the trained ones (see Table XVIII). Furthermore

Table XVII. Training: Light–Heat/Dark–Cold; Preference
Testing: Light–Cold/Dark–Cold; Training from 10^{35} to 12^{15a}

| Time | Preference test | | "Fright reaction" | |
	Light–cold	Dark–cold	Light	Dark
0–5′	0′07″	4′53″		
5–10′	0′21″	4′39″		
10–15′	0′25″	4′35″		
15–20′	1′24″	3′36″		
20–25′	1′11″	3′49″		
25–30′	0′49″	4′11″		
30–35′	1′37″	3′23″		
35–40′	1′32″	3′28″		
40–45′	2′15″	2′45″		
45–50′	1′04″	3′56″		
50–55′	1′56″	3′04″		
55–60′	2′46″	2′14″		

[a]Dots under "Fright reaction" indicate place where subjects showed an aversion or avoidance response.

Table XVIII. Time Spent in Lighted Side by Naive Animals Placed Into a Well in Which "Training" of Other Animals Had Been Carried Out (Grabowski, 1939)

	Animal				
	1	2	3	4	M
0– 5'	0'31"	0'00"	0'16"	0'12"	0'15"
5–10'	1'30"	0'00"	0'34"	1'24"	0'52"
10–15'	1'20"	0'16"	0'44"	1'25"	0'56"
15–20'	2'40"	0'59"	0'48"	2'01"	1'37"
20–25'	—	1'03"	1'03"	2'16"	1'36"
25–30'	—	1'34"	1'46"	—	1'40"
30–35'	—	2'01"	1'25"	—	1'43"
35–40'	—	1'47"	1'54"	—	1'51"
40–45'	—	2'15"	1'59"	—	2'07"
45–50'	—	—	2'09"	—	2'09"
50–55'	—	—	1'35"	—	1'35"
55–60'	—	—	2'36"	—	2'36"

if this type of learning were possible in paramecia, then a reversed stimulus condition should work, i.e., dark–heat and light–cold halves. Both Grabowski and Bramstedt failed to observe this reversal. A further control test placed subjects in a well that had one half heated and the other half cooled but with both sides lighted. During testing, subjects were exposed to a light–cold and dark–cold choice and were found to avoid the previously heated half. These and other controls served to cast considerable doubt on Bramstedt's assumption of a light–heat avoidance learning. Additional evidence that the purported associative conditioning demonstrations were due to changes in the medium was provided by Sgonina (1939). When a culture drop (without subjects) was shocked, it was found that when subjects were later placed in the drop they avoided the lighted half. A similar argument was made by Grabowski (1939) against the avoidance learning reported by Tschakhotine (1938). Tschakhotine had observed that *P. caudatum* avoided a microbeam of ultraviolet light in a culture drop and would continue to avoid the area where the beam was after it had been turned off. It is highly possible that the physicochemical changes in the irradiated area were sufficient to drive paramecia off.

The ensuing debate in the late 1930s over protozoan learning involved a number of investigators, including Alverdes and Koehler, and led to a number of exchanges in the literature as well as to attempts to replicate (see Table XIX). Although it is possible to point to procedural differences that could possibly account for replication failures, the earlier studies are characterized by a lack of systematic study of variables that could account for laboratory differences. Alverdes (1937a,b, 1939) commented on the fact that

Table XIX. Reference Summary for Avoidance-Conditioning Controversy
in Protozoans

Citation	Reported success (+) or failure (−)	
A. Basic demonstrations		
	+	−
Smith (1908)		x
Bramstedt (1935)	x	
Wawrzyńczyk (1937b)	x	
Soest (1937)	x	
Tschakhotine (1938)	x	
Grabowski (1939)		x
Sgonina (1939)		x
Diebschlag (1940)	x	
Dembowski (1950)		x
Best (1954)		x
Dabrowska (1956)		x
Chailakhian (1957)		x
Mirsky and Katz (1958)		x
Machemer (1966 a, b)		x
Bergström (1968 a, b; 1969)	x	
Applewhite et al. (1971)		x
B. Commentary and debate		
Alverdes (1937 a,b, 1939)		
Koehler (1939)		
Wichterman (1953)		
Kellogg (1958)		
Thorpe (1963)		
Poskocil (1966)		
Corning (1971a)		

the more distant an animal is from the level of man, the more difficult it is to
define the appropriate experimental conditions in which the organism can
display its true capacities. The exchanges and discussions between Bramstedt
(a student of Alverdes), Grabowski (a student of Koehler), and others failed
to come up with any clear reasons for failures to replicate—there were those
who could demonstrate associative learning and those who could not.

In an effort to define factors that were operative in the Bramstedt–
Grabowski studies, Diebschlag (1940) conducted a replication of some of
the critical parameters used by both investigators. With respect to Grabowski's adoption of 50°C (Bramstedt used 42°C) as a negative stimulus, it
was found that this temperature produces changes in the medium that can
affect behavior. A culture drop (without animals) was subjected to Grabowski's "training" condition for 2 hr; i.e., the hot side was 50°C and the
cold side 15°C. The temperatures were then equalized and the subjects placed
in the drop. Animals were observed to display negative reactions whether
they were in the lighted or darkened halves, and Diebschlag concluded that

the higher temperature used by Grabowski produced chemical changes in the drop and that at 42°C such changes are not observed. This is not a very strong argument, since the temperature gradient (15–42°C) of Bramstedt was sufficient to produce a gas difference (Dembowski, 1950). However, Diebschlag analyzed the effects of differing culture solutions and found the hay infusion an inadequate medium at high temperatures, while normal behavior could be maintained by using "Liebig's extract solution" (Dieb-schlag's citation for this is Schlieper, 1940). In this solution, the negative reactions of subjects were fewer, and they appeared to move about more freely. With the heated side held at 38°C, the hay infusion was found to be a safe medium.

Diebschlag replicated Bramstedt's experiment but altered both the temperature and the medium to minimize sensitization of subjects (the temperature was 38°C and Liebig's solution was used). He failed to confirm Grabowski's finding that when naive subjects that had been subjected to "training" were put in the drop they avoided light. After training, the light–dark boundary was moved, and the subjects would respond to the new boundary, not to the place where the old boundary had been. As for the sensitization criticisms of Koehler (1939) and Sgonina (1939), Diebschlag conducted two tests schematicized in Fig. 18. In the first (Fig. 18A), subjects were placed in a drop that was totally lighted but was divided into heated and cooled halves for a 2 hr period. The temperatures were then equalized and the animals tested for light–dark preferences. There were no preferences display-ed to either half. A similar experiment was accomplished in a drop that was totally dark (Fig. 18B). As for Sgonina's data that shock sensitizes animals to light, Diebschlag replicated this experiment and failed to confirm it. These findings are summarized in Table XX.

In an experimental apparatus similar in construction to that arrived at by Best (1954), Wawrzyńczyk (1937b) reported success in light–shock avoid-ance conditioning in *P. caudatum*. Subjects were placed in a long tube (6 cm by 3 mm) with electrodes at both ends and with a darkened center area.

Test

A.

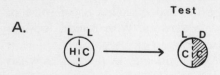

B.

Fig. 18. Schematic representation of Dieb-schlag's sensitization controls (Diebschlag, 1940).

Table XX. Replication of Sgonina's Experiment (Diebschlag, 1940)

Experimental series	Number of shocks	Duration of test (min)	Percent of Ss in Light	Dark	N	Test condition
1. Behavior of Ss before shocks	—	3	51	49	1867	light
2. Experimental treatments	30	33	45.3	54.7	389	light
(Ss shocked in drop)	30	33	52.3	47.7	367	light
	60	33	49.2	50.8	305	light
	30	33	52.9	47.1	925	dark
3. Control test (water was						
shocked with no animals;	60	35	52.0	48	394	light
animals were then put in	60	35	49.7	50.3	1055	light
and counts made after 5	60	35	50.2	49.8	1009	dark
min)						

Table XXI. Avoidance Conditioning Using Red and Blue Light in
Paramecium caudatum (Wawrzyńczyk, 1937b)

Light	Subject No. 1	2	3	4	5	6	7	8	9	10	Totals
Red (+)	66^a	52	73	66	90	85	89	109	73	85	788 = 89%
vs.											
Blue (−)	7	20	11	4	9	5	14	18	4	9	101 = 11%
Blue (+)	45	27	24	28	23	16	26	28	23	17	257 = 86%
vs.											
Red (−)	4	6	2	4	4	2	9	4	3	4	42	14%

[a]Data represent time in minutes during training that the subjects stayed in the red or blue
areas of the tube.

When the subjects moved from the central part, they were shocked; after
training, it was found that they remained in this portion of the tube approximately 80% of the time. The light avoidance persisted for 50 min. Wawrzyńczyk also used this apparatus to examine finer discriminations, obtaining
avoidance to blue when the central area was red and vice versa. Discrimination between green (+) and white light (−) and green (+) and red (−) was
also reported. He was unable to obtain any results using green and blue.
After extinction, the conditioned avoidance was rapidly reinstated after the
first few shocks. A sample of the data is presented in Table XXI.

As with most surges of positive protozoan conditioning data, the Wawrzyńczyk and Diebschlag studies were followed by a number of negative
reports. A replication of the light–shock conditioning data of Wawrzyńczyk
demonstrated that subjects begin to turn at the border of the positive and
negative areas in the tube but went on to show that when the filters are removed or displaced, subjects will respond to the *place* where the border was

during training (Dembowski, 1950). This control, similar in concept to that of Grabowski's, suggested to Dembowski that "we have to deal not with an acquired new reaction but with the occurrence of a new stimulus probably of chemotropic nature" (Dembowski, 1950, p. 11). Further tests such as reversing stimulus conditions and changing light filters showed that the subjects stayed in the previously nonshocked area regardless of the type and degree of impinging illumination. The negative findings of Dembowski were replicated by Dabrowska (1956) in *P. caudatum, Spirostomum ambiguum,* and *Stentor coeruleus.*

Mirsky and Katz (1958) attempted to clarify the factors operating in some of the earlier debates, in particular the Bramstedt–Grabowski studies. Their groups consisted of a light control used to test for a differential response in the well, a "conditioning" group exposed to the light–heat and dark–cold halves, and an oxygen control group. The latter group was exposed to a well in which the lighted half had previously been heated and the dark half cooled. Just before the subjects were introduced, the temperature was equalized. Unlike Grabowski, they found no preferences in the controls, nor did they find any evidence of light avoidance in the "conditioning" group. Incidental observations of a convection current in the well during differential heating and cooling appeared to rule out an oxygen gradient. It is possible that Grabowski managed to avoid these convection currents in his studies.

Another test of light–heat associative learning in paramecia was performed by Best (1954). Instead of the depression slide arrangement used by Bramstedt, the paramecia were placed in a capillary tube that had a heating coil positioned around one third of it (see Fig. 19). The light source was heat filtered and was perpendicular to the tube. Any portion of the tube could be darkened by covering it with opaque material.

The rationale on which the experimental method is based will merit a brief discussion. Consider a system such as the capillary divided into segmental regions such as those indicated schematically in Fig. 19. Let

L/H denote a region that is light and warm,
D/H denote a region that is dark and warm,
L/C denote a region that is light and cool,
D/C denote a region that is dark and cool.

If a paramecium, swimming in the tube, comes in contact with boundary 1, it will turn back into the D/C region. If it has become conditioned, then it will "remember" that light (to which it is supposedly ordinarily neutral) is associated with heat. If it contacts boundary 2 on its swim back from 1 before it has "forgotten" its experience at 1, then it will turn back from boundary 2. If a protozoan turns back from boundary 2 when it is in D/C but not when it is in L/C then it is apparent that protozoans will tend to collect in D/C and the number in L/C will be less than that anticipated on the basis of a random spatial distribution of the protozoans throughout the cool region. If, on the other hand, the distance interval between 1 and 2 is sufficiently large that by the time a paramecium swims

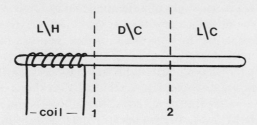

Fig. 19. Capillary tube arrangement of Best (1954) to test associative learning in paramecia.

from 1 to 2 it has forgotten what it "learned" at 1 then it will swim as freely past 2 from D/C to L/C as from L/C to D/C and one will observe a random distribution of paramecia between 1 and 2 for the protozoans, then by observing that distance of separation of 1 and 2 that causes the distribution to switch from random to nonrandom one can obtain an estimate of the duration of the "memory." (Best, 1954, pp. 91–92)

The results showed that light associated with heat did produce an avoidance of the L/C zone for a short time. However, it was also demonstrated that this light avoidance could be induced when the paramecia were exposed to a D/H zone rather than a L/H zone. Thus heat can be said to sensitize paramecia to light, or, as Best puts it, it induces a "photophobia." This "photophobia" was observed to last 30–45 sec.[10]

Recently, Bergström (1968a) has used the ciliate *Tetrahymena pyriformis* in attempts to demonstrate avoidance learning with light–shock pairings. The choice of *Tetrahymena* was based on the observation that this species had a better innate avoidance reaction than paramecia. Basically, the training paradigm was to expose animals to light–shock pairings and then to test them in a container where there were light and dark zones and no negative reinforcement. The experimental and control groups were as follows: group E, light–shock pairings; group S, shock only; group L, light only; group N, no light or shock. The groups E, S, and L were given 15 trials each with a 1 min intertrial interval. In the experimental group (E), the shock was administered 0.5 sec after the onset of light, and both light and shock were terminated simultaneously. The duration of the light was 2 sec. Each group contained approximately 1100–1200 Ss. After the treatments were given, a screen was placed over the rectangular container that had two holes in it and the light was turned on. This created two lighted wells in the animal's

[10]Machemer (1966a,b) has also reported a failure to condition *Stylonychia mytilus* and *Keronopsis rubra* using roughness and light–dark associations. He did demonstrate that shaking the culture drop could induce a sensitization to rough surfaces. We thank Dr. R. Eckert for bringing these studies to our attention.

container. The containers were photographed every 20 sec for 15 min. Thus an objective measure of the number of subjects in the lighted wells could be obtained. The results are presented in Table XXII. It can be seen that there were fewer animals in group E than controls in the light. A subsequent study (Bergström, 1968b) replicated this finding in cultures that had undergone a synchronous division. Later research also demonstrated that the retention of the avoidance survived cell division, that the avoidance persisted for almost 2 hr, and that extinction occurred (Bergström, 1969c).

These experiments present rather startling results, particularly in view

Table XXII. Quotients Between the Number of Animals in the Light Wells and the Total Number of Animals per Condition (Bergström, 1968a)

Time from start of the test period (min)	Proportion of animals in the lighted parts of the glass chamber			
	Group E	S	L	N
0	0.285	0.296	0.302	0.305
$\frac{1}{2}$	0.297	0.289	0.301	0.307
1	0.296	0.296	0.300	0.305
$1\frac{1}{2}$	0.276	0.297	0.303	0.307
2	0.276	0.294	0.312	0.309
$2\frac{1}{2}$	0.258	0.294	0.312	1.304
3	0.253	0.298	0.303	0.312
$3\frac{1}{2}$	0.237	0.295	0.297	0.302
4	0.237	0.305	0.301	0.309
$4\frac{1}{2}$	0.237	0.299	0.305	0.310
5	0.239	0.300	0.311	0.315
$5\frac{1}{2}$	0.243	0.294	0.297	0.306
6	0.234	0.304	0.309	0.313
$6\frac{1}{2}$	0.241	0.300	0.308	0.303
7	0.236	0.308	0.311	0.308
$7\frac{1}{2}$	0.239	0.316	0.319	0.303
8	0.240	0.305	0.324	0.318
$8\frac{1}{2}$	0.244	0.303	0.310	0.315
9	0.248	0.309	0.318	0.319
$9\frac{1}{2}$	0.250	0.209	0.327	0.311
10	0.251	0.306	0.334	0.317
$10\frac{1}{2}$	0.246	0.314	0.325	0.319
11	0.247	0.316	0.325	0.335
$11\frac{1}{2}$	0.245	0.310	0.319	0.319
12	0.239	0.311	0.326	0.333
$12\frac{1}{2}$	0.243	0.312	0.324	0.339
13	0.251	0.318	0.326	0.337
$13\frac{1}{2}$	0.239	0.308	0.330	0.329
14	0.250	0.308	0.318	0.326
$14\frac{1}{2}$	0.243	0.323	0.321	0.321
Total animals	5891	5608	6458	5380

of the somewhat classical avoidance-conditioning paradigm that was employed. The incorporation of a shock control group would appear to rule out a shock "sensitization" effect, but a more convincing control would be a random light and shock procedure. This would ensure equivalence between experimental and control groups with respect to the total amount of stimulus episodes. This control was included in a recent investigation (Applewhite *et al.*, 1971) that attempted a replication of Bergström's basic findings. The random group (subjects receiving unpaired light and shock) showed little evidence of light avoidance, but the "trained" group was also found to remain neutral with respect to light; *there were no significant differences between any of the experimental and control groups*. Thus there is something of a replication here, i.e., the inability of an independent laboratory to repeat an associative learning demonstration.

Other Demonstrations

Reactive Inhibition. Lepley and Rice (1952) conducted an experiment "half in jest" designed to measure reactive inhibition in paramecia. According to Hullian theory, the more a particular act is repeated, the less probable becomes a subsequent repetition of the act. The execution of a particular response generates a certain amount of inhibition for that response. When *Paramecium multimicronucleatum* individuals were placed in a T-maze, the probability of a left or right turn was 50%. If the animals were then placed in a T-maze with one or two turns before the T, the probability of choice at the T was altered. These data are summarized in Table XXIII and show that when a one-turn experience was given prior to the T the probability of an opposite choice was significantly changed ($p < 0.01$). When two turns were given prior to the T, the choice was still altered significantly, but it was less, perhaps, as the authors suggest, because of a partial negation of reactive inhibition by habit strength.

The results of Lepley and Rice were not replicated in subsequent research (Lachman and Havlena, 1962), but variations in the maze dimensions may account for this failure (Rabin and Hertzler, 1965). When the Lepley and Rice maze dimensions were used, their results were confirmed (see Table XXIIIB).

Food Selection and the Question of Learning. Investigations of the readiness of paramecia to selectively assimilate or reject various substances led Metalnikow (1913*a*,*b*, 1914, 1916) to conclude that learning was involved. For example, subjects were placed in a medium containing aluminum and 20 hr later were found to have few vacuoles. They were then placed in a fresh medium containing aluminum, and the avoidance of aluminum continued, although they would digest other nutritious materials. With carmine, the selective aversion lasted for 3 days. A more striking observation concerned

Table XXIII. Summary of "Reactive Inhibition" in Paramecia

Maze	N	Percent turning		p
		Left	Right	
A. Percentages of paramecia turning right and left and the significance of the deviations from chance expectancy (Lepley and Rice, 1952)				
T	613	50.08	49.92	n.s.
R-T	516	86.24	13.76	0.01
R-R-T	559	73.17	26.83	0.01
L-T	599	16.69	83.31	0.01
L-L-T	554	22.20	77.80	
B. Proportion of left and right turns (Rabin and Hertzler, 1965)				
T	100	51	49	0.70
R-T	100	61	39	0.05
R-R-T	100	60	40	0.05
L-T	100	38	62	0.025
L-L-T	100	38	62	0.025

the possibility of stimulus association (Metalnikow, 1912). When paramecia that had been exposed to carmine under a red light were tested under red light, they were found to form fewer vacuoles than animals exposed to yeast or sepia under the same lighting conditions.

Criticisms of these studies are summarized by Wichterman (1953) and Thorpe (1963). The major points of debate are over the possibility of a physiological depression or fatigue induced by the ingestion of nonnutritive or toxic material and over the possibility of size of the ingested particle being a factor rather than "learning." Bragg (1939) notes that if fatigue were involved then the animals should become fatigued with the continuous ingestion of bacteria. The more serious criticism is Bozler's (1924) observation that the size of the carmine particles may have been the factor in their eventual rejection and that if particles are small enough paramecia will take them in indiscriminantly. As Wichterman points out, these experiments are briefly reported and difficult to evaluate. It is clear that the issue of food selection remains unsettled.

Pattern Learning. In addition to his research on associative conditioning, Bramstedt (1935) attempted to demonstrate pattern learning in paramecia. Basically, the procedure was to place subjects in a container that was circular, square, or rectangular for a period of time. The swimming movements would eventually approximate the geometry of the container. Bramstedt reported that when the paramecia were transferred to a container of a different shape the initial behavior patterns were those of the original container.

Grabowski (1939) attributed Bramstedt's observation of a persisting swimming pattern to observer bias. He claimed that the pattern perception was a matter of the observer's preconceptions, just as it is possible to see figures in clouds and ink blots. Grabowski notes that one can pick out a circular or triangular pattern in naive subjects if one watches long enough.

VI. CONCLUSIONS

In reviewing the protozoan learning literature, it is clear that the findings are highly controversial with any training paradigm except habituation. The evidence at both the physiological and behavioral levels of analysis is compelling for a "learning not to respond" that can be differentiated from such alternatives as fatigue. From the early 1900s to the early 1970s, demonstration after demonstration has appeared without serious criticism. With improved technology with respect to measuring intracellular processes associated with changes in responsivity, there is a promise of a true "molecular biopsychological" breakthrough. Other techniques that could add power to the research have yet to be incorporated. For example, Jeon et al. (1970) have shown that in Amoeba proteus it is possible to assemble an animal from dissociated nuclei, cytoplasm, and membranes. This technique might possibly be used to localize experiential storage mechanisms or to discover interactions between components during the acquisition and retention phases of habituation.

Attempts at demonstrating simple forms of learning such as that of French (1940) have been only moderately convincing, and associative conditioning remains a highly questionable phenomenon in protozoans. It is true that a number of earlier studies such as that of Plavilstchikov (1928) have not as yet been replicated, but, based on the history of replication attempts, pessimism must prevail. As with many controversies, there are those who see it and those who do not. In many cases, it may be due to the a priori expectations or theoretical bents of the experimenter, In other cases, experimenters have attempted replications with a positive preexperimental bias and have failed [for example, Applewhite's attempt to replicate Bergström (1968a,b)]. Those who contemplate a "critical and conclusive experiment" are most likely to be disappointed, but those who enjoy controversy and a challenge are likely to be rewarded.

REFERENCES

Aaronson, S., 1963. Protozoan pharmacodynamics: The use of protozoa to study the cellular action of drugs. In Ludvik, J., Lom, J., and Vavra, J. (eds.), *Progress in Protozoology*, Academic Press, New York, pp. 175–176.
Allen, R. D., 1962. Amoeboid movement. *Sci. Am. 206*, 112–122.

Alverdes, F., 1937a. Das Lernvermögen der einzelligen Tiere. *Z. Tierpsychol., 1,* 35–38.

Alverdes, F., 1937b. Gewöhnung und Lernen in Verhalten der Tiere. *Zool. Anz., 120,* 90–95.

Alverdes, F., 1939. Zur Psychologie der niederen Tiere. *Z. Tierpsychol., 2,* 258–264.

Anderson, E., 1967. Cytoplasmic organelles and inclusions of protozoa. In Chen, T. T. (ed.), *Research in Protozoology,* New York, Pergamon Press, pp. 1–40.

Applewhite, P., 1968a. Non-local nature of habituation in a rotifer and protozoan. *Nature, 217,* 287–288.

Applewhite, P., 1968b. Temperature and habituation in a protozoan. *Nature, 219,* 91–92.

Applewhite, P., 1968c. Retention of habituation in a protozoan improved by low temperature. *Nature, 219,* 1265–1266.

Applewhite, P., 1970a. Habituation in *Spirostomum.* In Adler, J. (ed.), *10th International Congress for Microbiology,* in press.

Applewhite, P., 1970b. Protein synthesis during protozoan habituation learning. *Commun. Behav. Biol., 5,* 67–70.

Applewhite, P., 1971. Similarities in protozoan and flatworm habituation behavior. *Nature, 230,* 284–285.

Applewhite P., and Davis, S., 1969. Metallic ions and habituation in the protozoan *Spirostomum. Comp. Biochem. Physiol., 29,* 487–489.

Applewhite, P., and Gardner, F. T., 1968. RNA changes during protozoan habituation. *Nature, 220,* 1136–1137.

Applewhite, P., and Gardner, F. T., 1970. Protein and RNA synthesis during protozoan habituation after loss of macronuclei and cytoplasm. *Physiol. Behav., 5,* 377–380.

Applewhite, P., and Gardner, F. T., 1971. A theory of protozoan habituation. *Nature, 230,* 285–287.

Applewhite, P., and Morowitz, H. J., 1966. The micrometazoa as model systems for studying the physiology of memory. *Yale J. Biol. Med., 39,* 90–105.

Applewhite, P., and Morowitz, H. J., 1967. Memory and the microinvertebrates. In Corning, W. C., and Ratner, S. C. (eds.), *The Chemistry of Learning: Invertebrate Research,* Plenum Press, New York, pp. 329–340.

Applewhite, P., Gardner, F. T., and Lapan, E., 1969a. Physiology of habituation learning in a protozoan. *Trans. N.Y. Acad. Sci., 31,* 842–849.

Applewhite, P., Lapan, E., and Gardner, F. T., 1969b. Protozoan habituation learning after loss of macronuclei and cytoplasm. *Nature, 222,* 491–492.

Applewhite, P. B., Gardner, F., Foley, D., and Clendenin, M., 1971. Failure to condition *Tetrahymena. Scand. J. Psychol., 12,* 65–67.

Barnes, R. D., 1966. *Invertebrate Zoology,* Saunders, Philadelphia.

Bergström, S. R., 1968a. Induced avoidance behaviour in the protozoa *Tetrahymena. Scand. J. Psychol., 9,* 215–219.

Bergström, S. R., 1968b. Acquisition of an avoidance reaction to light in the protozoa *Tetrahymena. Scand. J. Psychol., 9,* 220–224.

Bergström, S. R., 1969a. Amount of induced avoidance behaviour to light in the protozoa *Tetrahymena* as a function of time after training and cell fission. *Scand. J. Psychol., 10,* 16–20.

Bergström, S. R., 1969b. Avoidance behaviour to light in the protozoa *Tetrahymena. Scand. J. Psychol., 10,* 81–88.

Bergström, S. R., 1969c. Induced avoidance behaviour to light in the protozoa *Tetrahymena.* Doctoral dissertation, University of Uppsala.

Best, J. B., 1954. The photosensitization of *Paramecium aurelia* by temperature shock. A study of a reported conditioned response in unicellular organisms. *J. Exptl. Zool., 126,* 87–100.

Bozler, E., 1924. Über die physikalische Erklärung der Schlundfadenströmungen, ein Beitrag zur Theorie der Protoplasmaströmerengen. *Z. Vergl. Physiol.*, *2*, 82–90.

Bragg, A. N., 1959. Selection of food by protozoa. *Turtox News. 17*, 41–44.

Bramstedt, F., 1935. Dressurversuche mit *Paramecium caudatum und Stylonychia mytilus. Z. Vergl. Physiol.*, *22*, 490–516.

Brown, F. (ed.), 1950. *Selected Invertebrate Types,* Wiley, New York.

Buchsbaum, R., 1948. *Animals Without Backbones,* University of Chicago Press, Chicago.

Bullock, T. H., and Horridge, G. A., 1965. *Structure and Function in the Nervous Systems of Invertebrates,* Freeman, San Francisco.

Bullock, T. H., and Quarton, G. C., 1966. Simple systems for the study of learning mechanisms. *Neurosci. Res. Program Bull., 4,* 105–233.

Buytendijk, F. J., 1919. Acquisition d'habitudes par des etres unicellulaires. *Arch. Néerl. Physiol., 3,* 455–468.

Byrne, W., 1970. *Molecular Approaches to Learning and Memory,* Academic Press, New York.

Chaïlakhian, L. M., 1957. On conditioned connections in protozoa and coelenterata. *Zh. Vyssh. Nervn. Deiatel., 7,* 765–774.

Corning, W. C., 1971*a*. Recent studies of learning and its biochemical correlates in protozoans and planarians. In Adám, G. (ed.), *The Biology of Memory,* Plenum Press, New York, pp. 101–119.

Corning, W. C., 1971*b*. Conditioning and "transfer of training" in a colonial ciliate: A summary of the work of N. N. Plavilstchikov. *J. Biol. Psychol., 13,* 39–41.

Corning, W. C., and Freed, S., 1968. Planarian behavior and biochemistry, *Nature, 219,* 1227–1230.

Corning, W. C., and Ratner, S. C., 1967. *The Chemistry of Learning: Invertebrate Research,* Plenum Press, New York.

Dabrowska, J., 1956. Tresura *Paramecium caudatum, Stentor coeruleus, Spirostomum ambiguum. Nr. Budźce Świetne Folia Biol. Polska Akad. Nauk., 4,* 77–81.

Danisch, F., 1921. Ueber Reizbiologie und Reizempfindlichkeit von *Vorticella nebulifera. Z. Allg. Physiol., 19,* 133–188.

Davis, M., and Wagner, A. R., 1969. Habituation of the startle response under an incremental sequence of stimulus intensities. *J. Comp. Physiol. Psychol., 67,* 486.

Day, L. M., and Bentley, M., 1911. A note on learning in *Paramecium. J. Anim. Behav., 1,* 67–73.

Dembowski, J., 1950. On conditioned reactions of *Paramecium caudatum* towards light. *Acta Biol. Exptl., 15,* 5–17.

Diebschlag, E., 1940. Über die Lernfahigkeit von *Paramecium caudatum. Zool. Anz., 11,* 17–271.

Diller, W. F., 1956. Nuclear behavior in the ciliated protozoa. *Bios, 27,* 217–234.

Eisenstein, E. M., 1967. The use of invertebrate systems for studies on the bases of learning and memory. In Quarton, G. C., Malnechuk, T., and Schmitt, F. (eds), *The Neurosciences,* Rockefeller Univ. Press, New York, pp. 653–665.

French, J. W., 1940. Trial-and-error learning in *Paramecium. J. Exptl. Psychol., 26,* 609–613.

Gardner, F. T., and Applewhite, P. B., 1970*a*. Protein and RNA inhibitors and protozoan habituation. *Psychopharmacologia, 16,* 430–433.

Gardner, F. T., and Applewhite, P. B., 1970*b*. Temperature separation of acquisition and retention in protozoan habituation. *Physiol. Behav., 5,* 713–714.

Gelber, B., 1952. Investigations of the behavior of *Paramecium aurelia:* I. Modification of behavior after training with reinforcement. *J. Comp. Physiol. Psychol., 45,* 58–65.

Gelber, B., 1954. Investigations of the behavior of *Paramecium aurelia:* IV. The effect of different training schedules on both young and aging cultures. *Am. Psychologist, 9,* 374.

Gelber, B., 1956a. Investigations of the behavior of *Paramecium aurelia:* II. Modification of a response in successive generations of both mating types. *J. Comp. Physiol. Psychol., 49,* 590–593.

Gelber, B., 1956b. Investigations of the behavior of *Paramecium aurelia:* III. The effect of the presence and absence of light on the occurrence of a response. *J. Genet. Psychol., 88,* 31–36.

Gelber, B., 1957a. Investigations of the behavior of *Paramecium aurelia:* VI. Reinforcement with three values of training. *Am. Psychologist, 12,* 428.

Gelber, B., 1957b. Food or training in paramecium? *Science, 126,* 1350–1341.

Gelber, B., 1958a. Retention in *Paramecium aurelia. J. Comp. Physiol. Psychol., 51,* 110–115.

Gelber, B., 1958b. Extinction in *Paramecium aurelia. Am. Psychologist, 13,* 405.

Gelber, B., 1962a. Acquisition in *Paramecium aurelia* during spaced training. *Psychol. Rec., 12,* 165–177.

Gelber, B., 1962b. Reminiscence and the trend of retention in *Paramecium aurelia. Psychol. Rec., 12,* 179–192.

Gelber, B., 1965. Studies of the behaviour of *Paramecium aurelia. Anim. Behav., 13,* Suppl. 1, 21–29.

Gelber, B., and Rasch, E., 1956. Investigations of the behavior of *Paramecium aurelia.* V. The effects of autogamy. *J. Comp. Physiol. Psychol., 49,* 594–599.

Grabowski, U., 1939. Experimentelle Untersuchungen ueber das angebliche Lernvermoegen von *Paramecium. Z. Tierpsychol., 2,* 265–282.

Gurowitz, E. M., 1969. *The Molecular Basis of Memory,* Prentice-Hall, Englewood Cliffs, N. J.

Hamilton, T. C., Blair, H. J., and Eisenstein, E. M., 1971. Variety of modifiable behaviors in the protozoan, *Spirostomum ambiguum. Biophys. Soc. Abst.,* 204a.

Harden, C. M., unpublished manuscript. Behavior modification of *Stentor coeruleus.*

Hegner, R. W., 1933. *Invertebrate Zoology,* Macmillan, New York.

Herrick, C. J., 1924. *Neurological Foundations of Animal Behavior,* Holt, New York.

Hyman, L., 1940. *The Invertebrates: Protozoa Through Ctenophora,* McGraw-Hill, New York.

Jennings, H. S., 1901. Studies on reactions to stimuli in unicellular organisms. IX. On the behavior of fixed infusoria (*Stentor* and *Vorticella*) with special reference to modifiability of protozoan reactions. *Am. J. Physiol., 8,* 23–60.

Jennings, H. S., 1906. *The Behavior of Lower Organisms,* Columbia University Press, New York.

Jensen, D. D., 1955. A critical examination of learning in paramecia, M. A. thesis, University of Nebraska.

Jensen, D. D., 1957a. Experiments on "learning" in paramecia. *Science, 125,* 191–192.

Jensen, D. D., 1957b. More on "learning" in paramecia. *Science, 126,* 1341–1342.

Jensen, D. D., 1959. A theory of the behavior of *Paramecium aurelia* and behavioral effects of feeding, fission, and ultraviolet microbeam irradiation. *Behaviour, 15,* 82–122.

Jensen, D. D., 1965. Paramecia, planaria, and pseudo-learning. *Anim. Behav., 13,* Suppl. 1, 9–20.

Jeon, K. W., Lorch, I. J., and Danielli, J. F., 1970. Reassembly of living cells from dissociated components. *Science, 167,* 1626–1627.

Katz, M., and Deterline, W. A., 1958. Apparent learning in the paramecium. *J. Comp. Physiol. Psychol., 51,* 243–247.

Kellogg, W. N., 1958. Worms, dogs, and paramecia, *Science, 127.*

Kinastowski, W., 1963a. Der Einfluss der mechanischen Reize auf die Kontraktilitat von *Spirostomum ambiguum. Acta Protozool., 1,* 201–222.

Kinastowski, W., 1963b. Das Problem "des Lernes" bei *Spirostomum ambiguum*. *Acta Protozool., 1,* 223–236.

Kindleman, P., Applewhite, P., and Morowitz, H. J., 1968. Capacitive detection of very small aquatic animals. *Rev. Sci. Instr., 39,* 121–123.

Kohler, O., 1939. Diskussion zu den Vortragen Alverdes-Bramstedt. *Verh. Deutsch. Zool. Ges.* 41.

Lachman, S. J., and Havlena, J. M., 1962. Reactive inhibition in the *Paramecium. J. Comp. Physiol. Psychol., 55,* 972–973.

Leedale, G. F., 1966. *Euglena:* A new look with the electron microscope. *Advan. Sci.,* May, 22–37.

Lepley, W., and Rice, G. E., 1952. Behavior variability in paramecia as a function of guided act sequences. *J. Comp. Physiol. Psychol., 45,* 283–286.

Machemer, H. von, 1966a. Versuche zur Frage nach der Dressierbarkeit hypotricher Ciliaten unter Einsatz hoher Individuenzahlen. *Z. Tierpsychol. 6,* 641–654.

Machemer, H. von, 1966b. Erschütterungsbedingte Sensibilisierung gegenüber rauhem Untergrund bei *Stylonychia mytilus. Arch. Protistenk., 109,* 245–256.

Manwell, R. D., 1961. *Introduction to Protozoology,* St. Martins Press, New York.

Mast, S. O., 1926. Structure, movement, locomotion and stimulation in *Amoeba. J. Morphol. Physiol., 41,* 347–425.

Mast, S. O., and Pusch, L. C., 1924. Modification of response in *Amoeba. Biol. Bull., 46,* 55–60.

McConnell, J. V., 1966. Comparative physiology: Learning in invertebrates. *Ann. Rev. Physiol., 28,* 107.

Metalnikow, S., 1912. Contributions à l'étude de la digestion intracellulaire chez les protozoaires. *Arch. Zool. Exp. Gén., 49,* 373–498.

Metalnikow, S., 1913a. Sur la faculté des infusoires d'apprendre à choisir la nourriture. *Compt. Rend. Soc. Biol. Paris, 74,* 701–703.

Metalnikow, S., 1913b. Comment les infusoires se comportent vis-à-vis des mélanges de diverses matières colorantes. *Compt. Rend. Soc. Biol. Paris, 74,* 704–705.

Metalnikow, S., 1914. Les infusoires, peuvent-ils apprendre à choiser leur nourriture? *Arch. Protistenk., 34,* 60–78.

Metalnikow, S., 1916. Les reflexes chez les protozoaires. *Compt. Rend. Soc. Biol. Paris, 79,* 80–82.

Mirsky, A. F., and Katz, M. S., 1958. Avoidance "conditioning" in paramecia. *Science, 127,* 1498–1499.

Naitoh, Y., 1966. Reversal response elicited in nonbeating cilia of paramecium by membrane depolarization. *Science, 154,* 660–662.

Naitoh, Y., and Eckert, R., 1969a. Ionic mechanisms controlling behavioral responses of paramecium to mechanical stimulation. *Science, 164,* 963–965.

Naitoh, Y., and Eckert, R., 1969b. Ciliary orientation: Controlled by cell membrane or by intracellular fibrils? *Science, 166,* 1633–1635.

Osborn, D., and Eisenstein, E. M., unpublished. The distribution of calcium and other elements in the ciliated protozoan, *Spirostomum ambiguum.*

Osborn, D., Hsung, J. C., and Eisenstein, E. M., in press. The involvement of calcium in contractility in the ciliated protozoan, *Spirostomum ambiguum. Commun. Behav. Biol.*

Osborn, D., Blair, H. J., Thomas, J., and Eisenstein, E. M., unpublished. The effects of mechanical and electrical stimulation on habituation in the ciliated protozoan, *Spirostomum ambiguum.*

Plavilstchikov, N. N., 1928. Observations sur l' excitabilité des infusoires. *Russ. Ark. Protist., 7,* 1–24.

Poskocil, A., 1966. If you're a paramecium, can you learn? A query. *Worm Runner's Digest,* *8,* 31–42.

Rabin, B. M., and Hertzler, D. R., 1965. Replications of two experiments on reactive inhibition in paramecia. *Worm Runner's Digest, 7,* 46–50.

Razran, G. H. S., 1933. Conditioned responses in animals other than dogs. *Psychol. Bull., 30,* 261–324.

Schlieper, C., 1940. *Praktikum der Zoophysiologie,* Jena.

Sgonina, K., 1939. Vergleichende Untersuchungen über die Sensibilisierung und den bedingten Reflex. *Z. Tierpsychol. 3,* 224–247.

Smith, S., 1908. Limits of educability in *Paramecium. J. Comp. Neurol. Psychol., 18,* 499–510.

Soest, H., 1937. Dressurversuche mit Ciliaten und rhabdocoelen Turbellarien. *Z. Vergl. Physiol., 24,* 720–748.

Sonneborn, T. M., 1950. *Paramecium* in modern biology. *Bios, 21,* 31.

Telfer, W. H., and Kennedy, D., 1965. *The Biology of Organisms,* Wiley, New York.

Thorpe, W. H., 1963. *Learning and Instinct in Animals,* Harvard University Press, Cambridge, Mass.

Tschakhotine, S., 1938. Réactions "conditionées" par microponction ultraviolette dans le comportement d'une cellule isolé *(Paramecium caudatum). Arch. Inst. Prophylac. Paris, 10,* 119–131 (not seen).

Ungar, G., 1970. *Molecular Mechanisms in Memory and Learning,* Plenum Press, New York.

Warden, C. J., Jenkins, T. N., and Warner, L. H., 1940. *Comparative Psychology,* Vol. II: *Plants and Invertebrates,* Ronald Press, New York.

Wawrzyńczyk, S., 1937*a.* Badania nad pamiecia *Spirostomum ambiguum major. Acta Biol. Exptl., 11,* 57–77.

Wawrzyńczyk, S., 1937*b.* Reakcje *Paramecium caudatum* na bodzce swietlne. *Trav. Soc. Sci. Wilno, 12,* 1–28.

Weiss, P., 1961. The concept of perpetual neuronal growth and proximodistal substance convection. In Kety, S. S., and Elkes, J. (eds.), *Regional Neurochemistry,* Pergamon Press, New York, pp. 220–242.

Weiss, P., 1969. "Panta' Rhei"—and so flow our nerves. *Am. Scientist, 57,* 287–305.

Wichterman, R., 1953. *The Biology of Paramecium,* Blakiston, New York.

Wood, D. C., 1970*a.* Parametric studies of the response decrement produced by mechanical stimuli in the protozoan, *Stentor coeruleus. J. Neurobiol. 1,* 345–360.

Wood, D. C., 1970*b.* Electrophysiological studies of the protozoan, *Stentor coeruleus. J. Neurobiol., 1,* 363–377.

Wood, D. C., 1970*c.* Electrophysiological correlates of the response decrement produced by mechanical stimuli in the protozoan, *Stentor coeruleus,* in press.

Chapter 3

BEHAVIORAL MODIFICATIONS IN COELENTERATES

NORMAN B. RUSHFORTH

Department of Biology and Department of Biometry
Case Western Reserve University
Cleveland, Ohio

I. INTRODUCTION

Coelenterates have proved to be perplexing animals to the behavioral physiologist. Their structural simplicity, diffuse nervous organization, and the early observations of their simple, stereotyped activities suggested that they were fruitful preparations for the study of the physiological mechanisms underlying behavior. Indeed, Pantin in his famous Croonian Lecture, *The Elementary Nervous System,* predicted that in the coelenterates we are nearer to "a complete analysis of the structural units on which behavior is based" than in any other animals (Pantin, 1952, p. 147). Such a belief was based on the extensive studies of the contraction reflex responses of the sea anemones *Metridium* and *Calliactis.* However, such simple reflexes play only a limited part in coelenterate behavior. The behavioral simplicity of these animals has proved to be more apparent than real, for today, 20 years later, we seem further from a complete analysis than was previously suspected.

In this chapter, some behavioral modifications of coelenterates are discussed. Many of the studies reviewed were undertaken at the turn of the century during a period in which controversy existed as to relative importance of external and internal factors in controlling behavior. Pavlov's discovery of the conditioned reflex had stressed the role of the external stimulus in eliciting a specific response in higher animals, while Loeb in his studies of lower organisms developed a "tropism theory of animal conduct" (1918). Loeb maintained that behavior in large part consisted of movements forced by environmental stimuli acting on bilaterally symmetrical structures. These views were severely questioned, however, since they were thought to characterize behavior in far too mechanistic terms. They were criticized by Jennings, who emphasized the importance of endogenous factors in controlling behavior. In his studies of the behavior of lower organisms, Jennings (1906) focused on the changing activities induced at different times by the same stimulus. He stressed the role of spontaneity and internal changes within the animal in the control of its behavior.

These early studies of behavioral modifications should be viewed historically in the context of this controversy between the so-called mechanistic and vitalistic viewpoints. To a large extent, the observations are anecdotal and are unsatisfactory as evidence of learning in the coelenterates, since the results are open to a variety of interpretations. These published reports do not meet the more rigorous standards of experimental design and statistical analysis required today. Data are frequently lacking on the number of animals studied and the use of controls as well as important criteria for the stimulus parameters and regimes. Thus some of the studies have proved difficult to repeat, and attempts to confirm the results have been unsuccessful. However, several of these observations do provide indications that the behavior of coelenterates can be modified in various ways by repeated experiences.

Are coelenterates capable of learning? Evidence appears strongest for the demonstration of habituation, while reports of conditioning are fragmentary and less convincing. Efforts have been made to study the physiological bases of habituation in *Hydra* and to contrast it with other forms of behavioral inhibition in the animal. Some progress has been made using electrophysiological recording techniques, which have been applied to coelenterates during the past decade. However, further insights into the underlying mechanisms of habituation must await the successful use of intracellular recording methods on such preparations.

As a preface to reviewing studies of behavioral modifications, a brief discussion of the nervous system and coelenterate behavioral physiology will be given. How far have we progressed during the past decade in understanding the mechanisms underlying coelenterate behavior? Ultrastructural studies have unearthed complexity in the nervous system not formerly apparent by use of light microscopy. Electrophysiological methods have provided us with electrical correlates of spontaneous and evoked activity and, coupled with histological evidence, have demonstrated the existence of epithelial cell conduction. Now a major question is what the respective roles of the nervous system and of nonnervous mechanisms are in the control of coelenterate behavior.

II. COELENTERATE NERVOUS SYSTEMS

The phylum Coelenterata includes marine forms such as jellyfish, corals, and sea anemones together with the freshwater *Hydra*. This group constitutes the most simply constructed animals with nervous systems. While sponges (phylum Porifera) are more simply organized metozoans, current histological and behavioral evidence indicates that they lack nerve cells (Bullock and Horridge, 1965).

Coelenterates are made up primarily of two concentric epithelial cell layers, an outer *epidermis* (or ectodermal layer) and an inner *gastrodermis* (or endodermal layer). These two cell layers are separated by an acellular layer, the *mesoglea*. The principal structural cells of the epithelial layers are *epitheliomuscular cells*. Such cells have long contractile bases which lie against and are partially imbedded in the mesoglea. In *Hydra*, for example, the muscular processes in the epidermis are longitudinal, while those in the gastrodermis have a circular orientation. These form two antagonistic muscle systems responsible for shortening and elongating the body column. In the epidermis, there is a diffuse two-dimensional network of nerve cells running among the contractile bases of the muscle cells (Fig. 1). Such cells are bipolar and multipolar neurons having relatively long processes from the cell body (McConnell, 1932; Burnett and Diehl, 1964; Davis *et al.*, 1968). They are more concentrated at the base of the tentacles, in the hypostome, and at the base of the body column than in other regions of the ectodermal layer (Noda, 1969). Such nerve cells, however, are sparsely distributed within the gastrodermis of the hydra (Semal van Gensen, 1952). In other coelenterates, anemones and jellyfish, nerve nets can exist in both epithelial layers, and sometimes two nerve nets having independent activity may jointly occur in the same tissue layer. In addition to nerve nets, jellyfishes have localized concentrations of nerve cells in marginal nerve rings or marginal ganglia.

There is a characteristic absence of specialized sensory receptors in most coelenterates. However, sensory nerve cells frequently are abundant between the epitheliomuscular cells in the outer epithelial layer. Investigations with light microscopy indicate that these neurosensory cells often contain sensory hairs (Burnett and Diehl, 1964; Mackie, 1960), which ultrastructural studies show to be modified cilia (Lentz and Barrnett, 1965). It has been postulated that such structures have a mechanoreceptive function (Jha and Mackie, 1967).

Studies with the electron microscope have revealed that coelenterate

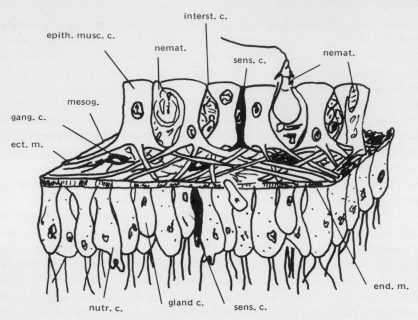

Fig. 1. Diagram of the organization of the epithelia and nervous system of *Hydra*. The ectoderm is above; the endoderm below (after Bullock and Horridge, 1965). *Ect. m.*, ectodermal muscle fiber layer; *end. m.*, endodermal muscle fiber layer; *epith. musc. c.*, epithelial muscular cell; *gang. c.*, ganglion cell; *gland. c.*, gland cell; *interst. c.*, interstitial cell; *mesog.*, mesoglea; *nemat.*, nematocyst; *nutr. c.*, nutritive cell; *sens. c.*, primary sensory neuron.

neurons have features typical of nerve cells in other animals—mitochondria, golgi region, neurotubules, and several types of vesicles, including synaptic vesicles. However, the nerve cell bodies and axons lack the glial cells which sheath the neurons of higher animals (Horridge and MacKay, 1962). The first axon-to-axon synapses, discovered in the marginal jellyfish *Cyanea*, were symmetrical, with vesicles on each side of the synaptic cleft. Horridge and MacKay (1962) suggested that such synapses were electrically transmitting and that the vesicles had a trophic role in the maintenance of the synapses.

Asymmetrical synapses with vesicles only on one side of the synaptic junction as well as symmetric synapses have been found in the hydromedusan *Sarsia* (Jha and Mackie, 1967). Similar polarized synapses have been reported by Westfall and her coworkers in several coelenterates: a hydromedusan, *Gonionemus;* the sea anemone *Metridium;* and two species of *Hydra* (Westfall 1969, 1970; Westfall *et al.*, 1970). Such synapses are present at three types of cell junctions: *neuromuscular, interneural,* and *nematocytic.* Neuromuscular synapses occur between axons from the epidermal nerve net and the contractile bases of the epitheliomuscular cells. Interneural synapses are evident between different axons in the nerve net. In *Hydra,* such junctions are seen along the *neurite* (*en passant* synapses) as well as at the nerve terminal. Neuronematocyte synapses occur between neurites and the bases of developed *nematocytes* (the stinging cells each containing a nematocyst). In the tentacles of *Hydra,* individual nerve cells sometimes are observed to have synaptic contact with both a nematocyte and an epitheliomuscular cell. This is the first reported example in any animal of a single neuron with synapses to two different types of effector cells (Westfall *et al.,* 1971).

The three types of synapses appear to have similar features. They consist of several dense-cored or clear vesicles, 100–200 mμ in diameter, associated with two parallel, thickened, and electron-dense plasma membranes. The membranes are separated by a cleft approximately 10 mμ wide. Dense-cored vesicles were previously observed in some of the neurons of *Hydra* and were classified as neurosecretory granules (Lentz and Barrnett, 1965; Davis *et al.,* 1968). However, the synaptic configuration of some of these vesicles in this animal, as well as in other coelenterate preparations, suggests that they play a role in neural transmission. Westfall *et al.* (1971) suggest that the presence of both clear and dense-cored synaptic vesicles may indicate that there are two types of transmitter substance used in this simple nervous system. The synaptic vesicles with dense cores appear similar to those which in higher animals contain adrenergic neurotransmitters, although they are somewhat larger in size than the vesicles of vertebrate synapses. On the other hand, the clear synaptic vesicles at the neuronematocyte junction may implicate a neurotransmitter such as acetylcholine, since pharmacological studies suggest that a cholinergic mechanism is involved in nematocyst discharge (Lentz and Barrnett, 1961, 1962). However, the identification of a specific neurotransmitter substance in any coelenterate has yet to be established.

III. COELENTERATE BEHAVIORAL PHYSIOLOGY

A. Conducting Systems

A characteristic feature of coelenterates, both the sedentary polyps (e.g., anemones) and the free-swimming medusae (e.g., jellyfish) is the diffuseness of many of their conducting systems. In the Scyphomedusae, typically there are eight marginal bodies, any one of which is capable of initiating the normal rhythmic swimming contractions of the bell. After removal of these structures, the normal wave of contractions can be elicited by a single electrical stimulus applied to any point on the subumbrella surface of the bell. Romanes (1885) showed that complex patterns of incisions made in the bell were unable to halt the propagation of a contraction wave as long as a narrow bridge of tissue (about 1 mm in width) remained between the cuts. He concluded from this that the pathways of conduction are present throughout the subumbrella in a diffuse array. Such surgical procedures have been used on the body wall of the sessile polyp *Metridium,* giving similar results (Parker, 1919).

In addition to diffuse conducting systems, there are several examples of discrete conducting pathways. In Hydromedusae, Scyphomedusae, and anthozoan polyps, radially oriented pathways have been established, and there is evidence for three discrete and parallel conducting systems in the stalk of the hydroid *Tubularia* (Josephson, 1965a). Studies of the behavioral physiology of coelenterates have established the presence of multiple conducting systems in both polyps and medusae. In *Hydra,* there are at least two conducting systems (Passano and McCullough, 1964, 1965). Three conducting systems are found in the anemone *Calliactis parasitica:* a fast-conducting system, which can initiate both quick and slow muscle contractions, and two slow-conducting systems (McFarlane, 1969). Electrical activity in one of the slow-conducting systems (SS 1) is associated with detachment of the basal disk. However, no clear behavioral correlate has been discovered for the other slow-conducting system (SS 2). There is some evidence that SS 1 and SS 2 are located in the ectodermal and endodermal layers, respectively, but the conducting elements have not been identified. There are two conducting systems in the subumbrella of medusae (Horridge, 1955, 1956). One of these overlies the radial and circular muscles (the fast system) and activates the swimming pulsations, the other (the slow system) is diffuse and spreads over the whole epithelium. This diffuse nerve net is utilized in feeding behavior.

Conducting systems can be distinguished by differences in their mode of spatial propagation. *Through-conducting systems* transmit a single impulse throughout their pathways. On electrical stimulation, the spread is all-or-nothing and is independent of the stimulus parameters. Typically, these systems control the rapid, symmetrical muscular responses used in jellyfish swimming or polyp withdrawal. In contrast, *incrementing systems* frequently mediate variable, localized movements. In these systems, the distance of spread increases with increasing numbers and frequency of the stimuli. This property is attributed to *facilitation* of the junctions in the conducting systems. The first impulse arriving at a given junction does not cross it, but temporarily allows the junction to transmit subsequent impulses. A decrease in the intensity of muscle contraction with increasing distance from the point of stimulation can occur if muscles are activated by an incrementing conducting system. When bursts of impulses are induced by mechanical stimulation, muscles farther from the point of stimulation receive fewer pulses than those close to it and thus contract less strongly. This process is dramatically illustrated in the withdrawal response in colonies of hydroid polyps. Individuals near to the stimulated point strongly contract, while those peripheral to it withdraw only slightly (Josephson, 1961).

In addition to activating muscles, conducting systems frequently link together *pacemakers* (described later) and are the pathways used to modulate these centers of endogenous activity. For example, activity in the slow-conducting system may either excite or inhibit spontaneous swimming contractions in Scyphomedusae (Horridge, 1956). In the hydroid *Tubularia,* electrical stimuli trigger a conducting system which induces firing of pacemakers in the stalk. Electrical stimulation can also activate a second conducting system which inhibits these pacemakers (Josephson and Uhrich, 1969). In muscle activation, facilitation frequently occurs at the junction of the muscle and conducting system. Thus an evoked contraction increases in strength with the number and frequency of conducted electrical events. A more complex form of facilitation, termed *heterofacilitation* is seen in the scyphomedusan *Cassiopea* (Bullock and Horridge, 1965). Contractions of the bell induced by impulses in the fast-conducting system are enhanced in amplitude by previous activity in the slow-conducting system.

The conducting systems of coelenterates, therefore, are examples of simple forms of information processing. They transform temporally significant signals into varying patterns of anatomical spread and decrementing response. As mentioned previously, such systems may be broadly classified as through-conducting or incrementing, and several responses may be satisfactorily explained with their use. As yet, however, their assumed properties appear insufficient to account for other types of behavior, such as the various kinds of spreading observed for polyp retraction in differing coral species (Bullock and Horridge, 1965).

B. Pacemaker Activity

Spontaneous activity in the absence of changes in the external environment was well documented in Romanes' classical observation of jellyfish (1885). For many years, however, the generality of endogenous behavior in all coelenterate groups was questioned. In describing the behavior of sea anemones, for example, Parker (1919) stressed the importance of reflex responses. Calling on the work of Loeb and Nagel in addition to his own, he characterized polyp activities as simple responses to external stimulation. Such animals were thought of as having rather uniform internal states, so that spontaneity was absent from their behavior. It was much later that Batham and Pantin (1950), using time-lapse cinematography and slow kymograph recording, showed that in reality anemones were in a state of continual and varied activity, spontaneously produced, but too slow to be easily seen by the eye.

A characteristic feature of spontaneous behavior in coelenterates is the variability within the component events. In anemones, reciprocal contractions of the longitudinal and circular muscles result in alternate shortening and elongation of the body column. Sometimes this activity may show a more or less regular rhythm, but frequently it shows no trace of rhythmicity. In some cases, various parts of the body wall are coordinated in their contractions, while in others such coordination is quite absent. Spontaneous activity is found in muscle systems other than those in the column, and the undisturbed animal may pass through many different "phases" of activity often lasting several hours (Batham and Pantin, 1950). Even in the more overt swimming pulsations of medusae, a great variability is present in the intervals between successive contractions of the bell (Horridge, 1959). In *Hydra,* electrical impulses associated with column contractions have varying temporal features, occurring as widely spaced events or in bursts containing differing number of pulses (Passano and McCullough, 1964; Rushforth, 1971).

The structures initiating spontaneous events such as muscle contraction and electrical potentials are termed *pacemakers*. Their sites are widely scattered throughout the animals, and in excised tissue they function as autonomous systems. Pieces containing a marginal body, when removed from a jellyfish, beat at roughly the same frequency as the parent animal (Romanes, 1877). Spontaneous contractions occur in rings isolated from the body column of anemones (Hoyle, 1960). Segments cut from the tentacle of a hydra produce electrical impulses similar in temporal characteristics to those recorded from the entire tentacle structure (Rushforth and Burke, 1971). Such examples illustrate the diffuseness of potential pacemakers in coelenterate tissue. How are such pacemakers coordinated in the intact animal? A common arrangement is for groups of pacemaker units to be linked by conducting systems. All of the pacemakers in the group are reset by firing of any one of them. The same effect is achieved by electrical stimulation of their conducting system. A tightly coupled group of potential pacemakers together with its associated conducting system has been termed a *pacemaker system* (Josephson, 1965b). An example of a pacemaker system is the group of marginal bodies linked by the fast-conducting system in the medusan subumbrella epithelium. Firing of any one of the marginal bodies fires off all the others. Shifts in the leading pacemaker locus among the various marginal bodies result in a rhythm of the whole animal which is more regular than that of the individual units (Horridge, 1959).

While the individual pacemaker units are linked tightly by conducting elements to act as a single functional component, the different pacemaker systems also may interact. Such pacemaker interactions have been demonstrated most clearly in *Tubularia* (Josephson and Mackie, 1965). This hydroid contains several pacemaker systems loosely coupled to one another. One system (the *HP system*), located in the hydranth of the polyp, produces single pulses or bursts of pulses. These *hydranth pulses* (HPs) are associated with synchronous oral flexions of the tentacles followed by a peristaltic wave in the proboscis. A second pacemaker system (the *NP system*), located in the stalk of the animal, also produces pulses (NPs) as bursts or single events. Activity in this pacemaker system is reflected only indirectly in polyp behavior, through its interaction with the HP system. It is observed that when the NP system fires in a burst, it usually drives the HP system to fire concurrently. In addition, a single NP sometimes triggers an HP, while conversely an HP sometimes induces firing of the NP system (Fig. 2). Thus there is a two-way interaction between the pacemaker systems.

While variability in activity is intrinsic to coelenterate pacemaker systems even under constant environmental conditions, marked changes in endogenous output often result from imposed stimulation. There are many examples of modified behavior in response to stimuli of differing modalities. Probably the most complex and long-lasting effects are found in the slow-activity patterns of anemones (Batham and Pantin, 1950). Such changes in behavior are not simple reflex responses in which contraction follows directly upon stimulation. Rather, the stimuli appear to modulate spontaneous activity and channel it into new patterns. For example, low-frequency electrical stimulation for a period of minutes may cause a change in the pattern of activity which continues for several hours. Ingestion of food or temporary exposure to filtered food solution initiates a sequence of many phasic changes: expansion of the disk and elongation, followed by swaying, contraction and then distension of the body column, defecation, and shriveling (Batham and Pantin, 1950). Not all of these phases are exhibited by different animals, nor even by the same animal at different times. Nevertheless, all of them are found to

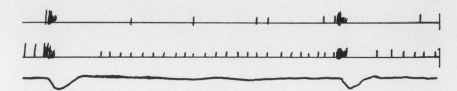

Fig. 2. Activity patterns recorded simultaneously from the hydranth pacemaker system (upper trace) and the neck pacemaker system (middle trace) of *Tubularia*. Coupling between these pacemaker systems leads to concurrent firing during bursts. Some single-pulse coupling is also evident; most of the single pulses from the hydranth are immediately followed by firing of the neck pacemaker system. The lower trace is from a transducer on the polyp neck recording neck contractions (indicated by a downward deflection). (After Josephson, 1968.)

occur spontaneously, so that they are potentially present in the animal and merely released by external stimuli. Thus the interplay of the stimuli and pacemakers results in the complex behavior observed in this animal.

C. Nonnervous Conduction

Until the mid-1960's coelenterate pacemakes and conducting elements were assumed to have a nervous origin. For example, the fast-conducting system of the medusan subumbrella epithelium was equated with a network of relatively large bipolar nerve cells overlying the radial and circular muscles. By analogy with functionally similar tracts in other invertebrate groups, this specialized system of neurons was termed the *giant fiber system* (Horridge, 1956). By simple cutting experiments, Horridge (1954) demonstrated that a single nerve cell was necessary and sufficient to conduct a contraction wave across a thin bridge of tissue from one part of the bell to another. Thus the conducting system in this case was shown conclusively to be nervous in nature. In other coelenterate preparations, however, evidence for nervous conduction is more indirect, and in some cases there is no histological evidence for nerve cells, even though behavioral responses suggest that contraction must occur. In addition, it is perplexing to find only a single nerve net when physiological studies indicate multiple conducting systems.

Nonnervous conduction was first found to occur in the epithelia of siphonophores. Using optical and electron microscopy, Mackie (1965) showed that nerve and muscle fibers are absent from large blocks of tissue in the exumbrella of *Hippopoduis*. However, electrical or mechanical stimulation of such regions results in a rolling movement of the bell margin and vellum some centimeters away. A small electrical impulse can be recorded at the surface of the epithelial tissue which is correlated with the conducted activity. Propagation of these potentials is unpolarized and nondecremental, with a conduction velocity of the order of 35–40 cm/sec.

In coelenterates, the major problem in identifying examples of nonnervous conduction lies in establishing histologically that nerves are absent. Selective nerve staining is notoriously difficult in such animals due to the complex folding of tissue layers and the presence of other cell types such as nematocysts and interstitial cells. The epithelium in the exumbrella of *Hippopodius*, however, is unusually simple tissue in which to demonstrate the absence of nerves. It consists of a thin layer of flattened epithelial cells, having similar shaped nuclei which are distributed in roughly geometric patterns. Phase microscopy of whole mounts reveals many details of these cells, such as their membrane and ciliary structures.

Since Mackie's original discovery, epithelial conduction has been investigated in several of the Hydromedusae (Mackie and Passano, 1968). A general picture of some of its properties is developing, and its relationship with nervous activity is beginning to be explored. Nonnervous conduction can occur independently in both the ectodermal and gastrodermal layers, whether the cells are epitheliomuscular or simple epithelial cells lacking contractile processes. Such conduction, like nerve net conduction, is all-or-none, unpolarized, and diffuse. It may have smaller conduction velocity than nervous conduction in the same preparation, but this is not a definitive criterion of classification (Mackie, 1970). Epithelial conduction tends to be associated with simple generalized responses such as contraction of large muscle sheets, whereas nervous activity appears to control complex, more localized movements. In medusae, pacemaker activity is always accompanied by nerve cells, but this association must be investigated more thoroughly for other coelenterates before pacemakers are equated with neurons.

IV. BEHAVIORAL MODIFICATIONS IN COELENTERATES

Several studies have been undertaken which show the range of behavioral modifications in coelenterates. Much of the earlier work focused on changes in simple motor activity as a result of prior stimulation, but recently some efforts have been made to demonstrate associative learning. The results of many of these studies are inconclusive, but the central problems still remain. What complexity of behavior is present in these animals with simple nervous organization? Are any of their responses parallel to those we classify as learning in higher organisms? What are the physiological bases for the modified behavior patterns we observe in this phylum?

A. Simple Behavioral Responses

Feeding Experiments

A series of investigations at the turn of the century showed that feeding reactions of anemones to weak stimuli wane as a result of repetition of the stimulation. Nagel (1894) showed that *Adamsia* at first readily accepts filter paper soaked in the dilute juice of crab meat. After the paper has been fed several times in alternation with the meat, the reaction to the paper becomes slower and finally ceases, while the meat continues to be accepted. This experiment was repeated by Parker (1896) using *Metridium* and gave similar results. He first used the right side of the tentacular zone of the animal and then repeated the procedure using the other side, "with a view of determining whether the condition brought about in the right lip had spread to the left one." However, there was no evidence of such an effect, and successive repetitions of the experiment showed independent activities of the two sides of the animal. This result was further substantiated by Allabach (1905), also using *Metridium,* in an experiment in which she fed six successive regions of the disk until each had rejected the paper. If the paper was removed with tweezers before it was engulfed, the same response was observed. Thus the selective rejection was not a result of accumulative effects of the paper in the digestive cavity.

In Parker's experiments, the trials were run on a single day, and there were no apparent residual effects observed during trials on subsequent days. Much longer-term effects were reported by Fleure and Walton (1907) using the anemones *Actinia* and *Telia.* In their experiments, pieces of filter paper were placed on specific tentacles every 24 hr. At first, the paper was carried to the mouth by the tentacles, swallowed, and subsequently ejected as inedible. After 2–5 days, the paper would not be swallowed, and after about another 2 days it would be rejected by the tentacles. At this stage, other tentacles not formerly used were tested. These tentacles accepted the filter paper at first, but began to reject it in fewer trials than did the tentacles originally used. This acquired response was lost after a period of 6–10 days without stimulation.

These feeding experiments are subject to some speculation, since they possibly indicate sensory adaptation or habituated responses. Allabach (1905) has suggested that the decrease in responsiveness might be due to the accumulation of mucus on the stimulated tentacle. Jennings (1905) attempted to repeat the experiments of Nagel and Parker, using the anemone *Aiptasia,* but obtained more variable results. He maintained that the reaction to the external stimulus depended on endogenous factors in the animal, particularly those depending on its state of metabolism. In his experiments with *Aiptasia,* Jennings most frequently found that rejection of the filter paper occurred at the same time as rejection of the food itself and concluded that both took place when the animals were satiated. Gee (1913), using the shore anemone *Cribrina,* confirmed the conclusions of the two earlier studies and rejected Jennings' interpretations. He found that tentacles of one side of the anemone could be fed to refusal, yet the tentacles of the other side of the disk would accept food. This condition could be achieved whether or not the meat was swallowed. He suggested that muscular fatigue plays little role in the modified response but that the lowered responsiveness results from mucous secretion on the stimulated tentacles. The results of Fleure and Walton's experiments are more provocative, since they suggest an acquired response of considerable duration for a coelenterate. Unfortunately, however, when Ross (1965*a*) tried to repeat the experiments using *Actinia* he found that original acceptance by the tentacles of filter paper dipped in food extracts did not always occur. Thus he abandoned his efforts at replication in the absence of initially reliable response.

Acquired Periodic Behavior

As mentioned previously, the coelenterates as a group are characterized by a variety of spontaneous behavior patterns. Many of these activities are modulated by external stimuli, which if they are periodic entrain a rhythmic response. If the animals are removed from the stimuli and placed under constant environmental conditions, frequently the rhythmic activity persists. This has been observed with both diurnal and tidal rhythmic responses. It has been suggested that the periodic behavior is acquired or imprinted as a result of the repeated stimulation (see Thorpe, 1963).

Bohn (1906, 1909, 1910b) observed that *Actinia* retracts when it is exposed to air by the falling tide and expands when it is covered again by water. Such rhythmic behavior was found to persist for 3–8 days when the animal was placed under a constant volume of water in an aquarium. Bohn and Pieron (1906) and particularly Pieron (1908a, 1910) claimed that in this anemone the responses start to occur a little in advance of the actual tidal changes, thus giving the impression of an "anticipatory reaction." This reaction is lost when the animal is placed under constant laboratory conditions but is regained again after a week or so of being reexposed to the tides.

A diurnal rhythm has been observed in *Eloactis* (Hargitt, 1907), *Actinia* (Bohn, 1906, 1907) and *Sagartia* (Pieron, 1908b). In *Metridium,* daily rhythmic contractions and expansions were shown to be dependent on light (Bohn, 1908a, 1910a; Parker, 1919) rather than on changes in the temperature or oxygen content of the water. Bohn (1908b, 1909) found that a normal cycle could be reversed experimentally by exchanging the light and dark periods, and he also maintained that controlled cycles could transplant the tidal rhythm. However, such effects were not substantiated by Parker (1919).

While both tidal and diurnal rhythms have been found in several anemones living in their natural habitats, the persistence of such rhythms under constant laboratory conditions has not invariably been established. Pieron (1908b), for example, was not able to substantiate Bohn's initial observations on the tidal rhythm in *Actinia.* Gee (1913) found no impressed diurnal or tidal rhythm in specimens of *Cribrina* removed to the laboratory. When placed under uniform illumination, they remained expanded for several days, whereas, when subjected to darkness they remained contracted for a similar period. Parker (1919) observed no persistent tidal rhythm in *Sagartia* and found that *Metridium* when placed in the dark remained expanded for over 36 hr.

In their extensive studies of spontaneous activity in *Metridium,* Batham and Pantin (1950) found that daily illumination initiates and controls regular daily phases of contraction. In observations made over a 20 day period, they noted that similar rhythmic activity occurred in animals placed in complete darkness and under constant environmental conditions. A subsequent new

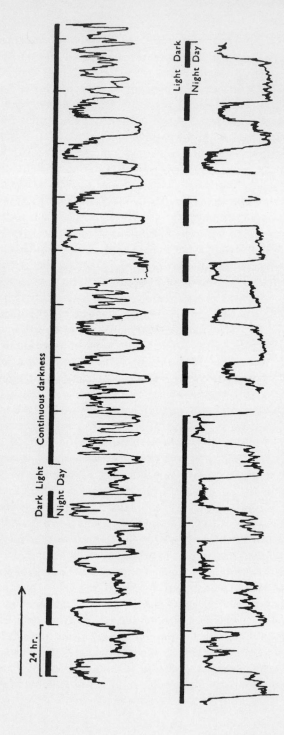

Fig. 3. Portion of 48-day kymograph record of activity of *Metridium*. Alternating 12 hr periods of light and dark at beginning correspond to day and night, at the end are the reverse of this. Note light-controlled rhythm. During the intermediate period of prolonged darkness, there develops an approximate phase rhythm unrelated to external diurnal or tidal rhythm. (After Batham and Pantin, 1950.)

light–dark regime was found to impose a new rhythm on the animal, causing a gradual phase shift in the activity cycle. During constant darkness, the pulses of contraction and expansion developed a rough regularity, with a period somewhat greater than 24 hr, and the phases became out of step with the previous periodic illumination (Fig. 3). Batham and Pantin (1950) also observed that the response started to begin slightly before the periodic changes in light intensity, recalling the claims of earlier workers that *Actinia* may "anticipate" tidal stimuli.

These early observations have ready interpretation in light of more recent work on biological rhythms in general (Bunning, 1967; Brown *et al.,* 1970) and on coelenterates in particular (Mori, 1960). Such periodic behavior is classified as a *circadian rhythm,* wherein a periodic stimulus causes endogenous rhythmic behavior to have the same period as the entraining cycle. Frequently the repeated stimulus is a light–dark cycle, but there are several examples of tidal rhythms (Enright, 1963). Some rhythmic phases are easily disturbed when the entraining stimulus is removed, such as the expansion and contraction rhythm of the anemone *Cymbactis* (Mori, 1948). On the other hand, other rhythms are relatively persistent (e.g., the expansion and contraction rhythm of the sea pen *Cavernularia* (Mori, 1948). The rhythms may be synchronized with different stimulus regimes providing the periods are within certain ranges relative to those of the naturally entraining stimulus. Since the endogenous behavior is frequently part of the continuum of activity rather than a discrete event, gradual increases in activity levels prior to onset of stimulation give the appearance of anticipatory events. The mechanisms underlying circadian rhythms are unknown, and controversy exists regarding the relative importance of extrinsic and endogenous factors in their control (Brown *et al.,* 1970). However, while such phenomena are certainly behavioral modifications, they are not generally considered as examples of learning.

Selective Extinction of Responses

There are several reports in the literature of behavioral modifications which may roughly be classified as the extinction of specific responses. Such examples have at various times been described as habit formation, manifestation of associative memory, and conditional reflexes. In no case have the physiological bases of such activities been investigated.

Jennings (1905) noted that when *Aiptasia* becomes extended its column often retains an irregular shape. He attributed this peculiar form to its life in irregular crevices beneath stones or in the hollows of coral reefs. He characterized the different irregular extensions of individual animals as *habits,* "the peculiarities of form being the structural correlates of habits." Jennings attempted to produce "new habits" in *Aiptasia* by repeatedly inducing it to extend in a new configuration. An animal attached to a plain horizontal glass surface was mechanically stimulated as it extended to the left side of the dish. This caused contraction of the animal, which was followed by its reextension to the same position. The procedure was continued for several trials, after which the animal was observed to bend over to the right. "Now when stimulated it contracted as before, but regularly reextended to the right. It seemed to have acquired a new habit—bending to the right instead of to the left" (Jennings, 1905, p. 461).

Bohn (1909) confirmed some observations on *Actinia* made by Van der Ghirst (1906) which were interpreted as demonstrating "associative memory." They found that if some polyps are taken from the underside of rocks and others from the upper surface and are placed between two plates of glass in an aquarium, the animals will take up roughly the positions they previously held. Polyps from the underside of the rocks adhere to the upper glass with their disks downward. Those from the upper surface attach to the lower glass with their disks directed upward. However, these original position "habits" may be eradicated and replaced by new ones after a period of 24–48 hr during which the animals are forced to assume new positions.

Zubkov and Polikarpov (1951) have observed the selective extinction of some responses in *Hydra*. They found that when animals were placed in a watch glass, those attached to the center of the dish and some distance from the water surface extended in any direction after a spontaneous contraction. Under constant illumination, these extensions occurred in the various directions in approximately equal frequencies. In contrast, in hydras situated near the surface of the water the extension behavior was quite different. After only a few contacts with the water surface, their movements toward it became highly restricted, and when the movements occurred they were of short duration. Movements in the opposite direction became relatively more frequent, and the extensions were more prolonged. This pattern of behavior was found to be established after about an·hour. If the water level in the glass was then raised, the selective extensions still persisted even though the physical restriction of the water surface had been removed. Usually it took 3–4 hr before a normal pattern of behavior was established, in which the extensions are equivalent in all directions. Zubkov and Polikarpov (1951) interpret this acquired behavior pattern as a conditional reflex, but it does not meet the criteria usually used for this type of learned response. Nevertheless, the observation is an interesting one and should be followed up with further studies.

Studies of Habituation

Several examples of habituation have been reported in coelenterates, and it has been studied most extensively in the freshwater *Hydra*. Jennings (1905) showed that an expanded specimen of the anemone *Aiptasia* initially contracts if a drop of water is allowed to fall onto the water surface just above the disk. If this stimulation is repeated with an interstimulus interval of less than 3 min, the animal does not contract after the first two or three drops. However, when the drops fall at intervals more than 5 min apart, the polyp contracts to each stimulus. Allabach (1905) observed a similar response to repeated mechanical stimulation in *Metridium*. A light stream of water directed with a pipette against the expanded disk causes the animal to contract. If such stimulation is repeated after reexpansion of the animal, the response is no longer elicited.

The contraction response of *Hydra viridis* to intermittent mechanical stimulation was first described by Wagner (1904). He observed that the animals contract at first when their culture dishes are tapped but reexpand if the stimulation is repeated at 1 sec intervals. Wagner found that if the interstimulus interval is increased so that the polyps expand after each contraction the animals continue to contract to each stimulus. He concluded from these observations that *Hydra* adapts to nonlocalized mechanical stimulus but that adaptation is dependent on the stimuli being rapidly repeated. He inferred that the recovery from the acclimatizing effect must be of short duration.

The effects of repeated mechanical stimulation on *Hydra pirardi* were investigated (Rushforth *et al.,* 1963; Rushforth, 1965*a*), and similar studies were extended to other species of *Hydra* (Rushforth, 1967, 1970).[1] In contrast to Wagner's results, it was observed that both *H. pirardi* and *H. viridis* adapt to nonlocalized mechanical stimulation even when the interval between stimuli is long enough to allow the animals to completely reexpand. In order to quantify the response, groups of hydras were placed in dishes on a rotary shaker. After the animals became attached, the dishes and their contents were rotated at a fixed rotation speed for a short predetermined time period and the numbers of hydras contracting were recorded. The rotation speeds could be systematically varied and were calibrated in revolutions per minute. The sensitivity of the hydras to this mechanical agitation is represented in the form of a response curve (Fig. 4a).

It was observed that if the hydras are exposed to repeated stimulation at a fixed rotation speed (e.g., a 3 sec period of agitation at 130 rpm every 30 sec), they become less responsive on continued exposure. The reduced sensitivity to the mechanical stimulus is reflected in a decrease in the proportion of animals contracting with an increasing number of stimulus trials (habituation curve, Fig. 4b). It may be demonstrated also by systematic lowering of the response curve from that observed before intermittent stimulation (Fig. 4c). The process of adaptation to the stimulus was not thought to be due to muscular fatigue, since light stimulation evokes contractions in hydras

[1]Supported in part by grants MH-10734 and GM-12302 from the National Institutes of Health.

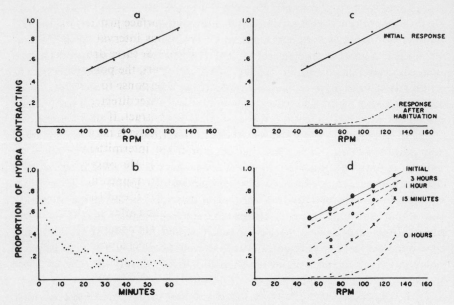

Fig. 4. (a) Sensitivity of *Hydra viridis* to mechanical stimulation: response curve. Each point represents the average proportion of hydras in five groups of ten animals that contracted within a 2 sec period that followed a 2 sec shaking period. (b) Effect of intermittent mechanical stimulation: habituation curve. Each point represents the average proportion of hydras that contracted within a 2 sec period that followed a 2 sec period of stimulation at 130 rpm every 30 sec, based on three groups of ten animals. (c) Effect of habituation on the sensitivity to mechanical stimulation: solid line, response curve of hydras prior to intermittent agitation; broken line, response curve of hydras after 6 hr of intermittent mechanical agitation. These curves are based on three groups of ten animals. (d) Decay of sensitivity to mechanical stimulation following habituation: response curves of hydras prior to intermittent agitation (solid line) and at various intervals (0 hr, 15 min, 1 hr, and 3 hr) following habituation during which the animals were not stimulated. (After Rushforth, 1967.)

unresponsive to mechanical agitation. Rather, the decrease in sensitivity to mechanical stimulation was considered to be an example of habituation. The animals show no evidence of habituation to repeated light stimulation over periods of several hours (Rushforth, *et al.,* 1963; Tardent and Frei, 1969).

The downward shifts in the response curves following habituation and their restoration to prestimulation levels provided a means of determining the length of retention of the habituated response. Specimens of *H. viridis* were habituated for a period of 6 hr after determination of the initial response curve. Response curves were determined immediately (0 hr), at 15 min intervals during the first hour, and then at hourly intervals, following the cessation of repeated stimulation (Fig. 4d). After approximately 4 hr, the response curve is not significantly different from the initial value (Rushforth, 1967). The modification in the contraction behavior of *H. pirardi* is retained for roughly a similar time period (Rushforth, 1965a). The effect of previous exposures to intermittent mechanical agitation on the subsequent response of the animal was also investigated. It was found that groups of *H. pirardi* habituated and then not stimulated for a 1 hr period habituate faster than control animals not previously agitated (Fig. 5). However, there appears to be no cumulative effect on the sensitivity to mechanical stimulation beyond that attained during the first habituating regime (Rushforth, 1967).

Thompson and Spencer (1966) have proposed several parametric relations which characterize habituation in intact organisms. The results of studies of the response of *Hydra* to repeated mechanical agitation may be summarized in terms of some of these criteria (Rushforth, 1965a, 1967). (a) Repeated applications of the mechanical stimulus result in a decreased response. The decrease is approximately a negative exponential function of the number of stimulus presentations. (b) If the stimulus is withheld, the original response recovers with time. (c) The shorter the interstimulus intervals, the more rapid is habituation. (d) Increasing the strength of the stimulus produces a slower rate of habituation. This result would not be expected if the process were one of fatigue. (e) The animals habituate faster on a second exposure to the initial habituation regime.

One parametric characteristic advocated by Thompson and Spencer— generalization of habituation to a stimulus in another part of the receptive

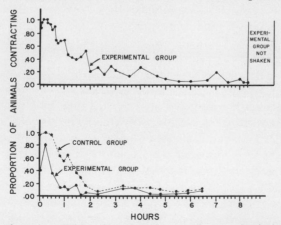

Fig. 5. Effect of previous habituation of *Hydra pirardi* to mechanical agitation on subsequent response to the stimulus: upper graph, habituation curve of an experimental group of ten hydras followed by a period of 1 hr during which the animals were not agitated; lower graph, habituation curves for the experimental group previously habituated and for a control group not previously shaken. (After Rushforth, 1965a.)

field—has not been demonstrated for the response of *Hydra* to mechanical agitation. However, stimulus generalization has not been found in other systems where the process is clearly one of habituation (Pinsker *et al.*, 1970). Other properties such as (a) dishabituation, (b) habituation of the dishabituatory stimulus with repeated presentations, and (c) prolongation of recovery following additional stimulation after the response has decremented to an asymptote remain to be established for the response of *Hydra* to repeated mechanical stimulation. However, some progress has been made in behavioral and electrophysiological studies in distinguishing the habituated response from other forms of inhibition in the animal.

B. Studies of the Physiological Bases of Habituation in *Hydra*

Studies of Behavioral Inhibition

The effects of repeated mechanical stimulation on two species of *Hydra* (*H. pirardi* and *H. viridis*) were found to conform with several of the criteria which have been used to define habituation operationally. However, it was found that a third species, *H. pseudoligactis,* does not habituate to similar regimes of mechanical agitation (Rushforth, 1970). In addition, it was observed that a decrease in sensitivity to mechanical stimulation could be induced for animals of all three species using both photic and chemical stimuli (Rushforth, 1965*b*, 1967, 1970). These forms of inhibition of mechanically induced contractions were studied in order to compare such responses with that produced by repeated mechanical stimulation.

The responsiveness of dark-adapted *Hydra* to mechanical agitation is markedly decreased by simultaneously exposing them to light (e.g., *H. viridis,* Fig. 6a). The duration of dark adaptation determines the extent of inhibition by light of contractions to mechanical stimulation, as does the wavelength of the light stimulus. The degree of inhibition increases gradually with increasing numbers of hours of dark adaptation up to roughly 24 hr (Fig. 6c). Studies of the spectral sensitivity of the response showed that inhibition of mechanically induced contractions is greatest with blue light (400–475 mμ, Fig. 6d). After dark-adapted animals have been exposed to light for 30 min, they have regained much of their former responsiveness to mechanical agitation (Fig. 6b), and after 2 hr the response is completely restored (Rushforth, 1967).

The contractions of *Hydra* induced by mechanical agitation are also suppressed when the animal is exposed to feeding stimuli: live *Artemia salina,* extracts of *Artemia,* or the tripeptide reduced glutathione (GSH). In the case of reduced glutathione, the sensitivity to mechanical stimulation is systematically lowered as the concentration is increased over the range 10^{-7} to 10^{-5} M (Rushforth, 1965*b*). In addition, the duration of decreased sensitivity is dependent on the GSH concentration. At 1×10^{-7} M GSH, restoration of the sensitivity to mechanical agitation is complete within an hour, but adaptation to 1×10^{-5} M GSH is much longer, requiring 6–7 hr for completion (Rushforth, 1967). Experiments using analogues of GSH showed that the molecular configuration of reduced glutathione is rather specific in its inhibitory effects on mechanically induced contractions. However, the sulfhydryl group is not essential for inhibition, since the S-methyl analogue is active. The substitution of large groups of the sulfhydryl group gives analogues which do not inhibit contractions. Thus neither S-acetyl nor oxidized glutathione blocks contractions stimulated by mechanical agitation. When these analogues are introduced into the medium, they compete with GSH, reducing its inhibitory effects (Rushforth, 1965b). The results of these investigations indicate that the action of GSH in suppressing contractions appears to have similar properties to those found by other workers studying the feeding response in *Hydra* (Loomis, 1955; Cliffe and Waley, 1958; Lenhoff and Bovaird, 1961).

Electrophysiological Correlates of Inhibited Responses

The responses of *Hydra* to external stimuli, mechanical agitation, light, and reduced glutathione are all modifications of the spontaneous contractions of the body column normally exhibited by undisturbed polyps. A series of electrophysiological studies has demonstrated several types of electrical potentials which are endogenously produced and which are correlated with overt behavior in the animal (Passano and McCullough, 1962, 1963, 1964, 1965; Josephson, 1967; Shibley, 1969; Rushforth, 1967, 1971; Rushforth and Burke, 1971). Accumulating evidence from such studies is beginning to provide some insight into the physiological bases of spontaneous contractions and their modifications by external stimuli.

Spontaneous Electrical Activity. Three major pulse types have been recorded from spontaneously active *Hydra* by use of external electrodes. (a) Large and rather slow triphasic pulses (up to 30–40 mv and lasting 200–500 msec) are associated in a one-to-one fashion with symmetrical contractions of the body column. These pulses (CPs) are produced by the *contraction-pulse pacemaker system*. They frequently originate in the hypostomal region and are conducted without decrement in the column at a velocity of 3–8 cm/sec. The temporal patterns of spontaneous CPs are species dependent; those in *H. pirardi* and *H. viridis* frequently occur in bursts, while those in *H. pseudoligactis* are produced primarily as single, widely spaced events. (b) Rhythmic potentials (RPs) are produced by the *rhythmic-potential pacemaker system,* whose primary locus resides in the base of the *Hydra*. These impulses consist of an initial relatively short monophasic potential followed by a slow compound pulse 80–150 msec later. The amplitudes of these pulses are generally two orders of magnitude less than CPs. The impulses are initiated

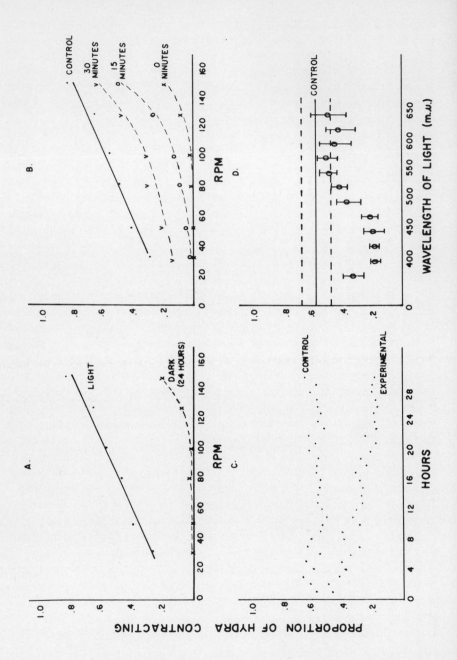

Fig. 6. Effect of light on the sensitivity of dark-adapted *Hydra viridis* to mechanical stimulation. (a) Solid line (light), response curve of 30 hydras tested in ambient light; broken line (dark 24 hr), response curve of 30 hydras dark-adapted for 24 hr and tested in ambient light. (b) Response curves of 150 hydras dark-adapted for 24 hr: solid line (control), tested in the dark; broken lines (0 min), immediately on exposure to ambient light, and after 15 and 30 min, respectively, in ambient light. (c) Effect of the length of dark adaptation on the sensitivity of hydras to mechanical stimulation. Each point represents the average proportion of a group of 60 hydras contracting to five shaking pulses at 130 rpm; the pulses were 1 min apart and of 5 sec duration. The experimental values were given by groups of hydras dark-adapted for different time periods and tested in ambient light, whereas the control values were given by groups of light-adapted hydras tested at the same times. (d) Sensitivity of dark-adapted hydras to mechanical stimulation when exposed to light of different wavelengths. Mean proportions and 95% confidence intervals are plotted for groups of 30 hydras dark-adapted for 24 hr, contracting to mechanical stimulation (five successive pulses of 5 sec of shaking every minute at 130 rpm) when exposed to monochromatic light. The control values (solid line, the mean; broken lines, the upper and lower 95% confidence limits) were determined for a group of 30 hydras dark-adapted for 24 hr and tested in the dark. (After Rushforth, 1967.)

endogenously in a more or less regular, rhythmic manner, but their frequency is highly variable even in unstimulated preparations. RPs are associated with contractions of the circular muscles in the gastrodermis. Thus CPs and RPs activate antagonistic muscle systems in *Hydra,* CPs triggering longitudinal muscle contractions producing shortening of the column, while RPs trigger the circular muscles giving rise to its elongation. In external recordings made at the base of intact hydras, *tentacle contraction pulses* (TCPs) are observed as small potentials (approximately 50 μv). These potentials are associated with symmetrical contractions of the longitudinal muscles of individual tentacles. In electrical recordings of endogenous TCPs made from the base of tentacles excised from *Hydra,* the impulses consist of large potentials up to 10 mv in amplitude and of approximately 200 msec duration. In all three species, TCPs occur spontaneoulsy only in bursts in these isolated tentacle preparations. The production of TCPs and CPs probably results from similar mechanisms, since these potentials have similar shape, amplitude, and temporal characteristics and are correlated with similar behavioral events.

Effects of External Stimuli on Electrical Activity. Contractions of the body column and associated CPs can be induced by short pulses of mechanical stimulation. In the case of *H. pirardi* and *H. viridis,* such stimuli produce single CPs (Fig. 7C) which supplant the normal spontaneous bursts of CPs characteristic of the animal. With repeated stimulation, however, many of the stimuli fail to induce CP firing, as the animal habituates to intermittent mechanical agitation. The endogenously produced bursts of CPs are gradually restored even though the hydra is repeatedly stimulated. In contrast, *H. pseudoligactis,* which normally spontaneously produces single CPs and not CP bursts, does not habituate to similar regimes of mechanical agitation. For example, when *H. pseudoligactis* was stimulated for a 2 sec period every minute it was observed that each pulse of agitation induced a single CP and an associated contraction of the body column. When the animal was stimulated in this manner for 3 hr, only about 5 % of the stimulus pulses did not elicit a CP, and at the end of the period there was no decrease in responsiveness to mechanical agitation.

When specimens of *H. pseudoligactis* are exposed to intermittent light stimulation, RP frequencies are enhanced in 2 min periods of direct photic stimulation, while bursts of CPs are induced during interposed periods of ambient illumination (Fig. 8). If such preparations are also exposed to repeated mechanical stimulation, light inhibits mechanically induced contractions and the associated CPs. During light stimulation, the RP system is strongly activated, and mechanically induced CPs occur in the 2 min periods of ambient light (Fig. 9).

The tripeptide GSH inhibits CPs associated with spontaneous column contractions or those induced by external stimuli (Rushforth, 1967, 1971).

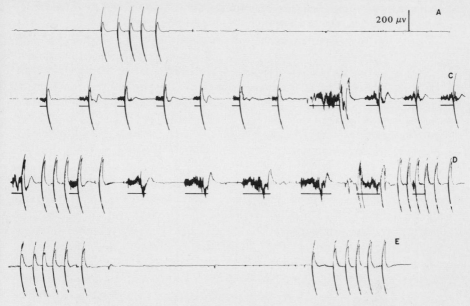

Fig. 7. Effect of repeated mechanical stimulation on the CP system in *Hydra pirardi*. A, Burst of spontaneous CPs in an unstimulated animal. C, Single CPs evoked by pulses of water-borne disturbances (unbroken line below the electrical record denotes the length of stimulation). D, Habituation of CP system to pulses of agitation with restoration of the normal CP burst pattern. E, Spontaneous pattern of CP bursts in unstimulated animal after cessation of mechanical agitation.

Figure 10 contrasts the electrical record of the unstimulated *H. pseudoligactis* spontaneously giving single CPs (C) with those of the same preparation exposed first to pulses of mechanical agitation at 1 min intervals (B), then exposed to 10^{-5} M GSH while continuing the mechanical stimulation (D). When GSH at this concentration is placed in the medium, both spontaneous and mechanically induced contractions are totally suppressed. After approximately an hour, the hydra fully adapts to GSH and is again responsive to mechanical stimulation (Fig. 10F).

Proposed Mechanisms for Inhibited Responses

Three examples of inhibition of mechanically induced CPs and column contractions in *Hydra* have been shown: (a) a reduction in responsiveness following repeated mechanical stimulation, (b) blockage by light stimulation in dark-adapted animals, and (c) suppression by GSH. While in no cases are the physiological mechanisms of such inhibition understood, habituation to mechanical agitation appears to be mediated by somewhat different processes

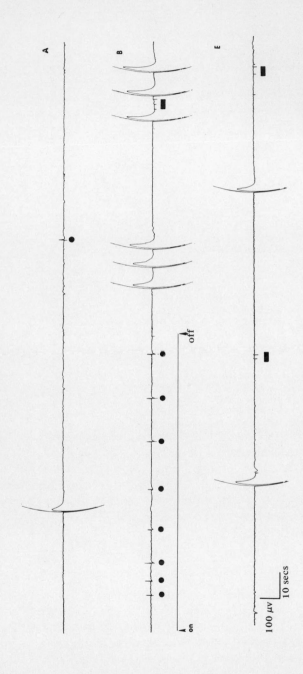

Fig. 8. Effect of light on RP and CP activity in *Hydra pseudoligactis*. A, Single CP and single RP in an unstimulated state. B, RPs evoked in a 2 min period of light stimulation (unbroken line, on to off), CPs and TCPs in a 2 min period of ambient illumination. E, CP singles and TCPs in the preparation 10 min after cessation of light stimulation. RPs are represented as closed circles, CP singles as closed circles, TCPs as closed rectangles.

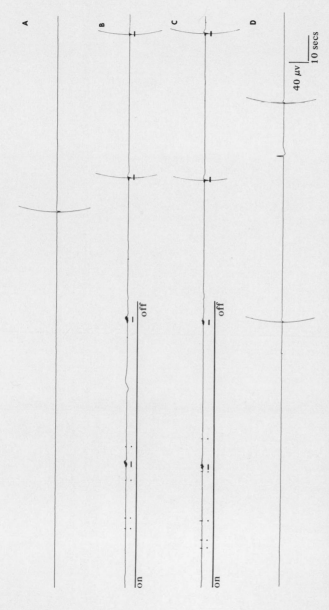

Fig. 9. Effects of light stimulation in blocking mechanically induced CPs in *Hydra pseudoligactis*. A, Spontaneous CP in unstimulated hydra. B and C, RPs evoked in a 2 min period of light stimulation (on to off) with no CPs elicited by pulses of mechanical agitation (unbroken line, and increased noise level in the record), while CPs are induced by mechanical agitation in following 2 min period of ambient illumination. D, Spontaneous CP singles in preparation after cessation of stimulation. RPs are represented as closed circles.

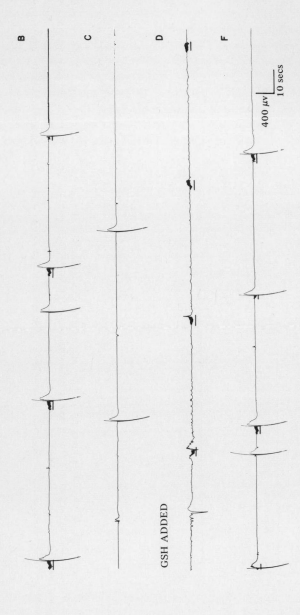

Fig. 10. Effects of GSH on mechanically induced CPs in *Hydra pseudoligactis*. B, CPs induced by pulses of mechanical agitation (unbroken line). C, Spontaneous CPs. D, Inhibition of mechanically induced CPs on exposure to 10^{-5} M GSH. F, Mechanically induced CPs in preparation after adaptation to GSH.

from those of inhibition by light and reduced glutathione. Such responses may be considered with reference to schematic models representing components of the CP and TCP pacemaker systems and the interactions among the various pacemaker systems in *Hydra* (Fig. 11).

Components of the CP and TCP Systems. On the basis of analyses of spontaneous pulse patterns and their modification by external stimuli, it has been proposed that the CP and TCP systems each possess two types of event generators, one producing single events and the other producing multiple events (Rushforth, 1971; Rushforth and Burke, 1971). In each system, the generators activate a conducting system which in turn triggers a pulse generator. The pulse generator produces CPs or TCPs in the respective systems (Fig. 11A).

The *multiple-event generator* has short-term positive feedback, leading to the production of bursts and the observed decrease of interval lengths in the initial phases of the burst. There is also slow rising inhibitory feedback to the

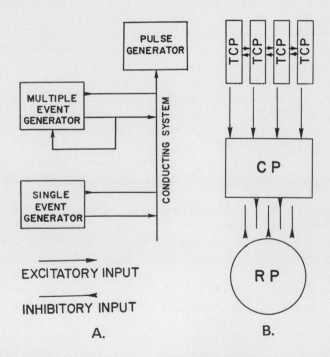

Fig. 11. A, Schematic model of the CP and TCP systems of *Hydra*. An arrow indicates excitatory input, a reverse arrow indicates inhibitory input. B, Schematic representation of the interactions among the various pacemaker systems in *Hydra*. See text for details.

multiple-event generator. This feedback probably is mediated via the conducting system, and the inhibitory effects of successive CPs are accumulative. Thus the increasing inhibition within a burst leads to increased intervals toward the end of the burst and eventually termination of the burst. The joint effects of the excitatory and inhibitory components give rise to the observed pattern of intervals within bursts; intervals are shortest in mid-burst (Rushforth, 1971; Rushforth and Burke, 1971).

The inhibitory effect of single pulses on the multiple-event generator, which acts via the conducting system, has been demonstrated by driving the CP system with single electrical stimuli. Triggering single CPs every 15–30 sec inhibits the production of CP bursts both in *H. pirardi* and *H. viridis* (Passano and McCullough, 1964; Rushforth and Burke, in preparation). Indirect evidence for the inhibition of CP bursts by single CPs was derived from an analysis of spontaneously produced inter-CP intervals in long-term recordings consisting of both single pulses and bursts. The interburst interval is significantly longer if it includes a single CP than if no CP is present in the interval. With increasing numbers of single CPs, the inter-CP burst interval is correspondingly increased in length (Rushforth, 1971).

There is also inhibitory feedback to the *single-event generator,* mediated by the conducting systems, which is evident when the system produces bursts. Such inhibition results in a decrease in the intervals between single CPs following a burst of CPs throughout the interburst interval. It suggests a gradual escape of the single-event generator from long-lasting inhibition following a CP burst. It also explains that whenever bursts of CPs are induced by external stimulation in *H. pseudoligactis* the normal production of single CPs is suppressed (Rushforth, 1971).

Interactions Among Pacemaker Systems. Much of the overt behavior of *Hydra* is coordinated through the interactions of the three principal pacemaker systems. Both behavioral studies and analyses of joint pulse patterns have provided evidence for interactions between (a) the CP and TCP systems and (b) the CP and RP systems. No RPs have been recorded from tentacles either attached to the polyp or isolated from it. Any interactions between TCPs and RPs appear indirect and do not play a major role in coordinating the animal's activities.

There is coupling between the CP and TCP systems so that TCP bursts tend to precede both single CPs and CP bursts. After a single CP in many preparations, and almost always after a CP burst, there is marked suppression TCP activity. Intervals between TCP bursts are significantly longer when CP firing takes place in the interval, and the elongation is much greater with a CP burst than with a single CP. Surgical removal of the tentacles gives rise to reduced CP frequency, but wounding of various parts of the animal does not have this effect (Rushforth, in preparation). By surgically creating a two-headed hydra and recording from each head, Passano and Kass-Simon(1969) found that distantly spaced tentacles stimulated each other to fire and that bursts of TCPs initiated CP bursts. Thus on the basis of these observations the following types of interactions are postulated: (a) excitation of the TCP system in tentacles by TCP firing in other tentacles, and (b) excitation of the CP system by TCP firing (Fig. 11B).

Both behavioral and electrophysiological evidence suggests that the CP and RP systems are mutually inhibitory (Fig. 11B). Direct evidence of RP inhibition of CPs comes from the finding that photic stimulation of the animal's base induces an increase in RP frequency accompanied by suppression of CP bursts (Passano and McCullough, 1965). Such stimulation has direct effects on the RP system, whose primary loci reside in the base of the hydra. Evidence of RP inhibition by CPs is given by the results of experiments in which *H. pirardi* specimens are jointly stimulated by light and GSH (Rushforth and Burke, in preparation). For this species, intermittent exposure to light results in increased frequencies of both CPs and RPs relative to interspersed periods of ambient illumination. However, it was found that animals exposed to 10^{-5} M GSH, which suppresses CP firing, have significantly greater RP frequencies during light stimulation than when they are exposed to light in the absence of GSH.

McCullough (1965) analyzed the joint firing patterns of the CP and RP systems. She suggested that RPs are generated in a cyclic manner and CP burst thresholds are attained when RP excitation has declined to low levels. The results of her analyses are consistent with the hypothesis that the CP and RP systems are mutually inhibitory. Probably the strongest evidence for this hypothesis stems from studies relating the production of the pulses with the movements of the body column. The impulses activate antagonistic muscle systems, CPs triggering longitudinal muscle contractions producing shortening of the column (Josephson, 1967), while RPs trigger the circular muscles which elongate it (Shibley, 1969).

Mode of Action of Mechanical, Photic, and Chemical Stimuli. Weak pulses of mechanical stimulation at first induce synchronous firing of single CPs, accompanied by a reduction in spontaneous CP burst frequencies in species such as *H. pirardi* and *H. viridis,* which endogenously produce bursts. As the polyps habituate to repeated stimulation, pulses of agitation cease to elicit firing of single CPs and the spontaneous patterns of CP bursts are restored. In *H. pseudoligactis,* a species which when unstimulated primarily produces single CPs, the response to mechanical agitation is long-lasting and there is not evidence of habituation. However, isolated tentacles of this species, which invariably produce TCP bursts, habituate to repeated mechanical stimulation (Rushforth and Burke, 1971). Pulses of agitation at first elicit firing of single TCPs, while TCP bursts are suppressed. As the preparation habituates, the mechanical stimulation no longer induces single TCPs and the spontaneous TCP burst pattern returns.

Since weak mechanical agitation, in the form of water-borne disturbances, does not affect the RP system (Passano and McCullough, 1965), it presumably acts directly on components of the CP and TCP systems. Such stimuli appear to trigger the single-event generator to produce single impulses. Activation of the conducting system in this process inhibits the multiple-event generator. As the preparation habituates to repeated stimuli, activity in the single-event generator declines, the conducting system is not activated, and the multiple-event generator escapes from its inhibitory effects. Recently, the inhibitory effects of mechanical stimulation on the multiple-event generating mechanisms have been further investigated (Rushforth and Burke, in preparation). Pulses of mechanical stimulation administered to *H. pirardi* after the first CP in a spontaneous CP burst were found to significantly reduce the number of pulses per burst.

The predominant effect of light in suppressing mechanically induced contractions appears to be on the RP system rather than on the other two pacemaker systems. Contractions to mechanical agitation may be inhibited by localized stimulation of the base of the hydra. With photic stimulation of the base or the whole animal, RP frequencies are markedly increased and the polyp elongates, often to one and a half times its dark-adapted length. In this extended state, the hydra is quite insensitive to mechanical stimulation. Secondary effects of light of the CP system occur, but these are somewhat complex, exhibiting both inhibitory and excitatory components (Passano and McCullough, 1964; Rushforth, 1967). Photic stimulation also induces the TCP system to fire in bursts when tentacles are attached to the parent hydra or excised from it (Rushforth, 1972; Rushforth and Burke, 1971). However, the primary mechanism for suppression of mechanically induced column contractions by light stimulation appears to result from excitation of the RP system, which in turn inhibits the production of CPs.

When hydras are exposed to 10^{-5} M GSH, the spontaneous contractions of the body column and tentacles and those induced by external stimuli are suppressed. The animal undergoes a relatively complex feeding response, consisting of sequences of tentacle movements and writhing activities which accompany mouth opening. Both single pulses and bursts of potentials produced by the CP and TCP systems are inhibited for periods of up to an hour. In isolated tentacles, TCP bursts are also blocked (Rushforth, 1971). Unlike photic stimulation, reduced glutathione does not directly affect the RP system, since spontaneous RP frequencies are unaltered by GSH (Passano and McCullough, 1965). Rather, the effects are on the components of the CP and TCP systems themselves. Possibly, GSH inhibits both the single-event and multiple-event generators, or it could operate by suppressing the pulse generator. However, this stimulus appears to have a different mode of action from those exhibited by light and repeated mechanical agitation.

C. Complex Behavioral Responses

Conditioning Experiments to Elicit Feeding Responses

Attempts have been made to condition coelenterates by linking stimuli with those normally eliciting components of complex feeding responses. In a short abstract, Hodgson and Hodgson (1966) report studies of habituation and conditioning in the sea anemone *Condylactis gigantea*. Various members of the species, living in different habitats, were observed to exhibit selective types of habituation to stimuli, depending on the specific stimuli present in the environment. The acquisition and loss of these modified behavior patterns were also demonstrated in the laboratory. Whole animals and isolated tentacles were found to have widely different responses to various amino acids, some of them exhibiting elements of feeding behavior. Pairing of feeding stimuli with electric shocks was shown to produce modifications of behavior which resembled conditioning. Such effects did not appear to result from sensitization, fatigue, or other general nonspecific mechanisms. Unfortunately, no further details of these provocative observations have been reported to date.

Tanaka (1966) has reported the establishment of a conditioned response in *Hydra japonica*. He first noted the response of the animal to stimulation from a jet of culture solution squirted onto the oral region. The hydra contracted to this form of mechanical stimulation. Then he observed that a jet of culture solution containing GSH when directed onto the polyp in a similar fashion produced a different response. The hydra did not contract but exhibited a considerable amount of tentacle writhing, a component of feeding behavior. This response, which he termed the *critical reaction,* was not observed in a group of 25 control hydras exposed to physical stress of the water jet alone (the conditioning stimulus, CS). Tanaka considered GSH to be the unconditioned stimulus (US) so that delivering it in the form of a jet of dilute GSH solution constituted a pairing of the CS and US. In a group of 22 experimental animals, critical reactions were observed in five members to the CS alone, after a variable number of joint applications of the CS and US. This result was interpreted as the establishment of a conditioned response in *Hydra*. However, in this experiment the GSH was not removed from the solution surrounding the polyp after each application of the CS and US. Thus the critical responses to subsequent CSs may have resulted from the effects of mechanical stimulation on hydras exposed to dilute solutions of GSH. Feeding responses have been observed when hydras adapted to GSH are mechanically stimulated (Burnett *et al.,* 1963), and therefore the behavior reported by Tanaka may not be a result of conditioning. Nevertheless, the observations are interesting and should be repeated with the use of more stringent controls.

Ross (1965a) performed conditioning experiments with *Metridium senile* by pairing electrical stimuli (CS) with food extracts (US). Such extracts induce mouth opening in the anemone, and low-frequency electrical pulses excite the nerve net without causing the animal to close up. Electrodes were placed on the margin of the base of small specimens of *Metridium*. In each trial, a drop of squid extract was placed on the disk while the animal was simultaneously electrically stimulated at a frequency of once per 5 sec. The extract usually elicited mouth opening, and the electrical stimuli were administered until the mouth closed. At the end of each trial, the sea water surrounding the animal was replaced. This operation often caused the polyp to close up so that the next trial could not be administered for about an hour or more.

After a sequence of ten trials, the animal was exposed to electrical stimuli in the absence of the extract, to determine if mouth opening was produced by this stimulus alone. Trials frequently were terminated when the response to the extract became sluggish. Sometimes the mouth would swell and remain partially open even when not stimulated, and frequently the tentacles would cluster around the mouth and obscure it. In these circumstances, the attempts to condition were stopped without definitive results. Thus in a group of 45 animals tested, 27 of them were discarded for one or more of these reasons. In seven of the animals, wide gaping of the mouth occurred to electrical stimuli alone after 20 trials, during which the stimuli were jointly administered. In three polyps, some mouth opening was observed, but the response was not consistent, and eight animals gave negative results. In the preparations which gave positive results, it was found that in the absence of reinforcement the response was extinguished after a few tests or after the passage of a few hours.

Ross interpreted these results as an indication that conditioning may take place in sea anemones. Further work with *Metridium* did not substantiate these early observations, but the conditions of stimulation were not identical. However, he did not pursue these experiments further but turned his attention to establishing a conditioned response in the swimming anemone *Stomphia coccinea* (Ross, 1965a,b).

Modifications of Swimming Responses

Most anemones are sessile animals whose movements are extremely slow. Except for the retraction and feeding responses, the behavior of these polyps consists of changes of shape and position too slow to be seen by the eye. However, there is a small group of anemones that exhibit swimming responses. Two different types of swimming are known. In *Gonactinia* and *Boloceroides,* contractions of the ectodermal muscles produce rowing movements of the tentacles, which propel these animals forward (Robson, 1966). The stimulus initiating such swimming activity appears to be mechanical in nature. In contrast, swimming in *Stomphia* consists of sharp bending movements of the body column which result from vigorous contractions of the endodermal musculature. In this animal, swimming is a response to specific chemical stimuli (Yentsch and Pierce, 1955). The swimming response of *Stomphia* to such stimuli has provided the most recent test of the learning capacity of a sea anemone.

In the early studies of swimming behavior in *Stomphia,* it was observed that the response was initiated by contact with either of two starfishes, *Hippasteria spinosa* or *Dermasterias imbricata* (Sund, 1958). Robson (1961*b*) extended these observations, finding that the anemone was responsive to a third starfish, *Hippasteria phrygiana,* and to the nudibranch *Aeolidia papillosa.* Since this nudibranch will feed on *Stomphia* and occurs in the same habitat, the swimming response has been regarded as an escape reaction to this predator. However, the response to the starfishes is difficult to understand, since they seem to affect the anemone in no other way. The general pattern of swimming behavior in *Stomphia* has been described as occurring in the following five phases after stimulation by *Dermasterias* (Robson, 1961*a*) (see Fig. 12):

(1) Initial response. Tentacles adhere to the starfish on contact. After several seconds they contract, followed at once by contraction of the sphincter and retractor muscles.

(2) Elongation. As the sphincter relaxes, a wave of elongation passes down the column. The tentacles and disk reappear and expand to their fullest extent.

(3) Detachment. At the same time the foot is released from the substratum, and unless it is very firmly attached it decreases in diameter. The centre of the pedal disk, initially convex, becomes delimited by a deepening circular groove, and is constricted to a conical projection.

(4) Swimming. A series of abrupt bending movements of the column ensues, which may cause the animal to "swim." Although the frequency of the contractions falls off, they may continue for several minutes. The bending movements may be preceded by a vigorous whirling of the column about its base.

(5) Recovery. An inactive period follows, during which the anemone resumes its normal shape. During this time it may not respond to renewed stimulation. The surface of the pedal disk presently exhibits a marked facility of adhesion, and as it reattaches to the substratum the ordinary sessile position is regained. (Robson, 1961*a*, p.344)

This swimming response is initiated by chemostimulation from the starfish. Ward (1962) found that tissue homogenates from the aboral surface of *Dermasterias* were as effective as entire animals. Homogenates from all other tissue gave negative results. The substance which causes the swimming

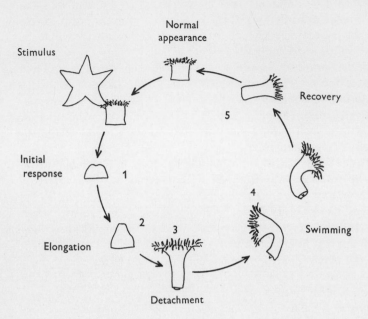

Fig. 12. Stages in the swimming behavior of *Stomphia coccinea* in response to the starfish *Dermasterias imbricata*. (After Robson, 1961*a*.)

response is dialyzable and thermostable, and qualitative tests indicated that it is primarily a carbohydrate. Chromatographic analysis showed that the active constituent is an aminopolysaccharide. The stimulus from the nudibranch *Aeolidia* is also chemical, but the active substance is not the same as that from starfish; Robson (1961*b*) found that it was not destroyed by boiling or dialysis and did not produce the rapid sphincter contraction induced by the starfish in the initial phase of swimming. In addition, the receptor sites for the two substances are different. Contact of *Stomphia*'s tentacles with the starfish initiates swimming. Similar stimulation by the nudibranch does not evoke the response, but the column of the anemone is responsive.

Several studies have shown that the swimming response of *Stomphia* may be activated by electrical stimuli (Yentsch and Pierce, 1955; Sund, 1958; Wilson, 1959; Robson, 1961*a*; Ross and Sutton, 1964*a*). Wilson (1959) claimed that there was a lowered threshold to the number of electrical shocks inducing swimming with successive trials, indicating a form of "learning" which he called *long-term facilitation*. However, such observations were not substantiated in later investigations, either of isolated rings of the body wall (Hoyle, 1960) or of intact *Stomphia* (Ross and Sutton, 1964*a*). The swimming response to electrical stimuli differs from that to *Dermasterias* in that the first few shocks cause the polyp to close completely. Further electrical stimuli are required for the anemone to detach and swim. A few stimuli (say, four or five at a frequency of 1/sec) may also induce delayed swimming without detachment, whereas the complete response occurs with more stimuli. Thus, unlike quick retraction, the swimming response to electrical stimuli is not an all-or-nothing event. Stronger stimuli must be applied to obtain swimming than are required to elicit the retraction response; stimuli of intermediate strength produce partial swimming movements (Ross and Sutton, 1964a).

The response to starfish is sometimes variable, and in different animals the behavior patterns may have different forms. Some *Stomphia* individuals exhibit swimming flexions before they detach, while in others the flexions always occur after detachment. Some anemones merely detach and give few flexions or none at all. Brief or very local contact with the starfish often produces delayed or incomplete responses. Longer or more massive stimulation is often required to give a full swimming activity. It is thought that chemoreceptors initiate impulses in nerve cells in the column which in turn excite a "pacemaker system" (Robson, 1961b) or "swimming control system" (Ross and Sutton, 1964a). Robson (1961b), on the basis of simple cutting experiments, postulated that the pacemaker system is localized in the median zone of the column. She further demonstrated the existence of widely dispersed multipolar nerve cells in the endodermal layer of the column of *Stomphia,* a cell type not located in the columns of sessile anemones such as *Metridium* and *Calliactis* (Robson, 1963). Thus it appears that the system controlling swimming is morphologically dispersed yet performs functions that are usually attributed to nerve cells aggregated in a specific location.

The parietobasilar muscles are chiefly involved in the anemone's detachment from the substratum. In the repeated flexions of the column during swimming, the parietobasilar and circular muscles function synergically. In response to starfish, there is contraction of the longitudinal muscles of the mesenteries and closure of the marginal sphincter. These events may be induced by electrical shocks (Sund, 1958). Hoyle (1960) has shown that electrical stimulation can cause slow movements of the parietal and circular muscles and can also evoke quick contractions of the parietobasilar muscles. These local rapid contractions occur to single stimuli if the electrodes are placed near the base, presumably close to the muscles.

Ross and Sutton (1964b) observed that the normal swimming response in *Stomphia* to objects rubbed on the aboral surface of *Dermasterias* was inhibited by flooding the tentacular crown with food extract. A localized inhibition of swimming was also found using the tentacles. Objects rubbed on *Dermasterias* caused swimming when brought into contact with a single tentacle. However, if the tentacle is first touched with a pipe cleaner dipped in food extract, it is unresponsive to *Dermasterias*. Other tentacles not exposed to food, when touched by a pipe cleaner rubbed on the starfish, cling to it, and a swimming response is induced. Ross and Sutton (1964*b*) also discovered a second type of inhibition; during swimming, the normal discharge of *Stomphia*'s nematocysts to food objects is suppressed. The disappearance of the nematocyst response coincides with detachment of the anemone from the substratum and remains until the reattachment of the basal disk.

Ross (1965*a*) utilized the specific chemical stimulus for swimming in *Stomphia* as the unconditioned stimulus, pairing it with conditioning stimuli in efforts to establish a conditioned response. The following three stimuli were used as conditioning stimuli: (a) A pipe cleaner previously dipped in food extract (from the clam *Pecten*) was touched on a single tentacle preceding contact of the tentacle with *Dermasterias*. (b) The starfish *Henricia* was exposed to the tentacles before their contact with *Dermasterias*. (c) Gentle pressure was applied simultaneously to the base of the *Stomphia* as the tentacles were exposed to the unconditioned stimulus. Each of these stimuli has an overt effect on the anemone. The tentacles cling to the pipe cleaner which has previously been dipped in food extract, and the swimming response to *Dermasterias* is inhibited. They also attach to the starfish *Henricia,* although it does not induce swimming behavior. Gentle pressure on the base causes partial retraction of the anemone, causing it to close. Using these stimuli, trials were performed at roughly hourly intervals and about 12 times per day. In about 20 consecutive treatments with *Dermasterias* alone, the swimming response was uniformly elicited. Often in subsequent trials, however, the swimming response was not fully established so that conditioning trials had to be terminated. Tests with the conditioning stimulus alone were applied after ten and 20 trials.

These tests did not provide evidence that *Stomphia* could be conditioned to any of three conditioning stimuli. But there was some indication of interaction between the US and CS in the case of the pressure stimulus. There was a weakening of the swimming response to *Dermasterias* which was interpreted as "conditioned inhibition" (Ross, 1965*a*). Further tests were undertaken to see if the response to the starfish could be suppressed by strong mechanical stimulation of the anemone's base. The starfish stimulus was applied to *Stomphia*'s tentacles, and 2 sec later the base was strongly stimulated. The mechanical stimulus was administered after the animal had fully opened up again. This sometimes required several hours so that the treatments ran over the course of 4 or 5 days. These trials were not performed as frequently as in the conditioning experiments, and different animals received widely differing numbers of treatments in the course of a single run.

In a group of eight experimental animals, all of which exhibited the swimming response to the starfish in five preconditioning trials, none of them swam to the first application of the starfish after the conditioning treatments. Each of the anemones responded to the *Dermasterias* on the first test by retracting, as if it had received mechanical stimulation. If the starfish stimulus was repeated in further trials, one by one the animals began to respond by swimming. It was found that animals which were given fewer conditioning trials in general required fewer subsequent exposures to the starfish before swimming behavior was restored. There was a rough correspondence between the number of conditioning treatments and the number of trials before swimming reappeared.

Ross (1965*a*) interpreted these results as evidence of conditioning in *Stomphia,* where after coupling the starfish stimulus (CS) with strong mechanical stimulation of the base of the anemone (US) the normal swimming response to the starfish was replaced by a retraction response. The relationship between the number of conditioning trials and the duration of the acquired response suggested a reinforcement–extinction relationship characteristic of true conditioning. Studies repeating and extending these investigations should be undertaken, since to date the phenomena constitute the most promising indications of associative learning in coelenterates.

V. SUMMARY AND CONCLUSIONS

The components which play a major role in controlling coelenterate behavior are the conducting systems and pacemaker units. Until recently, such components always have been equated with neuronal elements, frequently organized in simple, diffuse nerve nets. Only in the last decade have studies with the electron microscope revealed the complexity of interconnections in these nervous pathways, suggesting neural transmission similar to that in higher animals. Physiological studies have shown that in all of the coelenterate groups there are animals with multiple conducting systems. These systems activate musculature, couple different pacemakers, and modulate pacemaker and effector activity. A considerable advance in coelenterate physiology has been the discovery that epithelial cells are electrically excitable and can transmit behaviorally important signals. Nonnervous conduction is currently under investigation, since it may provide mechanisms of behavioral control which help to explain the inconsistency between the complex behavioral patterns and simple nervous organization seen in this phylum. Such mechanisms are of considerable interest, since they may be evolutionary forerunners of nervous functioning.

The most detailed description of a behavioral modification has been given for the response of *Hydra* to mechanical agitation. The body column of the polyp contracts to weak stimulation in the form of water-borne disturbances. The species *H. pirardi* and *H. viridis,* which spontaneously produce bursts of column contractions, adapt to repeated stimulation. Parameters of such adaptation satisfy many of the criteria used to operationally define habituation. The duration of this behavioral modification is longer than effects such as sensory adaptation and motor fatigue, whose time course of recovery is relatively short. The species *H. pseudoligactis,* which endogenously produces single contractions rather than contraction bursts, does not habituate to similar regimes of mechanical agitation. In all three species, a decrease in responsiveness to mechanical agitation may be induced by photic stimulation or exposure to reduced glutathione. The responses of *Hydra* to the various external stimuli are all modifications of the spontaneous contractions of the body column normally exhibited by unstimulated animals.

Electrophysiological recordings have been employed together with behavioral observations to investigate differences between habituation and the other forms of inhibition of mechanically induced contractions. On the basis of these studies, the following differences have been proposed (Rushforth, 1970): (a) Weak mechanical stimulation elicits single contractions of the body column (and associated single CPs) and inhibits spontaneous bursts of contractions (and CP bursts). In a model constructed for the components of the contraction pulse system, single-event generators trigger single CPs and multiple-event generators inhibit CP bursts. (b) Inhibition of mechanically induced contractions and associated CPs in dark-adapted hydras by light stimulation acts via excitation of the rhythmic-potential system (which produces RPs). This system is inhibitory to the contraction-pulse system. (c) Reduced glutathione (GSH) inhibits the contraction-pulse system from producing both single CPs and CP bursts, but has no direct influence on the rhythmic-potential system. Possibly, GSH inhibits both the single-event and multiple-event generators, or it might operate by suppressing the pulse generator of the CP system.

The cellular mechanisms underlying processes of habituation require the methods of intracellular recording for their elucidation. Such methods have not as yet been successfully applied to coelenterate systems. A major problem is the identification of the cellular elements in the production and propagation of the electrical events in *Hydra.* In their studies of the coordinating systems in this hydroid, Passano and McCullough (1964, 1965) have speculated as to the nature of the pacemaker units and conducting systems of both the RP and CP systems. They proposed that the pacemakers of the RP system are gastrodermal protoneurons, with the lead pacemakers located in the base. The conducting system was equated with the action system of the circularly arranged gastrodermal muscles.

Passano and McCullough (1965) proposed that nerve cells in the sub-hypostomal region of the epidermis are the pacemaker units of the CP system and that the conducting system is the epidermal nerve net. However, the size of the CPs in *Hydra* and that of similar pulses in other hydroids appear far too large to be the exclusive product of the small dispersed neurons in the nerve net. Josephson and Macklin (1967, 1969) showed that CPs are correlated with epithelial activity and can be recorded as a potential change occurring across the body wall of the whole animal. They proposed that the inner membranes of the ectodermal epitheliomuscular cells, which surround the contractile elements, are the sites of CP production. However, although the recorded CP represents an epithelial depolarization, the primary excitation may result from neuronal events. Conduction of CPs may take place via the nerve net or by the epitheliomuscular cells themselves. Nerve-free epithelia in coelenterates can propagate electrical impulses (Mackie and Passano, 1968), and septate desmosomes have been shown to provide low-resistance pathways between adjacent epithelial cells in other animals (Lowenstein and Kanno, 1964). Septate desmosomes have been discovered between adjacent epithelial cells of *Hydra* near the inner and outer surfaces of the column (Wood, 1959; Haynes *et al.,* 1968). Josephson and Macklin (1969) suggest that if CP conduction in *Hydra* is due to epithelial cells, it results from current flow through such junctions between adjacent cells.

Intracellular recording techniques have been used to study habituation in both vertebrate systems (Spencer *et al.,* 1966; Segundo *et al.,* 1967) and several invertebrate preparations (Kennedy and Mellon, 1964; Kupfermann *et al.,* 1969; Bruner and Kennedy, 1970). Two general classes of synaptic mechanisms have been proposed to account for habituation (Wickelgren, 1967). Theories advocating various types of fatigue at any or all of the synapses between the primary afferent endings and interneurons have been termed *synaptic depression theories.* Those theories proposing increased activity at synapses which inhibits excitatory pathways to motoneurons have been called *inhibitory build-up theories.* However, the processes of habituation and dishabituation may be displayed at a single neuromuscular junction and therefore do not require a complex central nervous network for their execution (Bruner and Kennedy, 1970).

Two factors have aided considerably the recent investigations of cellular mechanisms of habituation in invertebrates other than coelenterates: the use of intracellular recording techniques and the construction of idealized wiring diagrams for the neuronal components of the behavioral system. As pointed out by Kandel and Kupfermann (1970), mechanisms of habituation have been studied most successfully for the withdrawal reflexes where the wiring diagrams are quite simple. For *Hydra,* the contraction response to mechanical stimulation may be classified as a withdrawal reflex. Presumably, it has an adaptive advantage as an escape reaction to predators who transmit water-borne disturbances to the polyp. Habituation to repeated stimulation appears advantageous to hydras attached to rocks and plants submerged in a rapidly moving stream, for expansion of the column and tentacles provides a better food-collecting configuration than the constantly contracted state. The hypothetical wiring diagram for the withdrawal reflex in *Hydra* is complex, since the response is superimposed on spontaneously active systems that exhibit both single potentials and bursts of pulses. Mechanical stimuli may act via sensory nerve cells operating as simple stretch receptors, or such stimuli may have direct effects on the electrogenic properties of the epithelio-muscular cells themselves. The inability to make intracellular recordings from nerve and epitheliomuscular cells is a basic deterrent to further study of the mechanisms of habituation in *Hydra*. Nevertheless, the studies which have been reviewed show that habituation of the contraction response to mechanical stimulation is mediated by different processes from those giving rise to inhibition of mechanically induced contractions by external stimuli such as light and reduced glutatione.

The most promising indication of associative learning comes from studies utilizing the swimming response of the anemone *Stomphia*. Ross (1965a) paired the stimulus of pressure applied to the base of the polyp with that of chemostimulation from the starfish *Dermasterias*. There was some evidence that after linking these stimuli the normal swimming response to the starfish was inhibited and temporarily replaced by retraction, a response usually evoked by the mechanical stimulation. It is probable that the clear-cut demonstration of a conditioned response in a coelenterate will be made with similar strategies using various components of anemone behavior. Recent work has shown that these coelenterates may exhibit elaborate programs of activities consisting of several sequential events (Ross 1965b). For example, *Calliactis* transfers itself to an adjacent gastropod shell in a sequence of five distinct stages, while *Stomphia* undergoes equally complex behavior in its preferential settling response on shells of a mussel, after a swimming bout. The elucidation of mechanisms for each event affords a current challenge to the coelenterate physiologist.

Many of the studies of behavioral modifications were undertaken at the beginning of the present century. At that time, there was a widespread view that the behavior of lower organisms was composed of invariable reflexes, occurring always in the same way under the same external circumstances. Parker (1919), in his classic treatise *The Elementary Nervous System,* characterized the structural elements of these reflexes in coelenterates as simple receptor–effector systems—triads of receptor cells, protoneurons, effector cells. The work of Loeb, Nagel, and Parker led to a general conception of the sea anemone as "a delicately adjusted mechanism whose activities were made up of a combination of simple responses to immediate stimulation" (Parker, 1919, p. 172). This view, however, was criticized by Jennings as giving a much oversimplified picture. Jennings drew attention to the physiological state of the animal and the resulting effects of previous stimulation in determining its reactions. Several studies of behavioral modification—those of Jennings, Bohn, Pieron, and much later Batham and Pantin, drew attention to the role of spontaneity in coelenterate behavior. Today, we realize that endogenous activity in neural units is widespread, ranging from simple invertebrate ganglia to complex brain structures. Such activity underlies many components of behavior, since the internal events on which much of behavior is based are patterned in time, from the activity of a single pacemaker cell to that of the circadian rhythm (Aschoff, 1961). Now the prevailing belief is that both internal and external factors are important in patterns of behavior. The animal does not passively respond to an environmental stimulus but plays an active role in interacting with it.

REFERENCES

Allabach, L. F., 1905. Some points regarding the behavior of *Metridium*. *Biol. Bull., 10,* 35–43.

Aschoff, J., 1961. Exogenous and endogenous components in circadian rhythms. *Cold Spring Harbor Symp. Quant. Biol., 25,* 11–28.

Batham, E. J., and Pantin, C. F. A., 1950. Phases of activity in the sea-anemone *Metridium senile* (L) and their relation to external stimuli. *J. Exptl. Biol., 27,* 377–399.

Bohn, G., 1906. La persistance du rythme des marées chez l'*Actinia equina*. *Compt. Rend. Soc. Biol., 61,* 661–663.

Bohn, G., 1907. Le rythme nycthéméral chez les Actinies. *Compt. Rend. Soc. Biol., 62,* 473–476.

Bohn, G., 1908*a*. De l'influence de l'oxygène dissous sur les réactions des Actinies. *Compt. Rend. Soc. Biol., 64,* 1163–1166.

Bohn, G., 1908*b*. Les facteurs de la rétraction et de l'épanouissement des Actinies. *Compt. Rend. Soc. Biol., 64,* 1163–1166.

Bohn, G., 1909. Les rythmes vitaux chez les Actinies. *Compt. Rend. Ass. Franc. Av. Sci., 37,* 613–619.

Bohn, G., 1910*a*. Comparison entre les réactions des Actinies de la Méditerranée et celles de la Manche. *Compt. Rend. Soc. Biol., 68,* 253–255.

Bohn, 1910*b*. Les réactions des Actinies aux basses températures. *Compt. Rend. Soc. Biol., 68,* 964–966.

Bohn, G., and Pieron, H., 1906. Le rythme des marées et la phénomène de l'anticipation reflexe. *Compt. Rend. Soc. Biol., 64,* 660–661.

Brown, F. A., Hastings, J. W., and Palmer, J. D., 1970. *The Biological Clock—Two Views,* Academic Press, New York.

Bruner, J., and Kennedy, D., 1970. Habituation: Occurrence at a neuromuscular junction. *Science, 169,* 92–94.

Bullock, T. H., and Horridge, G. A., 1965. *Structure and Function of the Nervous System of Invertebrates,* Freeman, San Francisco.

Bunning, E., 1967. *The Physiological Clock,* Springer-Verlag, New York.

Burnett, A. L., and Diehl, N. A., 1964. The nervous system of *Hydra*. I. Types and distribution of nerve elements. *J. Exptl. Zool., 157,* 217–226.

Burnett, A. L., Davidson, R., and Wiernick, P., 1963. On the presence of a feeding hormone in the nematocyst of *Hydra pirardi*. *Biol. Bull., 125*, 226–233.

Cliffe, E. E., and Waley, S. G., 1958. Effect of analogues of glutathione on the feeding reaction of *Hydra*. *Nature, 182*, 804–805.

Davis, L. E., Burnett, A. L., and Haynes, J. F., 1968. Histological and ultrastructural study of the muscular and nervous system in *Hydra*. II. Nervous system. *J. Exptl. Zool., 167*, 295–332.

Enright, J. J., 1963. Endogenous tidal and lunar rhythms. *Proc. 16th Internat. Congr. Zool., 4*, 355–359.

Fleure, H. J., and Walton, C. L., 1907. Notes on the habits of some sea-anemones. *Zool. Ang., 31*, 212–230.

Gee, W., 1913. Modifiability in the behavior of the California shore-anemone *Cribrina xanthogrammic* Brandt. *Anim. Behav., 3*, 305–328.

Hargitt, C. W., 1907. Notes on the behavior of sea-anemones. *Biol. Bull., 12*, 174–284.

Haynes, J. F., Burnett, A. L., and Davis, L. E., 1968. Histological and ultrastructural study of the muscular and nervous systems in *Hydra*. I. The muscular system and the mesoglea. *J. Exptl. Zool., 167*, 283–294.

Hodgson, V. S., and Hodgson, E. S., 1966. Habituation and discrimination in sea anemones. *Am. Zoologist, 6*, 542.

Horridge, G. A., 1954. The nerves and muscles of medusae. I. Conduction in the nervous system of *Aurellia aurita* Lamarck. *J. Exptl. Biol., 31*, 594–600.

Horridge, G. A., 1955. The nerves and muscles of medusae. IV. Inhibition of *Aequorea forskalea*. *J. Exptl. Biol., 32*, 642–648.

Horridge, G. A., 1956. The nerves and muscles of medusae. V. Double innervation in Scyphozoa. *J. Exptl. Biol., 33*, 366–383.

Horridge, G. A. 1959. The nerves and muscles of medusae. VI. The rhythm. *J. Exptl. Biol., 36*, 72–91.

Horridge, G. A., and MacKay, B., 1962. Naked axons and symmetrical synapses in coelenterates. *Quart. J. Microscop. Sci., 103*, 531–541.

Hoyle, G., 1960. Neuromuscular activity in the swimming sea anemone. *Stomphia coccinea* (Muller). *J. Exptl. Biol., 37*, 671–688.

Jennings, H. S., 1905. Modifiability in behavior. I. Behavior of sea anemones. *J. Exptl. Zool., 2*, 447–473.

Jennings, H. S., 1906. *Behavior of the Lower Organisms*, Columbia University Press, New York.

Jha, R. K., and Mackie, G. O., 1967. The recognition, distribution and ultrastructure of hydrozoan nerve elements. *J. Morphol., 123*, 43–61.

Josephson, R. K., 1961. Colonial responses of hydroid polyps. *J. Exptl. Biol., 38*, 559–577.

Josephson, R. K., 1965a. Three parallel conducting systems in the stalk of a hydroid. *J. Exptl. Biol., 42*, 139–152.

Josephson, R. K., 1965b. The co-ordination of potential pacemakers in the hydroid *Tubularia*. *Am. Zoologist, 5*, 483–490.

Josephson, R. K., 1967. Conduction and contraction in the column of *Hydra*. *J. Exptl. Biol., 47*, 179–190.

Josephson, R. K., 1968. Functional components of systems controlling behavior in some primitive animals. In Mesarovic, M. D. (ed.), *Systems Theory and Biology*, Springer-Verlag, New York, pp. 246–260.

Josephson, R. K., and Mackie, G. O., 1965. Multiple pacemakers and the behavior of the hydroid *Tubularia*. *J. Exptl. Biol., 43*, 293–332.

Josephson, R. K., and Macklin, M., 1967. Transepithelial potentials in *Hydra*. *Science, 156*, 1629.

Josephson, R. K., and Macklin, M., 1969. Electrical properties of the body wall of *Hydra*. *J. Gen. Physiol., 53,* 638–665.

Josephson, R. K., and Uhrich, J., 1969. Inhibition of pacemaker systems in the hydroid *Tubularia. J. Exptl. Biol., 50,* 1–14.

Kandel, E. R., and Kupfermann, I., 1970. The functional organization of invertebrate ganglia. *Ann. Rev. Physiol., 32,* 193–268.

Kennedy, D., and Mellon, DeF. M., 1964. Synaptic activation and receptive fields in crayfish interneurons. *Comp. Biochem. Physiol., 13,* 275–300.

Kupfermann, I., Castellucci, V., Pinsker, H., and Kandel, E. R., 1969. Neuronal correlates of habituation and dishabituation of the gill withdrawal reflex in *Aplysia. Science, 167,* 1743–1745.

Lenhoff, H. M., and Bovaird, J., 1961. Action of glutamic acid and glutathione analogues on the *Hydra* glutathione-receptor. *Nature, 189,* 486–487.

Lentz, T. L., and Barrnett, R. J., 1961. Enzyme histochemistry of hydra. *J. Exptl. Zool., 147,* 125–150.

Lentz, T. L., and Barrnett, R. J., 1962. The effect of enzyme substrates and pharmacological agents on nematocyst discharge. *J. Exptl. Zool., 149,* 33–38.

Lentz, T., and Barrnett, R. J., 1965. Fine structure of the nervous system of *Hydra. Am. Zoologist, 5,* 341–356.

Loeb, J., 1918. *Forced Movements, Tropisms, and Animal Conduct,* Lippincott, Philadelphia.

Loomis, W. F., 1955. Glutathione control of the specific feeding reactions of *Hydra. Ann. N.Y. Acad. Sci., 62,* 209–228.

Lowenstein, W. R., and Kanno. Y., 1964. Studies on an epithelial (gland) cell junction. I. Modifications of surface membrane permeability. *J. Cell Biol., 22,* 565.

Mackie, G. O., 1960. The structure of the nervous system in *Velella. Quart. J. Microscop. Sci., 101,* 119–131.

Mackie, G. O., 1965. Conduction in the nerve-free epithelia of siphonophores. *Am. Zoologist, 5,* 439–453.

Mackie, G. O., 1970. Neuroid conduction and the evolution of conducting tissues. *Quart. Rev. Biol., 45,* 319–332.

Mackie, G. O., and Passano, L. M., 1968. Epithelial conduction of hydramedusae. *J. Gen. Physiol., 52,* 600–621.

McConnell, C. H., 1932. The development of the ectodermal nerve net in the buds of *Hydra. Quart. J. Microscop. Sci., 75,* 495–509.

McCullough, C. B., 1965. Pacemaker interaction in *Hydra. Am. Zoologist, 5,* 499–504.

McFarlane, F. D., 1969. Two slow conduction systems in the sea anemone *Calliactis parasitica. J. Exptl. Biol., 51,* 377–385.

Mori, S., 1948. Harmony between behavior rhythm and environmental rhythm. *Mem. Coll. Sci. Univ. Kyoto Ser. B, 19,* 71–74.

Mori, S., 1960. Influence of environmental and physiological factors on the daily rhythmic activity of a sea-pen. *Cold Spring Harbor Symp. Quant. Biol., 25,* 333–344.

Nagel, W. A., 1894. Experimentelle sinnesphysiologische Untersuchungen an Coelenteraten. *Arch. Ges. Physiol., 57,* 493–552.

Noda, K., 1969. On the nerve cells of *Hydra:* a light microscopic study. *Annot. Zool. Jap., 42,* 105–112.

Pantin, C. F. A., 1952. The elementary nervous system. *Proc. Roy. Soc. London Ser. B, 140,* 147–168.

Parker, G. H., 1896. The reactions of *Metridium* to food and other substances. *Bull. Mus. Harvard, 29,* 107–119.

Parker, G. H., 1919. *The Elementary Nervous System,* Lippincott, Philadelphia.

Passano, L. M., and Kass-Simon, G. 1969. Tentacle pulses: A new through conducted coordinating system in *Hydra*. *Am. Zoologist, 9,* 1113.

Passano, L. M., and McCullough, C. B., 1962. The light response and rhythmic potentials of *Hydra*. *Proc. Natl. Acad. Sci., 48,* 1376–1382.

Passano, L. M., and McCullough, C. B., 1963. Pacemaker hierarchies controlling the behaviour of hydras. *Nature, 199,* 1174–1175.

Passano, L. M., and McCullough, C. B., 1964. Coordinating systems and behavior in *Hydra*. 1. Pacemaker systems of the periodic contractions. *J. Exptl. Biol., 41,* 643–664.

Passano, L. M., and McCullough, C. B., 1965. Co-ordinating systems and behavior in *Hydra*. II. The rhythmic potential system. *J. Exptl. Biol., 42,* 205–231.

Pieron, H., 1908a. De l'influence réciproques des phénomènes respiratoires de du comportement chez certaines Actinies. *Compt. Rend. Acad. Sci. Paris, 147,* 1407–1410.

Pieron, H., 1908b. Des rythmes engendres par une variation périodique de la teneur en oxygène. *Compt. Rend. Soc. Biol., 64,* 1020–1022.

Pieron, H., 1910. L'étude expérimentale de l'anticipation adaptive. *Compt. Rend. Ass. Franc., 38,* 735–739.

Pinsker, H., Kupfermann, I., Castellucci, V., and Kandel, E. R., 1969. Habituation and dishabituation of the gill-withdrawal reflex in *Aplysia*. *Science, 167,* 1740–1742.

Robson, E. A., 1961a. Some observations on the swimming behaviour of the anemone *Stomphia coccinea*. *J. Exptl. Biol., 38,* 343–363.

Robson, E. A., 1961b. The swimming response and its pacemaker system in the anemone *Stomphia coccinea*. *J. Exptl. Biol., 38,* 685–694.

Robson, E. A., 1963. The nerve-net of a swimming anemone, *Stomphia coccinea. Quart. J. Microscop. Sci., 104,* 535–549.

Robson, E. A., 1966. Swimming in Actiniaria. In Reese, W. J. (ed.), *The Cnidaria and Their Evolution*, Academic Press, New York, pp. 333–360.

Romanes, G. J., 1885. *Jelly-fish, Star-fish and Sea-urchins*, D. Appleton and Co., New York.

Romanes, G. J., 1877. Further observations on the locomotor system of medusae. *Phil. Trans. Roy. Soc. London, 167,* 659–752.

Ross, D. M., 1965a. The behavior of sessile coelenterates in relation to some conditioning experiments. *Anim. Behav. Suppl., 1,* 43–55.

Ross, D. M., 1965b. Complex and modifiable behavior patterns in *Calliactis* and *Stomphia*. *Am. Zoologist, 5,* 573–580.

Ross, D. M., and Sutton, L., 1964a. The swimming response of the sea anemone *Stomphia coccinea* to electrical stimulation. *J. Exptl. Biol., 41,* 735–749.

Ross, D. M., and Sutton L., 1964b. Inhibition of the swimming response by food and nematocyst discharge during swimming in the sea anemone *Stomphia coccinea*. *J. Exptl. Biol., 41,* 751–757.

Rushforth, N. B., 1965a. Behavioral studies of the coelenterate *Hydra pirardi* Brien. *Anim. Behav. Suppl., 1,* 30–42.

Rushforth, N. B., 1965b. Inhibition of contraction responses of *Hydra*. *Am. Zoologist, 5,* 505–513.

Rushforth, N. B., 1967. Chemical and physical factors affecting behavior in *Hydra*: Interactions among factors affecting behavior in *Hydra*. In Corning, W. C., and Ratner, S. C. (eds.), *Chemistry of Learning*, Plenum Press, New York.

Rushforth, N. B., 1970. Electrophysiological correlate of habituation in *Hydra*. *Am. Zoologist, 10,* 505.

Rushforth, N. B., 1971. Behavioral and electrophysiological studies of *Hydra*. I. An analysis of contraction pulse patterns. *Biol. Bull., 140,* 255–273.

Rushforth, N. B., and Burke, D. S., 1971. Behavioral and electrophysiological studies of *Hydra*. II. Pacemaker activity of isolated tentacles. *Biol. Bull., 140,* 502–519.

Rushforth, N. B., Burnett, A. L., and Maynard, R., 1963. Behavior in *Hydra*. Contraction responses of *Hydra pirardi* to mechanical and light stimuli. *Science, 139,* 760–761.

Segundo, J. P., Takenaka, T., and Encado, H., 1967. Electrophysiology of bulbar reticular neurons. *J. Neurophysiol., 30,* 1194–1220.

Semal van Gansen, P., 1952. Note sur le système nerveux de l'hydre. *Bull. Acad. Roy. Belg. Cl. Sci., 38,* 718–735.

Shibley, G. A., 1969. Gastrodermal contractions correlated with rhythmic potentials and prelocomotor bursts in *Hydra*. *Am. Zoologist, 9,* 586.

Spencer, W. A., Thompson, R. F., and Neilson, D. R., Jr., 1966. Response decrement of the flexion reflex in the acute spinal cat and transient restoration by strong stimuli. *J. Neurophysiol., 29,* 221–239.

Sund, P. N., 1958. A study of the muscular anatomy and swimming behavior of the sea anemone *Stomphia coccinea*. *Quart. J. Microscop. Sci., 99,* 401–420.

Tanaka, J., 1966. A study of conditioned response in *Hydra*. *Ann. Anim. Psychol., 16,* 37–41.

Tardent, P., and Frei, E., 1969. Reaction patterns of dark- and light-adapted *Hydra* to light stimuli. *Experientia, 25,* 265–267.

Thompson, R. F., and Spencer, W. A., 1966. Habituation: A model for the study of the neuronal substrates of behavior. *Psychol. Rev., 73,* 16–43.

Thorpe, W. H., 1963. *Learning and Instinct in Animals,* Methuen, London.

Van der Ghirst, G., 1906. Quelques observations sur les Actinies. *Bull. Inst. Gen. Psychol., 6,* 267–275.

Wagner, G., 1904. On some movements and reactions of *Hydra*. *Quart. J. Microscop. Sci., 48,* 585–622.

Ward, J., 1962. A further investigation of the swimming response of *Stomphia coccinea*. *Am. Zoologist, 2,* 567.

Westfall, J. A., 1969. Ultrastructure of synapses in a primitive coelenterate. *J. Ultrastruct. Res., 32,* 237–246.

Westfall, J. A., 1970. Synapses in a sea anemone, *Metridium* (Anthozoa). *7th Congr. Internat. Microscop. Electron Grenoble, 717,* 718

Westfall, J. A., Yamataka, S., and Enos, P. D., 1970. Ultrastructure of synapses in *Hydra*. *J. Cell Biol., 47,* 266.

Westfall, J. A., Yamataka, S., and Enos, P. D., 1971, Ultrastructural evidence of polarized synapses in the nerve net of *Hydra*. *J. Cell Biol., 51,* 318–323.

Wickelgren, B. G., 1967. Habituation of spinal interneurons. *J. Neurophysiol., 30,* 1424–1438.

Wilson, D. M., 1959. Long-term facilitation in a swimming sea anemone, *J. Exptl. Biol., 36,* 526–532.

Wood, R. L., 1959. Intracellular attachment in the epithelium of *Hydra* as revealed by the electron microscope. *Biophys. Biochem. Cytol., 6,* 343–352.

Yentsch, C. S., and Pierce, D. S., 1955. A "swimming"anemone from Puget Sound. *Science, 122,* 1231–1233.

Zubkov, A., and Polikarpov, G. G., 1951. Conditioned reflex in coelenterates. *Advan. Mod. Biol. Moscow, 32,* 301–302.

Chapter 4

PLATYHELMINTHES: THE TURBELLARIANS[1]

W. C. CORNING

AND

S. KELLY

Division of Biopsychology
Department of Psychology
University of Waterloo
Waterloo, Ontario, Canada

I. INTRODUCTION

Interest in the phylogenetic development of learning capacities must lead to heavy emphasis on the flatworm, since it is the primitive turbellarian form Acoela which is widely accepted as representing the beginnings of bilateral existence. Of critical significance at this phylogenetic point are the appearance of the bilateral form and the consistently forward-directed anterior end, the concentration of nervous tissue and sensory apparatus in the anterior portion, and the appearance of refined organ systems. The importance of the planarians cannot be overestimated. As Hansen (1961) stresses, "Of the non-flatworm metazoan phyla . . . there is not one phylum, with the possible exception of the Cnidaria (Coelenterata), whose suspected affinities do not eventually lead us back to the Turbellaria." The presence of a well-developed nervous system complete with brain and longitudinal medullary cords connected by commissures, polarized synaptic conduction giving directed and rapid communication, and sensory and motor specializations provide the

[1]Preparation of this chapter was facilitated by Grant A 0351 from the National Research Council of Canada. Portions of this chapter have been published previously (Corning, 1971; Corning and Riccio, 1970).

comparative psychologist with a most interesting class of animals. It is no mere coincidence that it is in these animals that the first clear evidence of associative learning is obtained. These demonstrations were not without controversy in the initial stages, but the difficulties were perhaps more of a problem with semantics than with the ability of the planarian to perform in certain tasks.

Present-day interest by psychologists and psychobiologists in the planarian has also been stimulated by studies based on two critical properties of the planarian: the ability to regenerate and the ability of some species to readily cannibalize other worms. The findings that memory survives regeneration and that learned responses can be transferred via ingested tissue have generated considerable use of the planarian in research aimed at clarifying the molecular bases of memory. The controversy over these studies has yet to subside.

II. GENERAL CHARACTERISTICS

The phylum Platyhelminthes is composed of three classes of flatworms: Trematoda and Cestoda, which are both parasitic, and Turbellaria, which are free-living. Only the latter have been used in behavioral studies, and, accordingly, the present review will be restricted to this class.

Turbellarians are unsegmented flatworms found in marine, freshwater, and terrestrial habitats within a wide temperature range (Fig. 1). The major limitations on occurrence are extremes of acidity and alkalinity. All forms are dorsoventrally flattened and have a ciliated ectoderm. A midventral opening, the mouth, leads to a gastrovascular cavity which both digests and distributes food to the body. Both ingestion and excretion are mediated by the mouth. Excretory organs (nephridia) are present in all forms except order Acoela, which do not possess a gut. The nervous system varies across species but basically is characterized by the aggregation of cell bodies into two anterior cerebral ganglia and the extension of the central nervous system posteriorly via two or more nerve cords. There is also the appearance of glial cells along tracts and between cells. The nerve cords are connected by several commissures.

A. Cell Characteristics

Cell types found in Turbellaria are varied. The one-layered *epidermis* may be cuboidal in shape, or "insunk," the nuclei having descended into the mesenchyme. *Rhabdoids,* crystalline rod-shaped bodies secreted by epidermal or mesenchymal gland cells, may be present. The function of the rhabdoids

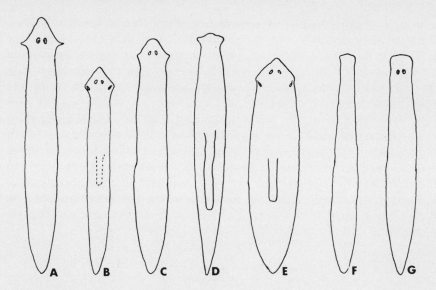

Fig. 1. Turbellarian forms. A, *Dugesia dorotocephala*. B, *Dugesia microbursalis*. C, *Dugesia tigrina*. D, *Polycelis coronata*. E, *Curtisia foremanii*. F, *Procotyla typhlops*. G, *Phagocata morgani*. (Modified from Pennak, 1953.)

is not definitely known—they may, on discharge, form the protective slime coat or be involved in cyst formation. The *subepidermis* is comprised of unicellular *glands* and *musculature*. The muscles appear as layers of circular, diagonal, and longitudinal fibers. A thin, transparent *basement membrane* serves as a point of attachment for both *subepidermal* and the more complex *mesenchymal muscles*. Muscle fibers appear as long, slender strands which may show primitive cross-striations. The contractile part of the muscle cell is often enclosed in a nucleated protoplasmic sheath which may send projections to the epidermis. A *submuscular plexus* exists but is not considered to be a nerve net since conduction will not occur around obstacles or a section (Bullock and Horridge, 1965).

Cells with long, irregular processes and fluid-filled interstices form the *mesenchyme*. The fibrillar appearance of mesenchymal cells is attributable to small granular particles, presumably the products of digestive activity. Excretion is carried out by the *protonephridia*, which have branching capillaries and *terminal flame bulbs;* the free ends of the latter extend into the mesenchyme for filtration of fluids. Undifferentiated cells capable of amoeboid movement are found in the intercellular spaces. Their function is not well understood, but they may be involved in regenerative processes.

The nervous system contains typical cell types that have been reviewed in detail by Bullock and Horridge (1965, pp. 546–549). Nerve cell inclusions are also typical, and uni-, multi-, and bipolar cells appear. The formation of an outer "rind" of cell bodies and an inner core of fibers complies with the basic invertebrate pattern. The *neuropile* in the brain contains fibers that range from 0.2 to 0.4 μ in diameter; they are unmyelinated (Best, 1967*b*; Morita and Best, 1966). *Neurosecretory* cells have also been found that contain 400–1100 Å *granules*. Evidence suggests that these granules migrate along neural processes to end bulbs (Morita and Best, 1965). Both *axoaxonic* and *axodendritic* synapses occur, and a presynaptic location of synaptic vesicles has been determined (Best, 1967*b*). In the anterior portion or "head" of the animal, the neural tissue and muscle are intimately connected, an arrangement that probably accounts for the fine degree and quickness of movement observed in this portion during food catching and orientation. Examples of nervous system organization are presented in Fig. 2.

B. Respiration

Since turbellarians have no respiratory structures, gaseous exchange occurs through the body surface via diffusion. The rate of O_2 consumption is constant unless the concentration of O_2 falls below 0.33 saturation. Calculations of the basal O_2 consumption in *Dugesia dorotocephala* give a rate of 0.2–0.3 $cm^3/g/hr$ (Hyman, 1919). The rate is higher in younger animals, during regeneration, in the initial stages of digestion, and in the later stages of starvation. An anterior–posterior gradient of O_2 consumption has also been shown (Hyman, 1923). CO_2 production shows much the same relationship to age and the anterior-posterior gradient as O_2, except that the gradient of CO_2 production is a U-shaped function with the middle area showing the least activity and the two extremes being the highest.

C. Digestion

Digestion in free-living flatworms is largely intracellular. Food is broken up by the muscular pharynx and proceeds to the highly diverging intestine, first filling up the anterior pouch and then moving to the two posterior rami which lie on either side of the pharyngeal cavity. Phagocytic cells lining the cavity begin engulfing food particles as well as fluids and form vacuoles. The food is converted over a period of 8 hr into protein balls; these are broken down over a period of 5 days. The fat present in the phagocytic cells disappears during the early stages of digestion, perhaps due to energy requirements for digestion. The protein of the food is thought to be stored as fat in the gastrodermis and to some extent in the mesenchyme. Extracellular digestion has been observed in some Turbellaria having a ciliated gastrodermis. The presence of large food particles may necessitate extracellular digestion prior to phagocytic activity.

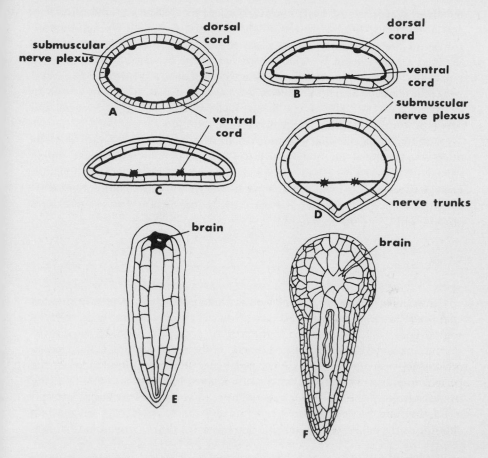

Fig. 2. Nervous system forms of turbellarians. A, Alloeocoel. B, Marine triclad. C, Fresh-water triclad. D, Land triclad. E, *Convoluta*. F, *Bothrioplana*. (Modified from Meglitsch, 1967.)

D. Reproduction

Reproduction may be sexual, asexual, or both. Asexual reproduction is usually triggered by temperature and nutritional conditions, while the sexual form is typically seasonal, occurring during cooler periods.

Turbellaria are hermaphroditic, having both male and female reproductive systems in the same animal. Copulation is usually mutual, self-fertilization being a rare exception. The male system consists of numerous pairs of testes with sperm ducts leading to a prepharyngeal copulatory organ. In the female system, pairs of ovaries are connected by oviducts to the atrium or genital chamber. After fertilization, the eggs and yolk cells formed by yolk glands are formed into a cocoon while still in the atrium. A few days later, the long slender thread by which the cocoons are attached to the substrate is extruded from the gonopore, followed by the cocoon. From one to several weeks (at temperatures of 18–20°C), the cocoon hatches, liberating one to 35 offspring. Immature planarians do not have well-developed sex organs and may reproduce by fission before achieving maturity.

Asexual reproduction, common in freshwater triclads, is achieved by transverse fission of the body at a point just behind the mouth; the gliding animal attaches the back half to the substrate while continuing to advance. Each half regenerates the missing parts. Morphological changes prior to fission are not pronounced. Further details concerning the regenerative process will be presented below.

E. Movement

In smaller flatworms, locomotion is made possible by ventrally situated cilia which beat against the thin layer of mucus that is constantly secreted by ventral tissue. Waves pass from anterior to posterior, sending the animal gliding across the substratum or underside of the water surface. Ciliary movement is not dependent on the central nervous system but is controlled by the peripheral plexus. Movement can also be effected by muscular action passing from the front of the body to the rear. The relative roles of muscular and ciliary action can be determined by paralyzing the muscles with magnesium chloride or the cilia with lithium chloride (Hyman, 1951). *Monotaxic* locomotion, in which the muscular waves pass across the entire width of the animal, is typical of planarians. However, in the marine forms, *ditaxic* locomotion is observed, in which muscular contractions alternate across two lateral margins. Swimming is rare in the Turbellaria and is generally elicited only when stimulation is applied during ditaxic movement. Both swimming and ditaxic locomotion are dependent on the presence of the brain and nerve cord activation of the muscle complex.

Relatively strong or aversive stimulation induces another kind of movement—crawling. The animal attaches its head to the substrate and pulls the posterior portion of its body by contraction of the longitudinal muscles. The circular muscles then contract to extend the head. This "humping" gait permits the animal to move rapidly.

The adhesive mucus produced at the margins of the body serves dual purposes: the slimy substance smoothes the substrate (a necessity for land forms), and by being drawn out into long threads it allows the animal to bridge gaps in the substrate or form a weblike food snare.

F. Receptors

Receptors consist of widely scattered single primary sense cells and the more highly specialized neurosensory organs composed of many cells that communicate information about light, pressure, orientation, water currents, and chemical compounds in the medium.

Tactile receptors are found over the entire body and may be concentrated on the auricles and along the lateral margins. A round nucleated cell body lying below the epidermis sends projections to the bristles. The axon extends to the submuscular plexus.

Ciliated pits and grooves of specialized epithelial tissue in the head region serve as chemoreceptors. Cells used for orientation toward food sources are found concentrated in the auricles. Gland cells are usually found in association with chemoreceptors.

Detection of water currents is made possible by four pairs of large cells lying just beneath the epidermis. Two pairs of rheoreceptors are anterior, one dorsal and one ventrolateral, and two pairs are situated on the lateral margins midway down the body. Each cell has many sensory bristles which extend well past the cilia. Axons terminate in the ventral cords.

Body orientation movements that depend on perception of gravity are served by statocysts located near the brain. The statocysts of Turbellaria consist of a vesicle enclosing small granular particles (statoliths). As the animal moves, these statoliths shift and make contact with different hairs of neurosensory cells.

Location, type, and complexity of eyes vary greatly in Turbellaria. In some cave-dwelling forms, they are entirely lacking. In others, they range from simple epidermal pigment spots to rather complex ocelli. In planarians, the cephalic eyes are composed of a light-transparent epidermis which overlies a cup formed of pigment cells and the photosensitive retinal cells. The pigment cells are thought to influence the directional photic response by blocking all light rays not parallel to the long axis of the retinal cells (Hyman, 1951; Taliaferro, 1920). The cell body of the retinal cell is situated in the mesenchyme and projects its distal processes into the eyecup and its proximal process into the central nervous system. Since a light-refractive apparatus does not exist, these eyes are only capable of light–dark discriminations. The presence of a phototaxic response in headless or blinded planarians suggests that photoreceptors other than the cephalic ocelli are present, although these have yet to be identified.

III. PHYLOGENY AND TAXONOMY

A. Phylum Platyhelminthes

The Platyhelminthes are bilaterally symmetrical, dorsoventrally flattened animals containing definite organ systems, including a central nervous system and gut. Three classes are recognized:

Class Turbellaria: free-living planarians.
Class Trematoda: the flukes.
Class Cestoda: tapeworms.

Since behavioral research has been confined to Turbellaria, only orders in this class will be listed:

Order Acoela: considered to be the most primitive. Members are marine and lack an intestine.
Order Rhabdocoela: found in fresh water.
   Example species:
   Stenostomum
Order Tricladida: freshwater, marine, and terrestrial inhabitants. These are the most widely used animals for biological and behavioral research.
   Example species:
   Dugesia dorotocephala.
   Dugesia tigrina.
   Dugesia lugubris.
   Cura foremanii.
   Dendrocoelum lacteum.
Order Polycladida: marine members.
   Example species:
   Leptoplana.

B. Evolution

The question of flatworm origins is still debated. Three distinct schools of thought persist. The Cnidaria are widely accepted as being the most primitive animals in Metazoa, and, accordingly, the Bilateria would all be descended from coelenterates. A second possibility is that the coelenterates developed from the bilaterally symmetrical turbellarians. This view is based on the biradial symmetry of anemones and on the fact that no cnidarian is as simple as the simplest flatworm. A third alternative is that turbellarians derived from protistans.

IV. SPECIALIZATIONS RELEVANT TO BEHAVIORAL RESEARCH

A. Summary of General Phylogenetic Advances

1. Bilateral symmetry which facilitates orientation; forward-directed anterior end.
2. Organ-grade level of construction.
3. Centralized nervous system and brain which facilitate integration of receptor–effector systems; also appearing are neurosecretory cells, synapses, polarized neurons, and the basic invertebrate nervous system organization.
4. Specialized sensory apparatus with an anterior concentration.

B. Regeneration

In a biological sense, regeneration represents a "retention" of sorts; former structural and functional systems are restored after their removal. In the planarian, the general strategy of tissue renewal has been well documented. Initially, the wound margin is reduced to a small size by contraction of the musculature. Cells at the margin, which interface with the surrounding medium, are transformed into mucoid cells serving a protective function. Gradually, the epithelium grows out from the lateral margins to cover the wound surface. Undifferentiated, embryonic-like cells known as *neoblasts* begin to accumulate at the wound site to form the regeneration blastema. The neoblasts may be local or may have migrated from other portions of the body. Formation of the blastema marks the end of the regressive phase of regeneration. The second phase is characterized by *morphallaxis,* the reworking of old tissue to form the developing structures, and *epigenesis,* involving cell growth, mitosis, and redifferentiation.

The stimulus setting off cell migration and differentiation in neoblasts is not known. Lender and Klein (1961) suggest that neurosecretions are responsible. They observed an increase of 100–300% in the number of neurosecretory cells in the brain during regeneration. As to the source of the neuroblasts, several possibilities exist. One theory is that these cells are totipotent; i.e., they are chronic embryonic cells providing a reservoir of cells used for tissue replacement. By following stained neoblasts, investigators have been able to locate them in the blastema (Lender and Gabriel, 1960*a,b*). However, the reliability of these stains has been questioned (Coward, 1969). Additionally, the appearance of neoblasts is not well correlated with the ability of certain species to regenerate. Species that are unable to regenerate a head from a posterior portion appear to possess the same number of neoblasts as those species that have the regenerative capability. However, there is no clear way of distinguishing between chronic embryonic cells, dedifferentiated cells, and recently divided ones (Coward, 1969).

Loss of specialization in extant cells could be a source of neoblasts (Coward, 1969; Hay, 1968; Hyman, 1951). It is known that other invertebrates and vertebrates regenerate by dedifferentiation of old tissue and that certain planarian structures (e.g., pharynx, nerve tissue, intestine, and epithelium) can be reproduced by morphallaxis. Kenk (1967) reports that in *D. tigrina* and *D. dorotocephala*, morphallaxis prevails, while in *C. foremanii* epigenesis is the rule.

The cells of the blastema are not autonomous. Sengel (1967) demonstrated that an interaction between the blastema and the old tissue exists before differentiation begins. The old tissue stipulates which structures will be formed (induction) and which will not (inhibition). The head end may be the major source of control over the pattern of morphology. The polarity observed in the regeneration of a head from the anterior cut and a tail from the posterior cut was first explained by Child in his theory of axial physiological gradients. Coward (1969) suggests that the head blastema may have more cells. By having a higher metabolic rate, the anterior end would receive more cells faster than a posterior cut. Thus a head blastema would be established first and would inhibit the formation of a second head, leaving a nonhead or posterior blastema at the other end of the animal. The mechanisms controlling subsequent differentiation are not clear. Inducers and inhibitory agents must be present. For example, a brain homogenate will induce eye formation. The brain also releases an inhibitory substance which diffuses throughout the planarian to prevent the duplication of cephalic structures. The potency of this substance decays with increasing distance from the head. Occasionally, this inhibition is inadequate, and bipolar planarian "freaks" result (Corning, 1967; Jenkins, 1963).

The rate and course of regeneration can be influenced in many ways. Agents that interfere with adequate oxidative metabolism and protein synthesis produce a temporary disorganization of tissue, while permanent and progressive disorganization is caused by agents interrupting purine metabolism (Henderson and Eakin, 1961). Starvation slows regeneration (Brøndsted, 1953), while size and extent of the wound have no effect (Brøndsted and Brøndsted, 1954). Regeneration and spontaneous fissioning rates can be accelerated by some agents, among them glutathione (Owen *et al.,* 1938, 1939), vitamin B (Brøndsted, 1955), and amino acids (Lecamp, 1942).

C. Cannibalism

Studies of learning and memory mechanisms have made use of both the regenerative and cannibalistic capacities of planarians. During regeneration, memory was apparently unhampered and was demonstrated by the regenerated offspring. This indicated a cellular basis of memory, and subsequent experiments asked whether the memory might be transferred via ingested cells.

Almost all species will cannibalize, some more readily than others. Generally, the preference is to eat different species that are smaller. However, controlled experiments usually require the use of the same species for donors and recipients. According to McConnell (1965), *D. dorotocephala* is the best cannibal and the probability of its cannibalizing other *D. dorotocephala* individuals will be increased if it is hungry and if it has had previous cannibalizing experience.

D. Receptor Properties

Planarians respond to a wide variety of stimuli. Brown (1962*a,b*, 1963) has found sensitivity to weak magnetic fields (0.17–10 gauss) with a seasonal and monthly cycle, to weak electrostatic fields, and to weak γ radiation with a semimonthly negative response. Planarians are positively rheotaxic. Responses to weak mechanical and thermal stimuli are positive, but avoidance or negative reactions are observed to more intense stimuli.

Light reception is mediated in planarians chiefly by the eyespots. Generally, planarians are negatively phototaxic, turning away from directed light and showing increased activity to nondirectional light (Jenkins, 1967). They demonstrate little sensitivity to the infrared end of the spectrum but show a strong aversive reaction to ultraviolet. Blinding animals by removing the eyespots does not obliterate the negative phototaxic response, according to some authors (Hyman, 1951; Parker and Burnett, 1900; Taliferro, 1920), but speed and precision of response are reduced. Uroporphyrin, a photoactive substance, is distributed throughout the body and may be the basis for light sensitivity in eyeless worms (Krugelis-Macrae, 1956). McConnell (1965) reports no evidence from his research on a phototaxis in eyeless animals.

Brown's electrophysiological and behavioral work on the photic response has necessitated some reevaluation of the results of learning experiments that have employed a light stimulus. Negative phototaxis may be augmented by light and by shock when they are temporally unrelated (Brown *et al.*, 1966*b*). Short planarians were found to be more light sensitive than tail regenerates (Brown *et al.*, 1966*a*). The latter finding, however, was not confirmed by Van Deventer and Ratner, (1964), who found the tail regenerates to be more sensitive. Exposure to ultraviolet light tends to reduce the photonegative response, but this is reversible ("photorestoration") by the application of white light (Brown, 1964; Brown, 1967*a*). Röhlich (1968) has found degeneration of the receptor apparatus in planarians kept in total darkness, a phenomenon that should be taken into account in studies attempting classical conditioning to light. Species differences with respect to light sensitivity have been documented by Reynierse (1967b). Electrophysiological studies on the photic response of the cephalic eyes have demonstrated that the ocellar potential generated by these receptors is dependent on the light intensity (Behrens, 1962; Brown, 1967b; Brown and Ogden, 1968). The ocellar potential amplitude and latency decrease with increasing light intensity. Temperature increases have the same effect. Of particular relevance to conditioning studies utilizing light as a CS is the finding that a brief (800 μsec) and intense (5000 lux) light flash produces electrophysiological changes that can last up to a minute.

Chemotaxis is positive in planarians in weak solutions and food juices. Some species can detect the presence of food juices up to a distance of 15 cm (Hyman, 1951). Pearl (1903) was able to elicit a positive chemotaxic response with weak solutions of acids, alkalies, and salts. Stronger concentrations produced negative reactions. Chemoreception is thought to be mediated by the auricles or lateral angles of the head, although in some species chemoreceptors appear to be spread over the body.

E. Slime

Slime production by planarians has been well documented in relation to such functions as transportation ("webs"), food capture, and protection from dehydration and adverse environmental conditions. Secretory cells have been identified that are similar in structure and function to mammalian mucoid cells (Best *et al.,* 1968). These goblet cells arise in the mesenchyme and migrate to ducts which lead to the surface through the epithelium.

The importance of slime secretion in planarian behavior has been emphasized by McConnell (1967b). Planarians display a clear preference for slimed areas and will learn better in a slimed environment. McConnell has seriously questioned the validity of any experiment which does not take into account the effects of slime. Light reactivity, for example, is lower when slime is present (McConnell, 1967b; Riccio and Corning, 1969). Under "nonstressful" conditions (darkened room, etc.), others have found no preference for a slimed portion of the environment (Riccio and Corning, 1969). However, in the presence of ultraviolet light, abnormal pH, and other deviant conditions the planarian demonstrates a clear slime preference. Functionally, this preference under threatening conditions would enhance the probability of worms locating each other and forming a protective "ball."

F. Galvanotaxis and Shock Responses

Investigation of galvanotaxic responses by several early researchers demonstrated that with medium-intensity currents the planarian shows a differential response based on its orientation in the electric field (Hyman and Bellamy, 1922; Pearl, 1903; Viaud, 1954). If the anterior end is oriented toward the positive pole during shock onset, the anterior end contracts and the animal turns its head so that it is oriented toward the cathode. Cephalic turning does not occur if the planarian is already cathodally oriented. Shafer and Corman (1963) supported these findings using current of short pulse duration and low intensity. Hyman states that all forms show cathodal orientation and may even swim or crawl toward the negative pole. Best and Elshtain (1966) suggest that differential responding is a function of the extent of neural excitation, which they relate to neuronal size and density. The use of the "dirty" waveform such as that produced by the Harvard Inductorium may be optimal as a shock source, since the probability of exciting neurons of different lengths is maximized.

V. LEARNING STUDIES

A. The Planarian Controversy—A Last Look[2]

Critical analysis of the available data and the published debates that stemmed from "the planarian controversy" compels a rephrasing of the question "Can planarians learn?" to "What can psychology learn from planarian research?" Not since the literature exchanges over protozoan learning in the 1930s and the Gelber–Jensen exchanges in the 1950s has there been so much heat generated over whether an invertebrate can demonstrate learning in what are mammalian paradigms. The controversy has centered around old issues, issues that are based on such things as sensitization and pseudoconditioning and whether these can be clearly segregated from the learning process. The penchant for a clear scientific semantics in psychology may have led to a premature senility in definitions. In spite of these problems of definition, demonstrations of learning capacities in planarians have shown that in most cases learning criteria have been met.

Focal Point of the Controversy

The basis of the recent controversy is the Thompson and McConnell (1955) study reporting classical conditioning in *D. dorotocephala* using light as the CS and shock as the US. As will be seen below, it was not this particular study that induced so much interest and debate over the planarian as

---

[2]"The planarian controversy" perhaps suffers from overreview. Many of the earlier criticisms have been adequately answered (Corning and Riccio, 1970; McConnell and Shelby, 1970). The present (and final) review of the controversy is included more for possible historical interest.

it was subsequent studies concerned with the biochemistry of learning that utilized the light–shock paradigm. Thompson and McConnell used three groups of animals to ask if light–shock pairings produced a different light responsivity than shock alone or light alone. Animals were trained individually, but trials were presented in three blocks of 50 each. A training trial consisted of a 3 sec period of light and a 1 sec burst of shock that overlapped the last second of the light period. All responses that were elicited during the first 2 sec of the light period were recorded. Controls consisted of a group that received shock trials with a light test trial interspersed after every fifth shock, a group that was presented light only on each trial, and a group that received neither light nor shock but was observed for 2 sec periods to determine the spontaneous response rate. The typical reaction to the shock stimulus was a longitudinal contraction, resulting in the animal balling up. As training progressed, two types of responses developed in the group being trained: (a) a longitudinal contraction and (b) a sharp turning of the anterior end. At the completion of the 150 trials, the group being trained displayed a significantly higher percentage of these responses (43.2%) than either the light control group (19.2%) or the spontaneous response group (17.2%). The shock control group showed no increase in light responsivity.[3]

Criticisms

The principal experimental results frequently cited as evidence against planarian learning were generated from a series of investigations by Halas and associates (Halas *et al.,* 1961, 1962; James and Halas, 1964). Essentially, this group differed in what measures constitute an adequate assessment of learning and which methodology best demonstrates it. In the Thompson and McConnell study, acquisition scores provided measures; the responses of individual subjects on every trial were recorded. In the initial study by Halas and associates (Halas *et al.,* 1961), no attempt was made to measure acquisition—subjects were group-trained in one container and tested individually under extinction conditions (CS only) in a different container. There was no way of knowing what sort of response modifications were taking place in the experimental and control groups during training. The reason for adopting these procedures was that individual conditioning required the constant attention of the experimenter to keep the animal gliding smoothly and oriented properly with respect to the electrodes. Group training was therefore adopted to keep the treatment of subjects uniform. Comparisons of procedures are summarized in Table I. The results of these experiments indicated that differences between experimental and control groups do not appear during extinction trials. It was also found that planarians exhibited extreme

[3]In terms of percentage responsivity to light, however, the shock group was high (37.3%) during the first 15 test trials and dropped to only 30.6% in the last 15 trials.

Table I. Classical Conditioning: Comparison of McConnell and Halas
Studies (Corning and Riccio, 1970)

Variables	Thompson and McConnell (1955)	Halas et al. (1961)	Halas et al. (1962)
Species	D. dorotocephala	D. tigrina	D. tigrina
Housing	20–24°C; background light "minimum"	20–24°C; "complete darkness"	20–24°C; "complete darkness"
Training mode and measures	Individual conditioning; acquisition scores	Mass training; tested 16–22 hr later under extinction conditions	Individual conditioning; acquisition scores
Trough	Semicircular: 3 by 1.5 by 30 cm	Rectangular (train): 3 by 1¾ by 2 inches Semicircular (test): 16 by 0.5 by 0.5 inches	V shape: 14 by 1 by 0.5 inches
Shock (US)	DC: 28 ma via inductorium	DC: 6 v via inductorium	DC: 6 v at 0.30 ma via inductorium
Light (CS)	Two 100 w bulbs 1 ft above trough	Two 100 w bulbs 1 ft above trough	Two 60 w bulbs 1 ft above trough
Intertrial interval	20 sec	30 sec	45 sec
Trial sequence	Three blocks of 50 trials each with 5 min between blocks	Same	Same

variability in their responsivity to light and that naive animals could display a very high response rate to light.

A second source of criticism stemmed from the research sponsored by Calvin (a Nobel chemist at Berkeley) presented at the MIT Neurosciences Meetings (Bennett and Calvin, 1964). The rather devastating conclusion that efforts to train planarians were "unrewarding" and "unsuccessful" received considerable attention and credibility coming from an eminent scientist. In short, several different attempts were made to classically condition planarians and to demonstrate learning in other tasks such as habituation to light, maze learning, and operant conditioning. While the data of Bennett and Calvin showed significant learning in classical conditioning and indicated a possibility of maze learning, they were not satisfied with the lack of control over the variability of behavior in the planarian and with the low levels of response achievement in the training situations. Their major points were that the factors responsible for producing learning in the planarian ought to be consistently described so that the biochemist can effectively use these animals and that training methodologies were inadequate to produce reliable training.

H. M. Brown's research at Utah (Brown, 1964, 1967a; Brown et al., 1966a,b) has also failed to support the contention that planarians can be conditioned. With light and shock as the stimuli, it was found that there were no differences between animals receiving paired training trials and control animals receiving spaced light and shock stimuli. Additionally, other studies showed that there was no habituation to light.

These various negative studies provided the basis for an attack by Jensen (1965) on the general validity of a learning interpretation for the planarian studies of McConnell and others. Jensen began with an overview of the paramecium learning controversy and an account of the confounding factors that might have been operative in certain studies and then applied "the strategy of the skeptic"[4] in an attempt to invalidate all planarian learning studies. Jensen criticized McConnell's view that in the training of planarians it is frequently necessary to try a number of different procedures until the right combination of factors is hit upon. According to Jensen, the continual variation of procedures will increase the probability of a success, "that with patience and time one will happen upon an experimental procedure which involves just the right confounding to produce pseudo-learning" (Jensen, 1965, p. 15). The assumption that the behavior change *will be* pseudolearning is of course a before-the-fact statement about what an animal can learn; the objection against varying procedures simply violates the history of progress in science.

Jensen also considered other forms of planarian learning, and, relying partially on Pearl's 1903 monograph, he proceeded to "explain" away all planarian learning demonstrations. Jensen noted Pearl's observation that strong mechanical stimulation produces negative reactions and jumped to the conclusion that shock ought to increase the responsivity of planarians to light. Pearl, however, found the opposite; i.e., continued strong stimulation activated the animal and *reduced* its sensitivity to stimulation. Best's maze-learning experiments (Best, 1963, 1965; Best and Rubinstein, 1962) were written off by Jensen because they had light as a cue. Again, Pearl was cited as support for this statement, but our careful examination of the 1903 monograph failed to yield any supportive data gathered by Pearl (Corning and Riccio, 1970). In the maze experiments of Best, animals were trained to go to either a lighted or a darkened side, and the learning curves for either cue were identical. Turning to Lee's (1963) operant conditioning experiment, Jensen again assumed hypothetical operations to account for the response modification. The Lee paradigm required planarians to move through a tunnel and break a photoelectric beam which in turn caused an intense overhead light to be turned off for 15 min. Jensen suggested that with this sort of contingency the planarian would come to rest near the tunnel since the absence of light results in less activity on the part of the worm. Thus,

[4] ". . . the strategy of the skeptic, who doubts any and every view, but will not assuage his doubt or selectively apportion it on the basis of data" (Jensen, 1964, p. 8).

having gone through the tunnel (turning off the overhead light), the animal would cease moving and would have a higher probability of repeating the response, since it would be resting near the hole. However, Best (1965) pointed out at the same conference that this did not occur—planarians did not stop when the stimulus went off but circled several times.

The objections of Jensen in themselves were not serious, since they could be refuted by data. What prolonged "the controversy" was the ready acceptance of Jensen's views by various reviewers and authors. Further details of this controversy have been presented previously (Corning and Riccio, 1970).

Rebuttal

Planarians Learn and Remember. One of the most convincing demonstrations of learning turns out to be one of the earliest, that of Van Oye in 1920. Basically, the Van Oye technique involves the location of food at the end of a wire lowered into the culture dish. The procedure has a number of advantages: there is minimal handling of subjects, unnatural stimuli such as light and shock are avoided, a natural response is used (food-seeking), and the assessments of behavior change are relatively more objective—the subjects are at the end of the rod or they are not. These procedures are portrayed in Fig. 3. The food is placed on the end of the wire, which is at first on the surface (making it easier to locate initially) and then is progressively lowered. Subjects eventually learn the route to food even when the wire is placed several centimeters below the surface. The experiments of Wells and associates have established the effectiveness of this training procedure and have included proper controls that mitigate against alternative explanations (Wells, 1967). In these experiments, the food was lowered in four stages until it was finally 32 mm below the surface. The controls were exposed to a baited rod on the bottom of the container. During the test sessions, both groups were exposed to a rod 32 mm below the surface. Because the planarians could be reacting to stimulus gradients in this situation rather than acquiring a specific maze habit, they were tested with a clean rod in a beaker containing water mixed with meat juice. Under these conditions, animals that had been trained were still found at the end of the clean rod in greater numbers than untrained controls. The food-seeking habit could also be brought under stimulus control. When light was associated with the training periods, it took on conditional properties; during the presence of light, greater numbers of trained animals were found at the end of the rod than controls.

The findings of Van Oye and Wells were replicated by McConnell (1967a) with some minor modifications that were found to facilitate acquisition. For example, location of food was improved by a colored stripe along the side of the beaker. Bennett and Calvin (1964) also attempted a replication but did not interpret their findings as indicating learning. Their general

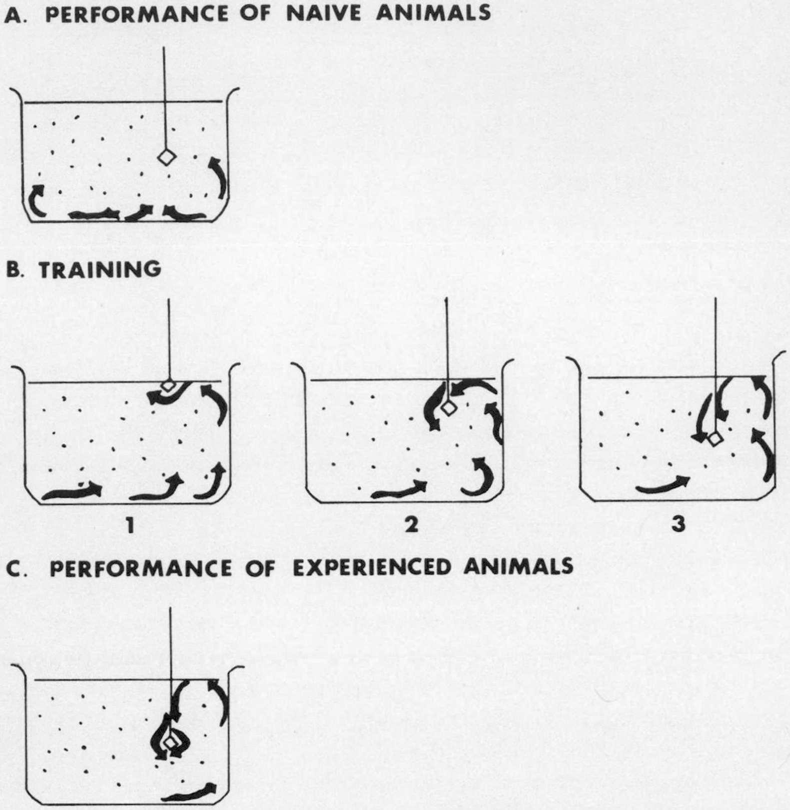

A. PERFORMANCE OF NAIVE ANIMALS

B. TRAINING

1   2   3

C. PERFORMANCE OF EXPERIENCED ANIMALS

Fig. 3. Schematic of the Van Oye training technique.

procedures were similar to those of Wells: in the experimental group the food was progressively lowered below the surface, while in the control group the baited rod remained at a fixed position below the surface. Because the experimental group did not find the food with shorter latencies or with increased frequency as trials progressed, it was concluded that learning did not occur. However, their data indicated some sort of response modification. While the controls found food in only 5% of the trials, it was reported that when food was presented on the surface for 45 min the trained subjects found food in 35% of 54 opportunities; when presented 0.5 cm below the surface, the subjects found food in 44% of the opportunities; at 1.0 cm, 30% found it; and at 2.0 cm, there was a 23% success. From these results, it would appear that the trained group was more likely to find the bait than controls, evidence that is more supportive than negative.

From the Van Oye study and subsequent replications, it is clear that the question of learning capacity is rather easily answered. The technique has been convincingly replicated in at least two independent laboratories, and Wells' group alone has run at least five separate experiments. We can now ask whether planarians can demonstrate any persistence of an acquired response. In Fig. 4, a schematic of an automatic mass-training apparatus that was devised at Brookhaven National Laboratories is presented (Corning, 1971; Corning and Freed, 1968; Freed, 1966). In this container, light (CS) and shock (US) were used to classically condition planarians. The temporal relationship between the CS and US is shown in Fig. 5 along with the general types of responses elicited by these stimuli. This particular apparatus has several advantages: (a) it permits several animals to be trained at the same time; (b) the porous clay ring prevents animals from climbing on the electrodes and yet does not impede current flow; (c) by keeping the animals in a central position, a more uniform exposure to the light and shock stimula-

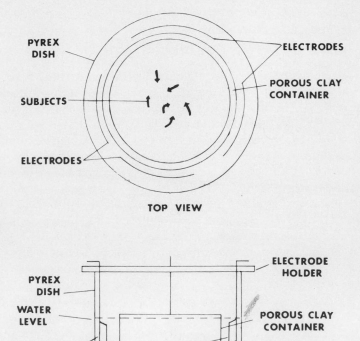

Fig. 4. Mass-training apparatus (Corning and Freed, 1968).

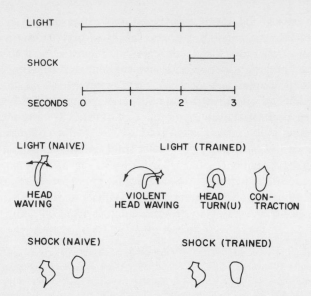

Fig. 5. Top: Relationships between light and shock in conditioning (Corning and Freed, 1968). Bottom: Types of responses elicited by light in planarians.

tion is achieved; (d) by randomly alternating the voltage drop between electrode pairs, the subjects are prevented from assuming an orientation in the field that would permit them to receive a less than adequate shock. Mass-conditioning studies using this apparatus have demonstrated a clear response modification to light that is dependent on light–shock pairing. With respect to the question of retention capacity in flatworms, Claudia Ferguson, working with Freed at Brookhaven National Laboratories, has obtained evidence that animals trained in this apparatus will demonstrate long-term retention (Freed, 1966). In Fig. 6 are shown the individual acquisition scores of six subjects. The ordinate represents the number of responses to light. The interval between the vertical lines at 380 trials indicates a 39 day break in the daily training sequence. From these data, it can be seen that after 39 days subjects were still at a high level of responsiveness. Occasional drops in performance, e.g., subject (d) in Fig. 6, were due to fissioning, etc.

Control Studies. Jacobson has perhaps provided one of the most effective rebuttals with respect to the issues of sensitization and pseudoconditioning[5]

[5](Comment by J. A. Dyal.) As Jacobson used the term, *pseudoconditioning* is to be construed in the broad sense, as any increment in the capacity of the CS to elicit the CR which is not based on the forward pairing of CS and US. Since the critical parameter is the forward pairing, he used a variety of control procedures to break up the forward CS–US sequence, including control groups that received random CS and US stimuli. The inadequacies of the CS only and the backward conditioning procedures as controls for Pavlovian excitation

raised by Jensen and Halas (Jacobson, 1967; Jacobson *et al.*, 1967). The initial research examined the effects of presenting planaria with paired and unpaired light and shock trials. Animals were trained in a group, but the performance of individual subjects was obtained by randomly selecting animals from groups and assessing them singly. The results of this study demonstrated that paired light–shock treatments (group CC) produce a different response modification than unpaired or randomly presented light and shock. These findings were confirmed by Corning and Freed (1968), who also assessed the base rate response to light alone during the conditioning sequence (see Fig. 7).

In subsequent research by the Jacobson group, more stringent controls were used to demonstrate conditioning. A "backward conditioning" group

---

conditions have been noted in Chapter 1. However, it would appear that for planarians the simultaneous and random procedure may generate about the same degree of excitatory conditioning (Vattano and Hullett, 1964). In any event, the Jacobson experiments strongly support the inference that substantially greater response increments occur in the forward pairing treatment than in any of the control procedures.

By the term *sensitization*, Jacobson (1967, p. 203) means alpha conditioning, which, as we have seen in Chapter 1, refers to the augmentation of the unconditioned response (R_1) to the CS as a result of the forward conditioning procedure. It will be recalled that the first prerequisite in asserting alpha conditioning of a new response is that the R_1 and UR be different responses; the CS should not elicit an R_1 which is indistinguishable in morphology from the UR elicited by the US. It turns out that Jacobson *et al.* (1967) fulfilled this criterion in one of their procedures but not in the other. They used a differential conditioning procedure with either light or vibration as the S+. The UR was a full-body contraction. Now it turned out that vibration also elicited a full-body contraction on an average of 15% of the adaptation trials prior to forward conditioning in this case (Jacobson himself notes that vibration is not a preferred CS for this very reason, 1967, p. 208). On the other hand, the preconditioning rate of contraction elicited by the light CS approached zero (two out of 90), and this level can reasonably be regarded as negligible. Thus the response increment which resulted from the forward pairing of the light–shock represents a possible demonstration of classical conditioning provided there is available an appropriate control group to show that the increment is indeed due to the forward pairing (a positive contingency) between CS and US. Jacobson's comparison group was a within-subject control, in which the performance differential between S+ and S− is taken as reflecting the difference between forward conditioning and a neutral unpaired condition. We have seen in Chapter 1 that this comparison provides an overestimation of the amount of true classical conditioning, since S− is not neutral but has conditioned inhibitory properties as a result of the negative contingency with US. Strictly speaking, then, differential conditioning does not, contrary to popular opinion, prove to be the best evaluation of true conditioning as compared with alpha conditioning. What is needed is a truly random control group.

With all of these factors taken into account, what can be said about the Jacobson *et al.* (1966 *a,b*, 1967) experiments? We would contend that they have to be evaluated as a whole, and to avoid the empirical nihilism inherent in the "strategy of the skeptic" we would conclude along with Jacobson that "It appears reasonable to assert that by the strictest criteria, classical conditioning has been conclusively demonstrated in planarians" (1967, p. 215).

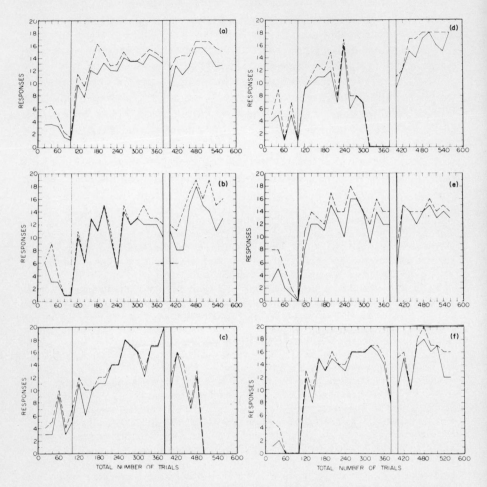

Fig. 6. Total (broken lines) and valid (solid lines) responses of six worms during training. The vertical line at 100 trials marks the end of the period of habituation to light. The interval between the vertical lines at 380 and 400 trials was 39 days. The sudden response decrements were caused by fissioning, e.g., subject (d) at 280 trials. (From Freed, 1966.)

received shock followed immediately by light; a "simultaneous condition-ing" group received the light and shock at the same time; a third control group received the CS only. The light responses obtained during test sessions are summarized in Fig. 8. It was also found that high-intensity vibration could be used as a CS, and subsequently Jacobson's group demonstrated that

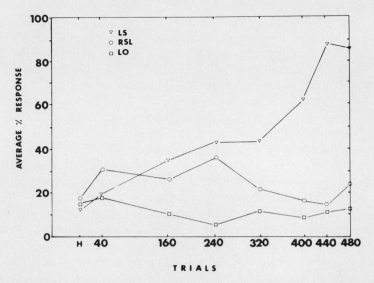

Fig. 7. Average percent response to light during 20 trial test sessions at various stages of conditioning. LS, Group receiving paired light–shock trials. RSL, Group receiving random light and shock. LO, Group receiving light only during trials. H, Habituation stage (Corning and Freed, 1968).

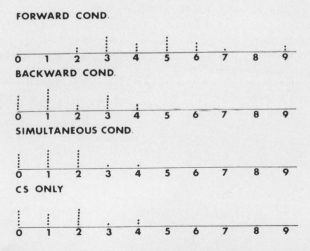

Fig. 8. Distribution of scores for the classical conditioning group and the three control groups. Each circle represents the number of contractions made by a given subject on the test series of 25 light presentations (Jacobson, 1967).

differential conditioning was possible. Both light and vibration were presented to subjects, but shock was paired with only one of the two stimuli. Two separate studies reported positive results (Jacobson *et al.,* 1967). In the first, subjects were trained individually, with shock being paired with either light or vibration. Differential responding occurred, with animals displaying a significantly higher response rate to the stimulus that was paired with the shock. Conditions were then reversed; the CS formerly paired with shock elicited decreasing response rates, demonstrating extinction, while the previously neutral stimulus began to elicit an increasing response rate.

In a second study, subjects were presented mass-training conditions with either light or vibration as the positive CS. Following training, subjects were given a random series of light or vibration trials individually, and, as can be seen from the data of Table II, a differential responsivity was attained.

A number of other investigations have confirmed that unpaired light and shock do not produce the response modification observed when they are paired and that a conditioned discrimination is possible. Baxter and Kimmel (1963) included an unpaired CS and UCS group and demonstrated that the responsiveness of the unpaired group to light decreased over trials while that of the paired group rose. A more extensive investigation by Vattano and Hullett (1964) arranged six different light and shock relationships summarized in Table III. During 25 post-treatment test trials, the forward conditioning groups showed a significantly greater number of contractions than the various controls. Some evidence of differential conditioning was also obtained (Kimmel and Harrell, 1964). A differential responsivity was obtained

Table II. Frequencies of Response to Light and to Vibration During Extinction Test Series for Light-Positive Group and for Vibration-Positive Group (Jacobson *et al.,* 1967)

S	Vibration (+)	Light (−)	S	Light (+)	Vibration (−)
V1	8	0	L1	7	2
V2	8	1	L2	6	2
V3	8	2	L3	6	4
V4	5	0	L4	5	3
V5	4	0	L5	4	2
V6	4	0	L6	3	1
V7	4	1	L7	5	4
V8	3	0	L8	4	3
V9	5	3	L9	4	3
V10	4	2	L10	3	2
V11	3	1	L11	4	4
V12	6	5	L12	1	1
V13	2	1	L13	5	6
V14	3	3	L14	5	6

Table III. Experimental Conditions (Vattano and Hullett, 1964)

Group	N	Duration of light (CS) (sec)	Duration of shock (UCS) (sec)	ISI (sec)	Intertrial interval (sec)	Type of conditioning
I	12	3	1	2	60	Forward
II	12	2	1	1	60	Forward
III	12	1	1	0	60	Simultaneous
IV	12	1	2	1	60	Backward
V	12	1	3	2	60	Backward
VI	12	2	2	Unpaired	60	Control

when vibration was positive and light was negative; in the reverse situation, no differential rates were observed. Griffard and Peirce (1964) reported success in training planarians to turn in one direction when a light was on and to orient in the opposite direction in the presence of vibration. Finally, Block and McConnell (1967) also demonstrated the capacity of the planarian to establish a differential conditioned response during acquisition and extinction measures and, in contrast to Kimmel and Harrell (1964), obtained a differential response rate when either light or vibration was the positive CS.

Critical Factors. Frequently, the failure of laboratories to replicate basic experimental results has been found to be due to the inappropriate control of certain variables in the experimental situation. A summary of these factors is provided in Table IV. Specific discussions of these may be found in reviews by McConnell (1967a,b) and by Corning and Riccio (1970).

Procedural Variations. One of the difficulties in initiating research on a relatively unstudied animal is that there is rarely any systematic study of all the variables that are operative and little agreement as to a standard set of experimental procedures. Accordingly, we tend to operate with techniques that work best in a particular laboratory. When there is disagreement or a failure to replicate, it is necessary that the different laboratories use similar methodologies and identical species. This is not an unreasonable demand, since "ratrunners" who use the albino have long been aware of strain differences that markedly affect results. It is quite surprising, therefore, that the negative studies have received so much credibility when they frequently failed to use the same species and in some cases did not even report which species were used (for example, Walker and Milton, 1966). In Tables I and V, we have summarized some of the procedural differences and similarities that occurred in the early, more controversial studies (Corning and Riccio, 1970).

B. Categories of Learning

Habituation

It is surprising that there are few systematic studies of habituation in the

Table IV. Summary of Critical Factors That May Influence Conditioning
Experiments

Reference	General findings
	A. Slime
McConnell and Mpitsos (1965) McConnell (1967a)	Planarians acquire a conditioned response to light more quickly when the trough is slimed; Ss also prefer the slimed part of their environment.
Riccio and Corning (1969)	Subjects tested in a trough coated with slime are less reactive to light than when tested under clean conditions. Larger Ss are somewhat less reactive than smaller Ss. Ss prefer a slimed portion of their environment when the pH of the medium is abnormal or when UV light is present; no preference is observed under normal conditions.
	B. Housing
Röhlich (1968)	Animals kept in total darkness for 21 days show eye degeneration.
Freed (1966)	Changes in the chemical content of the medium can affect performance.
McConnell (1967a)	General discussion of techniques of maintaining animals.
Van Deventer (1963)	Water contamination increases light sensitivity.
Van Deventer and Ratner (1964)	Increased temperature increases light sensitivity.
	C. Diurnal factors
Brown (1964)	Mobility is greatest at night; light reactivity is highest in the day.
Cohen (1965)	Maze learning is better at night.
	D. Species differences
McConnell (1967a)	Behavioral and conditionability differences exist between D. tigrina and D. dorotocephala.
Reynierse (1967a)	C. foremanii and D. tigrina are more photonegative than D. dorotocephala and P. gracilis. C. foremanii and D. tigrina are more sluggish and show a greater tendency to stop after activated.
Reynierse (1967b)	Species secrete specific attractants for other members of that species.
Ragland and Ragland (1965)	D. dorotocephala conditions better than D. tigrina.
	E. Shock parameters
Barnes and Katzung (1963)	Conditioning is achieved when the head is cathodally oriented but not when it is anodally oriented.
Best (1967a)	On anatomical grounds, a "dirty" shock waveform is best for conditioning.

Table V. Comparison of Studies Concerned with "Retention After Regeneration"

Variables	McConnell et al. (1959)	Van Deventer and Ratner (1964)	Corning and John (1961)	Brown (1964); Brown et al. (1966a,b)
Species	D. dorotocephala (10–24 mm)	D. tigrina (8–12 mm)	D. dorotocephala	D. tigrina
Housing	Aquarium water; minimum light	Melted snow[a]; total darkness	Pond water; indirect room lighting	Spring water; dim light
Trough	Semicircular: 12 by 0.5 inches	"Apparatus similar to McConnell"	Semicircular: 8 by 1 inches	Semicircular: inside cross-sectional area of 0.5 cm²
Light	Two 100 w bulbs 6 inches above trough		One 75 w bulb 8 inches above trough with 0.25 inch plastic shield between bulb and trough	Two 11 inch tungsten ribbon lamps operated at color temperature of 2870° K
Shock (US)	DC via inductorium	AC (30 ma)	DC via inductorium	Pulse generator; unipolar shock given[b]
Intertrial interval	84 sec	60 sec	60 sec	30 sec
Trials to criterion	23/25	23/25	34/40	10/15
Trial sequence	50/day	50/day	40/day	30/day
Regeneration time	28 days (bisected Ss)	23 days (bisected Ss)	14–18 days (bisected Ss)	14 days to 1 month (most Ss trisected, some bisected)

[a]In planarian research conducted by W. C. Corning at Michigan, it was found that melted snow was an inadequate medium for culturing planarians. Many died within a week after being placed in the water.

[b]Polarity alternated from trial to trial; e.g., on one trial the animal received a cathodal shock to the head, on the next trial it received an anodal shock, etc.

planarian considering the extent of the literature on lower phyla (see Chapters 2 and 3) and higher phyla. In efforts to ensure a stabilized laboratory housing condition and to reduce intersubject variability, we frequently assessed changes in light responsivity preexperimentally and during the experiment (Corning, Bluemfeld, and John, unpublished; Corning and Freed, 1968), and evidence of habituation to light during a daily session was obtained. Freed (1966), Togrol et al. (1966a), and Chapouthier (1967) have also reported that pretraining habituation trials resulted in a decreased light responsivity.

Earlier investigations by Walter (1908) and Dilk (1937) had provided evidence that habituation to mechanical and light stimuli was possible. Walter rotated the aquarium containing the animals and observed a response inhibition. This inhibitory reaction became habituated when the aquarium was rotated every minute. Walter also observed that when planarians crossed the border of two fields of differing light intensity there was a head-waving response. With successive crossings, this response waned. Mechanical stimulation delivered every 10–15 sec was found by Dilk to be effective for obtaining response decrements in *D. gonocephala*.

Westerman (1963) has provided more recent evidence of light habituation in *D. dorotocephala* and has reported data concerning the "somatic inheritance" of the response decrement. Animals were presented light trials of 3 sec duration with the intertrial interval varying between 30 and 60 sec. A total of 25 trials was given each day, and by the sixteenth to twentieth day the light reactivity had dropped from an initial level of 26 % to less than 5 %. Using a criterion of 50 trials (2 successive days) with no responses, Westerman found that the "tail drops" (planarians often spontaneously fission) or surgically removed tails of previously habituated whole animals habituated much more quickly when compared with controls. Controls consisted of tail drops or the cut tail sections of naive animals. Animals of another group were allowed to cannibalize two fully habituated planarians every fourth day for 20 days. When compared to controls that had eaten naive planarians, they demonstrated a more rapid response decrement to light. What is remarkable about this investigation is that retention of the habituation persisted after regeneration (9–14 days) and that, when tested 3 weeks and 7 weeks later, previously habituated animals still demonstrated a reduced light responsivity.

Stenostomum has been shown to habituate to mechanical stimulation (Applewhite and Morowitz, 1966, 1967). In the initial experiments, the depression slide containing the animals was mechanically vibrated by jarring the ring stand on which the slide was attached. The intensity of the shock could be controlled by measuring the degree of displacement with a micrometer. The criterion for habituation was no response to two successive stimuli. (The data for *Stenostomum* are summarized in Table VI.) Later research obtained habituation with modified techniques. Instead of the gross mechanical jarring of the depression slide, Applewhite (1971) used a 100 μ wire as a stimulus. The wire touched the side of the animal's body every 4 sec. Habituation criterion was no response for three consecutive stimulations. When animals were transected and tested for retention 30 sec later, savings were observed. The anterior and posterior portions habituated more quickly

Table VI. Nonlocal Habituation in *Stenostomum*[a] (Applewhite, 1971)

	Whole organism	Anterior	Posterior	Naive anterior	Naive posterior
Mean	6.6	2.5	2.3	7.5	6.4
S.D.	2.4	0.4	0.6	2.0	2.5

[a]Number of stimuli to habituate.

than the original whole organism. The transection did not desensitize *Stenostomum,* as cut, naive portions did not habituate any faster than the original, whole worm. Since some animals may have had more than one brain, Applewhite cut animals again to achieve brainless organisms. When these were then habituated, it was found that they habituated at the same rate as whole animals. To prove that the habituation process was not sensory fatigue, the animals were also habituated by touching only the anterior end. When habituation was completed, the posterior end was then stimulated, and the animals were still habituated. The reversal of this procedure produced the same transference effect. Thus, sensory fatigue can be ruled out as a factor in the response decrement. Further demonstrations of habituation and retention after regeneration have been provided by Togrol *et al.* (1966*a,b*) in *D. lugubris.*

Failure to obtain habituation in planarians has been reported in two laboratories (Brown, 1964; Bennett and Calvin, 1964). Brown obtained daily drops in light sensitivity between the first and second blocks of light trials (25 trials per block) but did not observe a consistent decrease over days. The Bennett and Calvin study used Westerman's criterion of 50 consecutive trials without a response and found that only four of 70 subjects habituated. In addition, analyses of the average percent response to light in all animals showed no decrease from the initial response rate of 16%.

Associative Learning: Earlier Investigations

Dark-adapted planarians will typically begin active movement when they are suddenly exposed to light. Hovey (1929) was able to reverse or inhibit this innate response by touching the animal at the anterior end each time it moved at the initiation of light. Thus Hovey "attempted to condition these worms against photokinesis by means of contact stimulation applied simultaneoulsy with the light," or establish what was termed associative hysteresis. Each animal was exposed to 25 trials, each trial consisting of 5 min of light exposure followed by 30 min of darkness.[6] As soon as the animal moved after the initiation of light, its anterior end was touched, causing

[6]The training procedures were not entirely independent of Hovey's dinner habits. The trial sequence deviated "between the 20th and 21st exposures, at which time the animals enjoyed peace for an hour while the experimenter ate dinner."

it to retract. This was repeated each time the animals tried to move. A summary of the number of touches over the experiment is presented in Fig. 9. Controls consisted of animals which were exposed to 20 trials of light but were not stimulated when they moved. On trial 21, training began and their behavior showed that light adaptation was not a confounding factor (see curve 3, Fig. 9). Injury could also be excluded as an explanation. Hovey deliberately injured several animals and demonstrated that their response to the training was no different from normal (curve 2, Fig. 9). Hovey also found that the persistence of the conditioned response was apparently dependent on the cephalic ganglia. When the ganglia were lesioned, the conditioned inhibition was lost. Hovey compared his findings with those of Pavlov and concluded that, like the vertebrate, the planarian brain is the seat of conditioning. Jacobson (1963) questioned this interpretation, noting that the ganglia are the critical mediators of sensory and motor capabilities.

Soest (1937) followed a basic paradigm that he had used with reported success in protozoans (reviewed in Chapter 2) to obtain evidence of aversive inhibitory conditioning in *Stenostomum*. As in the protozoan studies, the basic association was between a lighted or darkened half of a bowl and electric shock. Thus each time a subject moved into the incorrect area of the bowl, it was shocked. An example of the results obtained with a subject is presented

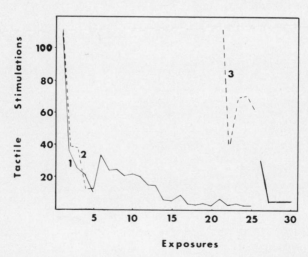

Exposures

Fig. 9. Inhibition of movement in planarians. Curve 1 represents the average number of tactile stimulations in 17 animals during each 5 min exposure to light. The intertrial interval was 30 min. The increase at trial 26 is due to a lapse of 10 hr. Curve 2 represents the behavior of ten animals that had the tips of their snouts cut off. Curve 3 represents the number of stimulations given to 12 subjects that had not been stimulated for the first 20 light exposures, indicating that light adaptation is not a factor in the response inhibition. (After Hovey, 1929.)

in Table VII. Soest notes that training to avoid the darkened half was more difficult. Dilk (1937) obtained similar results with *D. gonocephala*. Both of these early studies, however, lacked the critical controls for stimulus sensitization.

Classical Conditioning: Nondifferential

The basic techniques of the McConnell group at Michigan have been outlined in the prior section reviewing the controversy over planarian learning. Details of apparatus and discussions of the advantages of various procedures are available (McConnell, 1967*b;* McConnell *et al.,* 1960; Thompson and McConnell, 1955). Later studies of this group used somewhat different procedures. Instead of massed trials where each animal received all trials in 1 day, planarians were given only 50 trials per day (McConnell *et al.,* 1959), since the mass-training procedure enhances the probability of pseudoconditioning occurring (Cornwell, 1960). Later work by Jacobson (1967) stressed

Table VII. *Stenostomum*: **Light-Avoidance Training (Soest, 1937)**

Time since beginning of training (min)	Total number of contacts with light–dark border	Number of crossings at border without avoidance reaction	Number of avoidance responses at border	Avoidance responses (%)
1.5	3	3	0	0
3.5	6	3	0	0
6	9	3	0	0
9	12	3	0	0
12	15	3	0	0
15	18	2	1	33
18	21	3	0	0
21	24	3	0	0
23	27	3	0	0
25.5	30	0	3	100
27	33	0	3	100
29	36	1	2	67
31	39	0	3	100
33	42	0	3	100
34.5	45	1	2	67
36.5	48	0	3	100
(Extinction)				
38	51	1	2	67
40.5	54	2	1	33
43.5	57	2	1	33
46	60	3	0	0
48.5	63	3	0	0
50	66	3	0	0

the use of just the contraction as a reliable indication of a CR development. There has also been some confusion over what type of shock to use. Corning and Freed (1968), Best (1967a), and McConnell (1967b) all emphasized the effectiveness of the "dirty" waveform such as that put out by the Harvard Inductorium, and yet Jacobson's laboratory produced some of the cleanest experimental–control differentiations with AC current (Jacobson *et al.*, 1966a,b). As previously discussed, group training techniques have been developed which are quite effective in replicating the results obtained with individual training methods (Corning and Freed, 1968; Freed, 1966).

Repeated successes with classical conditioning paradigms appear. Applewhite and Morowitz (1967) report the following for *Stenostomum*:

> After 4 seconds of the light, an electric shock of 5 volts and 100 msec. duration in the biphasic mode was administered . . . this caused *Stenostomum* to stop moving. The light went off after the 5 seconds, then after 25 seconds it went on again and the procedure was repeated. . . . Fifteen animals were used and each was considered conditioned when it stopped moving four times in a row every time the light was flashed. . . . It took a mean of 27.3 stimuli (s.d. 12.3) to condition them this way. This response was extinguished with a mean of 4.2 (s.d. 1.6) stimuli by presenting the 150 watt light for 5 seconds every 30 seconds until the animal no longer responded to it. . . . Three control groups of 12 animals were run. In the first group, each animal was left in the chamber with only the ambient light on for 14 minutes, the mean time for conditioning to occur (27.3 stimuli multiplied by 0.5 minute). Then the 150-watt light was presented for 5 seconds every 30 seconds over a period of 14 minutes. For the second group, a shock of 5 volts for the 100 msec. duration was given every 30 seconds for 14 minutes. Then the 150-watt light was presented for 5 seconds every 30 seconds over a period of 14 minutes. The third group was given only the light every 30 seconds for 5 seconds over a period of 14 minutes. None of these groups showed any response to the light. (pp. 336–337)

A listing of other recent studies examining classical conditioning in planarians is provided in Table VIII.

Classical Conditioning: Differential

Differential conditioning in Platyhelminthes has been demonstrated in *D. tigrina, D. dorotocephala,* and *Phagocata gracilis.* One of the first reported successes was that by Griffard and Peirce (1964), who were able to train planarians *(P. gracilis)* to turn one way when a light was on and in the other direction when vibration stimulus was presented. The differential effects of cathodal and anodal shock were used to effect the directional turning—planarians usually turn away from the anode. Animals were presented a total of 200 training trials consisting of a 3 sec CS (light or vibration) overlapping with a 2.5 sec duration shock. One half of the subjects were run with light paired with an elicited right turn and the vibration paired with a left turn. The other half were run with conditions reversed. Ten test trials consisting of five

Table VIII. Positive and Negative Learning Studies in Planarians[a]

Positive	Negative
A. Habituation	
Applewhite and Morowitz (1967)	Bennett and Calvin (1964)
Applewhite (1971)	Brown (1964)
Dilk (1937)	
Freed (1966)	
Togrol *et al.* (1966*a,b*)	
Walter (1908)	
Westerman (1963)	
B. Classical conditioning: acquisition measures	
Applewhite and Morowitz (1967)	Bennett and Calvin (1964)[b]
Barnes and Katzung (1963)	Brown (1964)
Baxter and Kimmel (1963)	Brown (1967*a*)
Bennett and Calvin (1964)[b]	Brown *et al.* (1966*a*)
Brown and Parke (1967)	Brown *et al.* (1966*b*)
Chapouthier (1967)	Cummings and Moreland (1959)
Cherkashin and Sheiman (1967)	Halas *et al.* (1962*b*)
Cherkashin *et al.* (1967)	
Corning and John (1961)	
Cornwell (1961)	
Cornwell *et al.* (1961)	
Crawford (1967)	
Crawford and King (1966)	
Crawford *et al.* (1965)	
Crawford *et al.* (1966)	
Fantl and Nevin (1965)	
Freed (1966)	
Griffard (1963)[b]	
Guilliams and Harris (1971)	
Halas *et al.* (1962)[b]	
Hartrey *et al.* (1964)	
Hyden *et al.* (1969)	
Jacobson (1967)	
Jacobson *et al.* (1966)	
Jacobson *et al.* (1967)	
John (1964)	
Kimmel and Yaremko (1966)	
King *et al.* (1965)	
McConnell (1967*a*)	
McConnell and Mpitsos (1965)	
McConnell *et al.* (1959)	
McConnell *et al.* (1960)	
Thompson and McConnell (1955)	
Walker (1966)	
Walker and Milton (1966)	
Yaremko and Kimmel (1969)	

Table VIII. (Continued)

Positive	Negative
C. Classical conditioning: extinction measures	
Corning and John (1961)	Baxter and Kimmel (1963)
Crawford (1967)	Brown (1964)
Crawford and King (1966)	Brown (1967a)
Crawford et al. (1965)	Brown and Beck (1964)
Crawford et al. (1966)	Cornwell (1960)
Hullett and Homzie (1966)[b]	Crawford (1967)
Jacobson (1967)	Halas et al. (1961)
Jacobson et al. (1967)	James and Halas (1964)
Kimmel and Yaremko (1966)	
Vattano and Hullett (1964)	
Yaremko and Kimmel (1969)	
D. Differential conditioning	
Block and McConnell (1967)	Kimmel and Harrell (1964)[b]
Fantl and Nevin (1965)	Kimmel and Harrell (1966)[b]
Griffard and Peirce (1964)	
Jacobson (1967)	
Jacobson et al. (1967)	
Kimmel and Harrell (1964)[b]	
Kimmel and Harrell (1966)[b]	
E. Instrumental paradigms	
1. Instrumental avoidance	
Dilk (1937)	
Hovey (1929)	
Lacey (1971)	
Ragland and Ragland (1965)	
Soest (1937)	
Tushmalova (1967)	
2. Mazes	
Best (1963)	Bennett and Calvin (1964)
Best (1965)	
Best and Rubinstein (1962)	
Chapouthier (1968)	
Corning (1964)	
Corning (1966)	
Ernhart and Sherrick (1959)	
Haynes et al. (1965)	
Humpheries and McConnell (1964)	
Jacobson and Jacobson (1963)	
McConnell (1966b)	
Pickett et al. (1964)	
Roe (1963)	
Van Oye (1920)	
Wells (1967)	
Wells et al. (1966)	

Table VIII. (Continued)

Positive	Negative
3. Operant	
Best (1965)	Bennett and Calvin (1964)
Crawford and Skeen (1967)	
Lee (1963)	

[a]The columnar arrangement of positive and negative studies is not intended to argue for learning in planarians by providing a "numbers" comparison. It is provided to give the reader a general guide to the existing literature.

[b]Equivocal findings or interpretations:

Bennett and Calvin (1964): "Although it would be presumptuous to conclude that planarians cannot be trained, we do conclude that at the present time, the relevant factors necessary for reproducible and reliable training of planarians have not been described with sufficient precision to permit their use in studies of the biochemical bases of learning" (p. 4).

Griffard (1963): "In spite of the significance obtained between Group E and all others in both sets of data, the appearance of possible sensitization effects in Group SC renders a conclusion in favor of classical conditioning at least equivocal" (p. 600).

Halas *et al.* (1962): "The results agree in general with Thompson and McConnell's, but it is suggested that reflex sensitization is a more tenable interpretation of the data. . . . light is weak US . . . It would seem that what we may have here is a precursor or primitive type of conditioning" (p. 971).

Hullett and Homzie (1966): "Though the findings support the position of Halas . . . the present experiment does not preclude the possibility that worms can be classically conditioned. The fact that Group B-L_iS after 31 days of adaptation responded at a level significantly above that of the control indicates that there may be a classical conditioning component in this paradigm" (p. 230).

Kimmel and Harrell (1964): "Thus, while the differential responding that was found adds some support to the argument that conditioning *does* occur in the planarian, the possibility that sensitization and/or generalization were significantly involved cannot be ruled out" (p. 228).

Kimmel and Harrell (1966): "It does not seem appropriate at present to conclude that the planarian shows this type (differential) of conditioning, although it is possible that a longer training session might permit differentiation to occur" (p. 286).

light and five vibration stimulations were given every 50 trials. The frequency of correct responses obtained during these trials demonstrated a significant increase in discriminative performance.

A differential conditioning paradigm was attempted by Fantl and Nevin (1965) and Kimmel and Harrell (1964). In both studies, either light or vibration was used as CS+. The basic procedure was to present a random series of either light or vibration and to pair shock consistently with one of the stimuli. Fantl and Nevin observed differential response increases when either light or vibration was the CS+, while Kimmel and Harrell were able to achieve differential conditioning only when vibration was the CS+. A subsequent study using equipotential stimuli (light and vibration) also failed to demonstrate differential conditioning to light (Kimmel and Harrell, 1966). However, Jacobson's group (research reviewed in a previous section) was able to replicate the Fantl and Nevin study (Jacobson *et al.*, 1966*a*,*b*, 1967; Jacobson, 1967). At the completion of differential conditioning, Jacobson reversed the pairings, making the CS+ a CS−, and in subsequent trials obtained a reestablishment of differential responding to the new CS+. Block and McConnell (1967) have provided a more recent replication of differential responsivity during both acquisition and extinction. A list of differential conditioning studies is provided in Table VIII.

Instrumental Learning

Simple Escape Learning. With a sufficiently intense light stimulus, planarians display a clear photonegative reaction—they will crawl around the container until they reach a point where they are farthest from the light. Lacey (1971) used this response to gain evidence of escape learning in *D. dorotocephala*. The primary motivation was to avoid the possible sensitization of planarians by the use of shock. Thus the subjects were required to crawl to a particular place in a bowl to escape the light, and the latency of this escape response was measured. When they reached the goal area, the light was terminated. Over eight training trials, the animals showed a mean drop from 140 sec to approximately 117 sec. The improved escape time was found to persist after regeneration (19 days) in transected subjects.

Avoidance Conditioning. The earlier investigations of Soest (1937) and Hovey (1929) have been reviewed in a previous section. Recent instrumental paradigms have relied mainly on maze demonstrations. In general, there has been surprisingly little interest in the instrumental avoidance task with respect to Platyhelminthes. The Raglands (Ragland and Ragland, 1965) used light–shock pairings in attempts to condition avoidance responses in *D. tigrina* and *D. dorotocephala*. Light duration was 4 sec and the shock was delivered during the last second of the light period. If the animal responded during the first 3 sec, no shock was administered. In this training situation, *D. dorotocephala* showed an increased responsivity to light over 200 training trials, achieving a conditioning level of 65–70%, while *D. tigrina* failed to condition at all.

Soest's original procedure has been generally replicated with *Podoplana olivacea* (Tushmalova, 1967). Animals were placed in a container with lighted and darkened halves. When subjects swam into the incorrect side (either the light or dark portion), they received tactile stimulation, and an avoidance of a particular half was eventually obtained.

Maze Investigations. Interest in developing a maze-learning paradigm was partially motivated by the many criticisms over the light–shock classical conditioning procedure. A commonly used apparatus is a T-maze, where the primary motivation for correct traversal is return to the animal's container. One such maze is represented in Fig. 10A (Corning, 1964). The maze is positioned over the subject's bowl and moistened with water. The worm is picked up with a soft camel's hair brush and placed in the starting area of the maze, with care taken to orient the anterior end toward the choice point. When line CD or EF is reached, the choice is recorded. When the animal crawls back into its bowl, the maze can be positioned over another bowl or trials with that subject can be repeated. An incorrect choice is punished by poking the anterior end of the subject with the brush, a tactic that almost always results in a withdrawal response.

This maze is based on the design reported by Humpheries (1961) and has been moderately successful. In right–left discrimination, animals *(D. dorotocephala)* will significantly increase their correct performance but show response instability or even a reversal of choice in the later trials (Corning, 1964), a phenomenon replicated by Chapouthier (1968). Ernhart and Sherrick

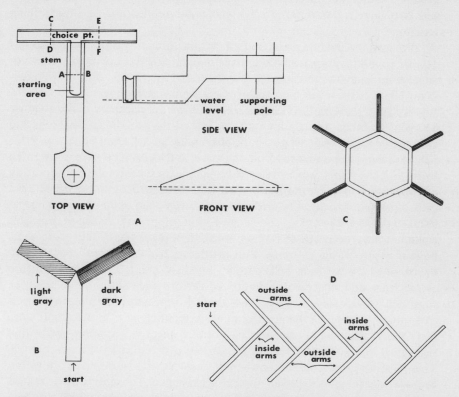

Fig. 10. Maze types. A, T-maze (Corning, 1964). B, Y-maze (McConnell, 1965). C, Hexagonal maze (Roe, 1963). D, Z-maze (Roe, 1963).

(1959), in one of the original maze investigations, found a double-unit T-maze quite successful in developing right–left discrimination in *D. tigrina*. Generally, it appears that 50–70% of the animals run in this type of maze can reach a criterion of nine out of ten correct responses within 100 trials. Humpheries and McConnell (1964) and McConnell (1967*a*) have detailed some of the critical factors that can influence maze performance and have listed the types of discriminations and methods of reinforcement. Factors deemed critical are the shape of the maze trough (semicircular and V-shaped are best), the number of trials per day (five trials may be the maximum), and the reinforcements for incorrect (touch, shock, bright light) and correct (water, food, darkness) choices.

Kiki Roe (1963) developed a hexagonal maze (Fig 10) to reduce handling time and also the interval between trials since the subject could run continuously. Using a black–white discrimination with electric shock as the aversive stimulus, 21 out of 30 subjects (*D. dorotocephala*) were able to achieve a rather stiff criterion of 12 consecutive correct choices. Humpheries and McConnell (1964) reported a successful replication of this training technique.

Perhaps the most ingenious of all "worm runners" is Jay Best, who has devised apparatus for automating maze running and has documented several interesting behaviors in *Cura foremanii* (Best, 1963, 1965; Best and Rubinstein, 1962). One maze consists of three wells interconnected by tunnels. Any well can serve as the starting area, and the animal must make a choice between the other two. By illuminating one of the remaining two wells and keeping the other dark, a light–dark discrimination task can be set up. Prior to a trial, the water is drained from the maze, and a correct choice is rewarded by refilling the maze. Subjects showed a significant increase in the proportion of correct responses for the first two trial sets (four to fifteen trials per set) but then displayed what Best termed a period of rejection. Animals would begin to make persistent incorrect choices and eventually would refuse to run. This could not have been due to fatigue, since when placed back into their home bowl they moved about freely. Best attributed this "rejection" behavior to the confined space of the wells and devised a second, less restricting version of the maze and observed an improved performance stability. It has been suggested previously (Corning, 1964) that the repeated presentation of trials to a subject may offset the positive reinforcement effect of a successful maze run. The subjects were placed in a generally negative situation and within that situation executed a response that led to a return to the home bowl. This positive experience of bowl reinforcement was short-lived, however, because the subjects received further treatments. The continued negative experience may enhance the probability of subjects attempting other behavioral modes to reduce this experience.

A fully automated operant paradigm has also been successful (Lee,

1963). The apparatus consists of a water-filled well that contains a tunnel at the bottom. An intense overhead light is turned on and when the animal moves through the tunnel, breaking a photoelectric beam, the light terminates for a period. The frequency of tunnel traversal was found to increase in experimental animals and to remain constant in yoked control subjects where tunnel entry was independent of light termination. Best (1965) has replicated this study and has discovered that performance can be related to lunar cycles. Further replications are reported by Crawford (Crawford, 1967; Crawford and Skeen, 1967), and a list of operant publications is included in Table VIII.

Physiological Bases of Associative Learning

Retention After Regeneration. In an attempt to localize the bases for retention of the conditioned response in planarians, *D. dorotocephala* specimens were conditioned, sectioned transversely, allowed to regenerate, and then reconditioned (McConnell *et al.,* 1959). An initial assumption might be that the anterior portion would demonstrate the greatest amount of savings during retraining, since it is this portion that contains the brain. However, it was found that both halves demonstrated equivalent savings. Control groups showed that regeneration did not sensitize animals to light and that the retention levels were similar to that observed in uncut animals left undisturbed for the regeneration period. Savings are also reported for second-generation animals (McConnell *et al.,* 1958). Planarians were transected, the anterior half was conditioned, and on completion of regeneration this portion was again transected and allowed to regenerate. During retraining, both halves demonstrated significant savings. These basic findings of McConnell's group were repeated at the University of Rochester (Corning and John, 1961; John, 1964). In these studies, all subjects were first habituated to light over a 3 day period and then trained in the light–shock classical conditioning paradigm to a criterion of 34/40 correct responses. Experimental groups were then sectioned and allowed to regenerate 14–18 days. Following regeneration, they were given another 3 days of light-only trials and then retrained. Thus the measures for retention were both the extinction data and acquisition scores. The response scores on both of these measures were equal for the regenerated anterior and posterior portions, and considerable savings were demonstrated. The regenerated subjects took approximately half as many trials to reach criterion, and they displayed higher light responsivity scores on the extinction trials. It appeared that the physiological consequences of conditioning were locked in a system that was relatively unaffected by the regeneration process. Other evidence of retention of a conditioned response after regeneration continues to appear (Cherkashin and Sheiman, 1967; Togrol *et al.,* 1966a,b).

Maze discriminations have also been shown to survive regeneration.

The first demonstration of this was in the Ernhart and Sherrick (1959) double-unit T-maze in which planarians were required to learn a position discrimination. Subjects were trained to a criterion of six consecutive correct choices, transected, and retrained after regeneration. The original animals took an average of 58 trials to achieve criterion, while the regenerated anterior and posterior portions took 29.5 and 28.6 trials to reach the same performance level. Humpheries and McConnell (1964) trained planarians in a black–white discrimination using anodal shock as punishment for an incorrect choice. Subjects were able to achieve a 73% correct response level after 150 trials. Tail drops (natural fissioning) that had occurred at the end of 30 trials and had been allowed to regenerate displayed a rather high (73%) correct response level when tested 2–3 weeks later. Tail drops that occurred after 60 trials exhibited a chance level of performance when tested after regeneration. These observations suggest that the negativity frequently observed in planarian maze behavior is also carried through regeneration. The results of a more systematic attempt to demonstrate retention of a position discrimination after regeneration are presented in Fig. 11 (Corning, 1966). *D. dorotocephala* individuals were required to select the right or left arm of an elevated T-maze in order to return to the home bowl. All subjects were first given ten free-choice or preference-testing trials in which they could choose either arm to escape. Thereafter, subjects were forced to select the least-preferred arm. Animals were trained until they achieved a criterion of nine out of ten correct trial runs, sectioned transversely, allowed to regenerate 18–26 days, and then retrained. After regeneration was completed, they were given an additional ten preference tests followed by retraining. Controls included animals that were placed in the maze and allowed to take either arm to escape and a naive group that were transected, regenerated, and then trained. The results of Fig. 11 demonstrate the retention in the regenerated portions of previously trained animals (I). During the second preference test, their original preference had been reversed, and during retraining the performance levels were considerably higher than those of the original subjects.

Equal retention or savings in the regenerated segments of a planarian is not always observed. Agoston (1960) reports that anterior portions of previously classically conditioned planarians demonstrated better savings than the posterior regenerates. Larger portions were also found to demonstrate more savings than animals derived from smaller pieces. In the hexagonal maze, Roe (1963) trained *D. dorotocephala* in a black–white discrimination and found that only the anterior portions showed any retention after regeneration. One suggestion was that a more complex form of learning such as the maze-discrimination task requires the cerebral ganglia for storage.

Other investigations have delineated variables that might offer alternate

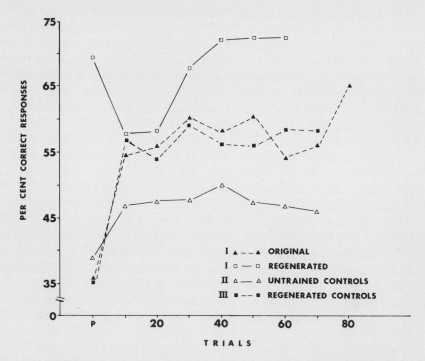

Fig. 11. Average percent correct response performance of planarians in a T-maze. All
*S*s were first given ten preference tests (P) in the maze. Group I subjects were trained to a
criterion of nine out of ten choices of the arm opposite to that selected during the preference
tests. Group II subjects were permitted to take either arm to escape the maze. Group III
contained newly regenerated subjects that were trained to criterion. It can be seen that the
regenerates of group I demonstrated retention of the habit during their postregeneration
preference tests and during retraining. The scores of the anterior and posterior regenerates
were the same and are combined in this figure (from Corning, 1966).

explanations for the regeneration findings. Van Deventer and Ratner (1964)
found that worm size is inversely related to light sensitivity, and, since transection and regeneration result in smaller animals, this might account for
the savings reported by others. A similar suggestion is made by Brown (1964),
who found that smaller worms *(D. tigrina)* are more reactive than larger
ones and that the origin of the worm along the axis is correlated with light
sensitivity. Anterior derivatives were more sensitive than middle or posterior
regenerates, and these differences persisted for up to a month. The question
is raised whether the regeneration time of other studies might have been too
short. Additionally, an examination of the relationship between light intensity and the origin of the regenerated animal indicated that at higher intensities (1000 ft-c) the anterior and posterior portions displayed equal response

levels, whereas at the lower intensities (500 ft-c) greater reactivity was observed in the anterior regenerates. Thus the combination of size decrease due to regeneration and the intensity of the CS could explain the equal retention reported by McConnell *et al.,* (1959) and by Corning and John (1961). However, this interpretation ignores the control groups included in the latter studies. Regeneration was found not to sensitize *D. dorotocephala* to light, nor was there a difference in light sensitivity in naive anterior and posterior regenerates. As discussed previously (Corning, 1967; Corning and Riccio, 1970), these negative studies have introduced considerable variation in procedure and do not constitute fair "replications." For example, Brown's use of a different species, alternating cathodal and anodal trials, and other variations could be critical.

The demonstration of retention after regeneration stimulated interest in using the planarian as a preparation to investigate memory mechanisms. Since retention in the planarian can survive regeneration, it is reasonable to assume that memory involves a physiological substrate that persists through this process and that the nature of these substrates can be ascertained by effecting changes in specific systems during regeneration. In earlier research on memory mechanisms, considerable interest was aroused in the role of ribonucleic acid (RNA). Such interest was stimulated by the theoretical paper of Holger Hyden (1959), in which he speculated about how information might reorder nucleotide sequences and subsequently affect a cell's firing probability. Since ribonuclease has a disruptive effect on planarian regeneration (Henderson and Eakin, 1961), the effects of less disruptive concentrations of the enzyme on conditioned response regeneration were examined (Corning and John, 1961). Conditioned and transected planarians were regenerated in pond-water solutions containing ribonuclease. The expectation was that if ribonuclease could affect CR persistence in regenerates its effect would be more pronounced on the posterior segment, since this portion would have to regenerate a dominant anterior end in the presence of the enzyme whereas the anterior segment would only have to grow a nondominant tail. The differential effect was obtained. The CR frequencies of anterior regenerates derived from previously conditioned worms were unaffected by enzyme treatment; they were comparable to retention levels of anterior portions regenerated in pond water and of uncut controls. In contrast, the posterior portions that had regenerated in ribonuclease exhibited the response of a naive animal—there was little indication that they had been trained, suggesting that the enzyme had interfered with the processes by which the posterior section transmitted experience to the regenerating anterior tissue. Various control groups demonstrated that the effect of ribonuclease on conditioned retention was only observed in regenerating posterior segments and not in whole worms and that neither regeneration nor the presence of the enzyme sensitized flatworms to light.

Research at the Academy of Sciences in Leningrad has confirmed certain aspects of the regeneration findings (Cherkashin and Sheiman, 1966, 1967; Cherkashin *et al.*, 1967). If the cerebral ganglia were removed in already conditioned *Ijimio tenuis*, the subjects returned to a naive response level. Regeneration of the cerebral ganglia was accompanied by a return of the conditioned response, a finding that essentially confirms the earlier work of McConnell. Memory appears to spread throughout the nervous system and is reestablished in newly developed tissue. Other experiments by these investigators examined the effects of various enzymes and inhibitors on retention. If whole worms were placed in a ribonuclease solution during conditioning, the animals did not learn; if already conditioned animals were put in the enzyme, the conditioned response was suppressed but not destroyed, and on removal of ribonuclease it returned. Other agents such as deoxyribonuclease, 8-azaguanine, bromine uracil, and aurantine had no effect on retention. When already conditioned planarians were transected and placed in a solution of ribonuclease, there was a loss of the conditioned response in both anterior and posterior portions. Tushmalova (1967) has found that ribonuclease can reversibly inhibit an avoidance discrimination in *Podoplana olivacea* but also found that the enzyme reduced motor activity and concluded that RNA regulated cell excitability rather than being critical for learning *per se*.

Recently, Lacey (1971) has examined whether there is a time dependency in the effects of either ribonuclease or deoxyribonuclease on a simple escape response in *D. dorotocephala*. Previously trained tail sections were placed in ribonuclease or deoxyribonuclease immediately after sectioning or 24 hr later. After regeneration (19 days), the group that was exposed to the ribonuclease immediately and the groups that were exposed to both enzymes 1 day later showed no retention when compared to controls.

A rather unique approach to the disruption of potential mnemonic substrates was adopted by Brown (1964, 1967a). Planarians that had been subjected to light–shock treatments were exposed to ultraviolet radiation at a wavelength of 260 nm, which damages nucleic acids. Responses to light were diminished as a consequence of the exposure, but photorestorative treatment (which partially reverses the ultraviolet effects) brought about an increase in light reactivity. The failure to obtain clear evidence of conditioning in these studies did not permit the full effectiveness of this technique to be explored.

Biochemical Correlates. While the regeneration studies had implied a biochemical basis of learning in planarians (Corning and John, 1961; McConnell *et al.*, 1959), it was the demonstration that a transfer of memory could be effected either via cannibalism or via RNA injections that stimulated the most interest in the planarian as a preparation to study memory molecules. In fact, the "memory transfer" strategy has been expanded to crabs,

fish, starfish, and rats, and the results with this technique have led to considerable debate occupying extensive space in journals and books. Some
consider RNA to be the transfer molecule, while others prefer small proteins.

In order to obtain a more complete picture of the nucleic acid changes
that might accompany conditioning, we carried out whole-body analyses of
worms at several stages of training (Corning and Freed, 1968). Analyses
were carried out after 160, 340, and 460 training trials. We performed whole-
body analyses because both the cannibalistic transfer studies and retention
after regeneration studies indicated a nonlocal storage of information in the
planarian. Two groups of D. tigrina were used: group I received the paired
light–shock classical conditioning treatment, and group II was presented with
random light and shock. Because the Ss of group I began to contract more
frequently as training progressed (both to the light and shock), the Ss of
group II were given an equivalent amount of contraction experience by
tapping their dish an appropriate number of times. Estimates of nucleic acid
turnover were obtained by using P^{32}. Previous studies had shown that when
radioactive inorganic phosphorus was placed in the culture medium there
was rapid incorporation. Electrophoretic and chromatographic separations
established a nucleotide location of the tracer in the nucleic acid fractions.
Extraction procedures were based on those described by Scott et al. (1956).
To keep exposure to the isotope at a minimum, the Ss to be loaded were not
placed in a P^{32} medium until 3 days before each of the stages. Samples
containing 18–24 worms were removed from each of the main groups and
placed in separate bowls containing P^{32} (20 μc/ml of pond water). During
the next 3 days, half of each sample received its usual experimental treatments
(A samples), while the other half was placed in the training bowls but received
no stimulation (B samples). No P^{32} was present in the water contained in
the training bowls. At the end of the 3 days, the samples were killed by immersing them in liquid nitrogen. The main groups and the samples received
their experimental treatments simultaneously. Further procedural details
have been presented elsewhere (Corning and Freed, 1968; Freed, 1966).

We found that there were no significant differences between groups with
respect to RNA and DNA quantities and RNA/DNA ratios. However, the
P^{32} data did yield interesting differences between the trained and control
groups. When RNA/DNA specific activity ratios were calculated, the conditioned Ss of the A samples showed significantly lower specific activity
ratios at 160 and 340 trial stages when compared to all other groups including
the conditioned B samples. At the terminal stage (460 trials), the ratios of the
various samples were again similar and not significantly different from those
observed at the beginning of training. These data are summarized in Fig. 12.

These findings suggest that the biochemical consequences of different
modes of stimulus presentation (paired light–shock vs. random light and

shock) occur before there is any dramatic behavioral change. They also show that the changes in specific activity ratios produced by paired light and shock are transient; i.e., the LS-B samples did not demonstrate the differentiation observed in the A samples.

Analyses of single cells in trained and pseudoconditioned planarians by H. Hyden, E. R. John, and coworkers have failed to yield any clear-cut differences in RNA base ratios (Hyden *et al.,* 1969). However, these analyses were made at the completion of training and did not include any attempt at estimating turnover rates. There is some evidence that training produces differences in amino acid content (Crawford *et al.,* 1965), but follow-up research has failed to establish this effect (King *et al.,* 1965).

Memory Transfer. The planarian was the first modern preparation to be used in a "memory transfer" experiment. The oft-cited study of McConnell (1962) that demonstrated an increase in light reactivity in planarians that had cannibalized previously conditioned subjects was immediately replicated by John (1964) and created a stir that has yet to settle. The original experiment was not done in jest:

> Planarians are interesting experimental animals with which to work, not only because they can learn and regenerate vigorously, but also because they lack the immune reaction found in most higher organisms. The head of one animal can fairly readily be grafted onto the body of another. . . . We reasoned that if heads could be transferred from one worm to another, why couldn't the chemicals involved in memory formation also be taken from a trained animal and "grafted"

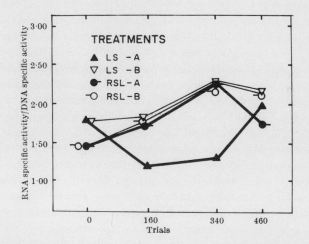

Fig. 12. RNA/DNA specific activity ratios at four stages of conditioning in planarians. LS, Groups receiving paired light and shock treatment. RSL, Groups receiving random light and shock. The A samples were presented their usual stimulus conditions during the isotope loading period, while the B samples were not (Corning and Freed, 1968).

somehow onto another animal? At first we attempted to graft trained heads onto untrained bodies, but our grafting techniques were less than adequate. Then it occurred to us that hungry planarians (of the species we were using) are cannibalistic; why couldn't we train a group of donor planarians, then cut them in pieces and feed them to hungry, untrained cannibals? (McConnell and Shelby, 1970, p. 84)

McConnell has pointed out that almost all who attempt a transfer experiment with planarians succeed. In many cases, the transfer effect is nonspecific; i.e., it appears to be a sensitization effect rather than a specific transfer of acquired information. However, it is argued that if acquired behavior can be transferred then it is reasonable to assume that other forms of behavioral change (those due to "sensitization," etc.) can also be transferred and that studies failing to replicate have not trained but have sensitized the donors (McConnell and Shelby, 1970). Reviews of the "memory transfer" literature and debate are readily available, and extensive discussion of the phenomenon will not be included in the present chapter (see Adam, 1971; Byrne, 1970; Corning and Ratner, 1967; Dyal, 1971; Fjerdingstad, 1971; Gurowitz, 1969; McConnell and Shelby, 1970; Ungar, 1970). To best demonstrate the potential usefulness of the transfer technique in planarians, studies by Jacobson and McConnell are exemplary.

Jacobson's group has been able to demonstrate that the transfer of training effect is specific to a group receiving paired light and shock and does not occur in animals experiencing unpaired treatments (Jacobson *et al.,* 1966a). RNA was extracted from animals that had paired light and shock treatment, animals that had random light and shock, and animals that were naive. The extract was injected into 75 planarians that had been habituated to light. One day after the injection, all subjects were given test trials consisting of light only. The results demonstrate quite convincingly that the performance of animals receiving the RNA from trained animals was different from that of animals receiving the RNA from the pseudoconditioned group and from the naive group. If the RNA was degraded and then injected, the differences did not appear. If the RNA was extracted from trained animals that had undergone extinction treatment, the transfer effect was as high as in those subjects receiving RNA from trained but unextinguished animals.

An acquired preference in a T-maze has also been shown to be transferable (McConnell, 1966b; McConnell and Shelby, 1970). Donor animals were trained to go to a light-gray or dark-gray arm of the maze. Group I cannibals were trained to go to the same arm as the donor animals were trained to choose; group II were trained to go to the opposite arm as the donors; group III digested a "conflict"—they cannibalized animals trained to go to the light-gray arm and animals trained to go to the dark-gray arm; and group IV was fed untrained donors. The results summarized in Table IX strongly suggest an experiential transfer. It should be stressed, however, that others

Table IX. Mean Trials to Criterion in a T-Maze for Four Groups of Planarians
Fed Different Types of Donor Animals (McConnell and Shelby, 1970)

	Mean trials to criterion
Group I (+ group)	113.8
Group II (− group)	166.3
Group III (± group)	263.8
Group IV (0 group)	228.8

have failed to achieve this degree of specificity. Wells and associates have
failed to do so using the Van Oye maze task and a T-maze (Haynes *et al.,*
1965; Pickett *et al.,* 1964; Wells, discussion in Corning and Ratner, 1967, p.
324). Exogenous yeast RNA was found to improve performance in the T-
maze.

VI. CONCLUSIONS

With the advent of a brain, defined sensory systems, and complex neural
cords, the capacity for complex learning clearly appears in Platyhelminthes.
There have been confirmed demonstrations of habituation, classical conditioning, differential classical conditioning, avoidance learning, and operant
training. The debate over learning in Turbellaria appears to have subsided
with increasingly sophisticated demonstrations that apparently answer most
of the early objections. Much of the debate stemmed from *a priori* notions
about what level of animal can learn. These objections can be easily dispensed
with. The failure of some investigators to replicate basic findings can be attributed to procedural deviations and to an inadequate knowledge of factors
that influence planarian behavior. The planarian learning literature lays bare
the problems inherent in psychological learning theory and has caused many
to reconsider the role of processes such as sensitization in response modification.

The use of the planarian in research on memory mechanisms appears to
be decreasing as researchers discover special advantages of other heretofore
unexplored invertebrate and vertebrate preparations. However, the lasting
and significant contributions of planarian research would appear to be (a)
the underscoring of advantages that certain "simple systems" have for studying certain processes, (b) the focusing of attention on invertebrate capacities
and mechanisms, and (c) the stimulation of interest in biochemical mechanisms of behavior. The flatworm may not achieve the position in psychology
that the fruit fly gained in genetics and the bacterium achieved in molecular
biology, but it certainly has stimulated interest in researching animals other
than the rat.

REFERENCES

Adam, G., 1971. *Biology of Memory*, Plenum Press, New York.
Agoston, E., 1960. Learning and regeneration in the planarian. *Worm Runner's Digest, 2,* 53–55.
Applewhite, P., 1971. Similarities in protozoan and flatworm behavior. Unpublished manuscript.
Applewhite, P. B., and Morowitz, H. J., 1966. The micrometazoa as model systems for studying the physiology of memory. *Yale J. Biol. Med., 39,* 90–105.
Applewhite, P. B., and Morowitz, H. J., 1967. Memory and the microinvertebrates. In Corning, W. C., and Ratner, S. C. (eds.), *Chemistry of Learning: Invertebrate Research,* Plenum Press, New York, pp. 329–340.
Barnes, C. C., and Katzung, B. G., 1963. Stimulus polarity and conditioning in planaria. *Science, 141,* 728–730.
Baxter, R., and Kimmel, H. D., 1963. Conditioning and extinction in the planarian. *Am. J. Psychol., 76,* 665–669.
Behrens, M. E., 1962. The electrical response of the planarian photo-receptors. *Comp. Biochem. Physiol., 5,* 129–138.
Bennett, E., and Calvin, M., 1964. Failure to train planarians reliably. *Neurosci. Res. Progr. Bull. 2,* 3–24.
Best, J. B., 1963. Protopsychology. *Sci. Am., 208,* 55–62.
Best, J. B., 1965. Behavior of planaria in instrumental learning paradigms. *Anim. Behav., 13,* Suppl. 1, 69–75.
Best, J. B., 1967a. Major factors in classical conditioning of planarians: Stimulus waveform and neural geometry. In Corning, W. C., and Ratner, S. C. (eds.), *Chemistry of Learning: Invertebrate Research,* Plenum Press, New York, pp. 255–270.
Best, J. B., 1967b. The neuroanatomy of the planarian brain and some functional implications. In Corning, W. C., and Ratner, S. C. (eds.), *Chemistry of Learning: Invertebrate Research,* Plenum Press, New York, pp. 144–163.
Best, J. B., and Elshtain, E., 1966. Biophysics of unconditioned response elicitation in planarians by electric shock. *Worm Runner's Digest, 8(1),* 8–24; *Science, 151,* 727–728.
Best, J. B., and Rubinstein, I., 1962. Maze learning and associated behavior in planaria. *J. Comp. Physiol. Psychol., 55,* 560–566.
Best, J. B., Morita, M., and Noel, J., 1968. Fine ultrastructure and function of planarian goblet cells. *J. Ultrastruct. Res., 24,* 385–387.
Block, R. A., and McConnell, J. V., 1967. Classically conditioned discrimination in the planarian, *Dugesia dorotocephala. Nature, 215,* 1465–1466.
Brøndsted, H. V., 1953. Rate of regeneration in planarians after starvation. *J. Embryol. Exptl. Morphol., 1,* 43–47.
Brøndsted, H., and Brøndsted, H. V., 1954. Size of fragment and rate of regeneration in planarians. *J. Embryol. Exptl. Morphol., 2,* 49–54.
Brøndsted, H. V., 1955. Planarian regeneration. *Biol. Rev., 30,* 65–126.
Brown, F. A., 1962a. Response of the planarian, *Dugesia*, to very weak horizontal electrostatic fields. *Biol. Bull., 123,* 282–294.
Brown, F. A., 1962b. Response of the planarian, *Dugesia*, and the protozoan, *Paramecium*, to very weak horizontal magnetic fields. *Biol. Bull., 123,* 164–181.
Brown, F. A., 1963. An orientational response to weak gamma radiation. *Biol. Bull., 125(2),* 206–225.
Brown, F. A., and Parke, Y. H., 1967. Association-formation between photic and subtle geophysical stimulus patterns—a new biological concept. *Biol. Bull., 132,* 311–319.

Brown, H. M., 1964. Experimental procedures and state of nucleic acids as factors contributing to "learning" phenomena in planaria. Unpublished doctoral dissertation, University of Utah.

Brown, H. M., 1967a. Effects of ultraviolet and photorestorative light on the phototaxic behavior of planaria. In Corning, W. C., and Ratner, S. C. (eds.), *Chemistry of Learning: Invertebrate Research*, Plenum Press, New York, pp. 295–309.

Brown, H. M., 1967b. Some characteristics of the light-evoked electrical response of the planarian eyecup. In Corning, W. C., and Ratner, S. C. (eds.), *Chemistry of Learning: Invertebrate Research*, Plenum Press, New York, pp. 164–165.

Brown, H. M., and Beck, E. C., 1964. Does learning in planaria survive regeneration? *Fed. Proc., 23,* 254.

Brown, H. M., and Ogden, T. E., 1968. The electrical response of the planarian ocellus. *J. Gen. Physiol., 51,* 237–253.

Brown, H. M., Dustman, R. E., and Beck, E. C., 1966a. Experimental procedures that modify light response frequency of regenerated planaria. *Physiol. Behav., 1,* 245–249.

Brown, H. M., Dustman, R. E., and Beck, E. C., 1966b. Sensitization in planaria. *Physiol. Behav., 1,* 305–308.

Bullock, T. H., and Horridge, A., 1965. *Structure and Function in the Nervous System of Invertebrates,* Freeman, San Francisco.

Byrne, W. K., 1970. *Molecular Approaches to Learning and Memory,* Academic Press, New York.

Chapouthier, G., 1967. Conditioning in the European planarian, *Dendrocoelum lacteum:* The effects of prolonged conditioning. *J. Biol. Psychol., 9,* 23–30.

Chapouthier, G., 1968. Relations entre deux réactions des Planaires face à une discrimination droite-gauche. *Compt. Rend. Acad. Sci. Paris, 266,* 905–907.

Cherkashin, A. N., and Sheiman, I. M., 1966. The use of simple biological models in memory mechanisms. Paper presented at International Psychology Congress, Moscow.

Cherkashin, A. N., and Sheiman, I. M., 1967. Conditioning in planarians and RNA content. *J. Biol. Psychol., 9,* 5–11.

Cherkashin, A. N., Sheiman, I. M., and Bogorovskaya, G. I., 1967. Conditioned reflexes in planarians and regeneration experiments. *Neurosci. Transl., 1,* 12.

Cohen, J., 1965. Diurnal cycles and maze learning in planarians. *Worm Runner's Digest, 7,* 20–24.

Corning, W. C., 1964. Evidence of right–left discrimination in planarians. *J. Psychol., 58,* 131–139.

Corning, W. C., 1966. Retention of a position discrimination after regeneration in planarians. *Psychon. Sci., 5,* 17–18.

Corning, W. C., 1967. Regeneration and retention of acquired information. In Corning, W. C., and Ratner, S. C. (eds.), *Chemistry of Learning: Invertebrate Research,* Plenum Press, New York, pp. 281–294.

Corning, W. C., 1971. Recent learning demonstrations and some biochemical correlates in planarians and protozoans. In Adam, G. (ed.), *Biology of Memory,* Plenum Press, New York, pp. 101–119.

Corning, W. C., and Freed, S., 1968. Planarian behavior and biochemistry. *Nature, 219,* 1227–1230.

Corning, W. C., and John, E. R., 1961. Effect of ribonuclease on retention of conditioned response in regenerated planarians. *Science, 134,* 1363–1364.

Corning, W. C., and Ratner, S. C., 1967. *Chemistry of Learning: Invertebrate Research,* Plenum Press, New York.

Corning, W. C., and Riccio, D., 1970. The planarian controversy. In Byrne, W. (ed.), *Molecular Approaches to Learning and Memory,* Academic Press, New York, pp. 107–150.

Cornwell, P., 1960. Classical conditioning with massed trials in the planarian. *Worm Runner's Digest, 2,* 34–39.

Cornwell, P., 1961. An attempted replication of studies by Halas *et al.* and by Thompson and McConnell. *Worm Runner's Digest, 3,* 91–98.

Cornwell, P., Cornwell, G., and Clay, M., 1961. Retention of a conditioned response following regeneration in the planarian. *Worm Runner's Digest, 3,* 34–38.

Coward, S., 1969. Regeneration in planarians: Some unresolved problems and questions. *J. Biol. Psychol., 10,* 15–19.

Crawford, F. T., and Skeen, L. C., 1967. Operant responding in the planarian: A replication study. *Psychol. Rep., 20,* 1023–1027.

Crawford, T., 1967. Behavioral modification of planarians. In Corning, W. C., and Ratner, S. C. (eds.), *Chemistry of Learning: Invertebrate Research,* Plenum Press, New York, 234–250.

Crawford, T., and King, L., 1966. Spontaneous recovery of a classically conditioned response in the planarian. *Psychon. Sci., 6,* 427–428.

Crawford, T., King, F., and Siebert, L., 1965. Amino acid analysis of planarians following conditioning. *Psychon. Sci., 2,* 49–50.

Crawford, T., Livingston, P., and King, F., 1966. Distribution of practice in the classical conditioning of planarians. *Psychon. Sci., 4,* 29–30.

Cummings, S. B., and Moreland, C. C., 1959. Sensitization *vs.* conditioning in planaria: Some methodological considerations. *Am. Psychologist, 14,* 410.

Dilk, F., 1937. Ausbildung von Assoziationen bei *Planaria gonocephala. Z. Vergl. Physiol., 25,* 47–82.

Dyal, J. A., 1971. Transfer of behavioral bias: Reality and specificity. In Fjerdingstad, E. (ed.), *Chemical Transfer of Learned Information,* North-Holland Publishing Co., Amsterdam.

Ernhart, E. N., and Sherrick, C., 1959. Retention of a maze habit following regeneration in planaria *(D. maculata).* Paper presented at meeting of Midwestern Psychological Association, St. Louis.

Fantl, S., and Nevin, J. A., 1965. Classical discrimination in planarians. *Worm Runner's Digest, 7,* 32–34.

Fjerdingstad, E. (ed.), 1971. *Chemical Transfer of Learned Information,* North-Holland Publishing Co., Amsterdam.

Freed, S., 1966. *Endogenous Biochemistry of Planarians Correlated with Learning Experiments at Brookhaven National Laboratory,* BNL Report No. 981 (T-414). (Available from the Clearinghouse for Federal Scientific and Technical Information, National Bureau of Standards, U.S. Department of Commerce, Springfield, Virginia.)

Griffard, C. D., 1963. Classical conditioning of the planarian *Phagocata gracilis* to water flow. *J. Comp. Physiol. Psychol., 56,* 597–600.

Griffard, C. D., and Peirce, J. T., 1964. Conditioned discrimination in the planarian. *Science, 144,* 1472–1473.

Guilliams, C. I., and Harris, C., 1971. Accelerated conditioning of contracted immobilized planarians. *J. Biol. Psychol., 12,* 27–33.

Gurowitz, E. M., 1969. *The Molecular Basis of Memory,* Prentice-Hall, Englewood Cliffs, N. J.

Halas, E. S., James, R. L., and Stone, L. A., 1961. Types of responses elicited in planaria by light. *J. Comp. Physiol. Psychol., 54,* 302–305.

Halas, E. S., James, R. L., and Knutson, C., 1962. An attempt at classical conditioning in the planarian. *J. Comp. Physiol. Psychol., 55,* 969–971.

Hansen, E. D., 1961. *Animal Diversity,* Prentice-Hall, Englewood Cliffs, N. J.

Hartrey, A. L., Keith-Lee, P., and Morton, W. D., 1964. Planaria: Memory transfer through cannibalism reexamined. *Science, 146,* 274–275.

Hay, E. D., 1968. Dedifferentiation and metaplasia in vertebrate and invertebrate regeneration. In Ursprung, H. (ed.), *The Stability of the Dedifferentiated State,* Springer-Verlag, New York.

Haynes, S. E., Jennings, L. B., and Wells, P. H., 1965. Planaria learning: Nontransfer and non-facilitation in a Van Oye maze. *Am. Zoologist, 4,* 424.

Henderson, T. R., and Eakin, R. E., 1961. Irreversible alteration of differentiated tissues in planaria by purine analogues. *J. Exptl. Zool., 146(3),* 253–263.

Hovey, H. B., 1929. Associative hysteresis in marine flatworms. *Physiol. Zool., 2,* 322–333.

Hullett, J. W., and Homzie, M. J., 1966. Sensitization effect in the classical conditioning of *Dugesia dorotocephala. J. Comp. Physiol. Psychol., 62,* 227–230.

Humpheries, B., 1961. Maze learning in planaria. *Worm Runner's Digest, 3,* 114–116.

Humpheries, B., and McConnell, J. V., 1964. Factors affecting maze learning in planarians. *Worm Runner's Digest, 6,* 52–59.

Hyden, H., 1959. Biochemical changes in glial cells and nerve cells at varying activity. *Proc. IV Internat. Congr. Biochem., 3,* 64–89.

Hyden, H., Egyhazi, E., John, E. R., and Bartlett, F., 1969. RNA base ratio changes in planaria during conditioning. *J. Neurochem., 16,* 813–821.

Hyman, L. H., 1919. Physiological studies in planaria. III. Oxygen consumption in relation to age (size) differences. *Biol. Bull., 37,* 388–403.

Hyman, L. H., 1923. Physiological studies on planaria. V. Oxygen consumption of pieces with respect to length, level and time after section. *J. Exptl. Zool., 37,* 47–68.

Hyman, L. H., 1951. *The Invertebrates,* Vol. 2: *Platyhelminthes and Rhynchocoela,* Mc-Graw-Hill, New York.

Hyman, L. H., and Bellamy, A. W., 1922. Studies on the correlation between metabolic gradients, electrical currents and galvanotaxis. I. *Biol. Bull., 43,* 313–347.

Jacobson, A. L., 1963. Learning in flatworms and annelids. *Psychol. Bull., 60,* 74–94.

Jacobson, A. L., 1965. Learning in planarians: Current status. *Anim. Behav., 13,* Suppl. 1, 76–81.

Jacobson, A. L., 1967. Classical conditioning and the planarian. In Corning, W. C., and Ratner, S. C. (eds.), *Chemistry of Learning: Invertebrate Research,* Plenum Press, New York, pp. 195–216.

Jacobson, A. L., and Jacobson, R., 1963. Maze learning in planaria—a case history. *Worm Runner's Digest, 5,* 69.

Jacobson, A. L., Fried, C., and Horowitz, S. D., 1966a. Planarians and Memory. I. Transfer of learning by injection of RNA. *Nature, 209,* 599–601.

Jacobson, A. L., Fried, C., and Horowitz, S. D., 1966b. Planarians and memory. II. Influence of prior extinction on RNA transfer effect. *Nature, 209,* 599–601.

Jacobson, A. L., Fried, C., and Horowitz, S. D., 1967. Classical conditioning, pseudo-conditioning, or sensitization in the planarian. *J. Comp. Physiol. Psychol., 64,* 73–79.

James, R. L., and Halas, E. S., 1964. No difference in extinction behavior in planaria following various types and amounts of training. *Psychol. Rec. 14,* 1–11.

Jenkins, M. M., 1963. Bipolar planarians in a stock culture. *Science, 142(3596):* 1187.

Jenkins, M. M., 1967. Aspects of planarian biology and behavior. In Corning, W. C., and Ratner, S. C. (eds.), *Chemistry of Learning,* Plenum Press, New York, pp. 116–143.

Jensen, D. D., 1964. Paramecia, planaria, and pseudolearning. Draft of paper presented at symposium "Learning and Related Phenomena in Invertebrates," Cambridge University.

Jensen, D. D., 1965. Paramecia, planaria, and pseudolearning. *Anim. Behav. 13*, Suppl. 1, 9–20.

John, E. R., 1964. Studies on learning and retention in planaria. In Brazier, M. A. (ed.), *Brain Function*, Vol. 2, University of California Press, pp. 161–182.

Kenk, R., 1967. Discussion on the biochemistry of memory. In Corning, W. C., and Ratner, S. C. (eds.), *Chemistry of Learning: Invertebrate Research*, Plenum Press, New York, p. 323.

Kimmel, H. D., and Harrell, V. L., 1964. Differential conditioning in the planarian. *Psychon. Sci., 1*, 227–228.

Kimmel, H. D., and Harrell, V. L., 1966. Further study of differential conditioning in the planarian. *Psychon. Sci., 5*, 285–286.

Kimmel, H. D., and Yaremko, R. M., 1966. Effect of partial reinforcement on acquisition and extinction of classical conditioning in the planarian. *J. Comp. Physiol. Psychol., 61*, 299–301.

King, F. J., Crawford, F. T., and Klingman, R. L., 1965. A further study of amino acid analysis and conditioning of planarians. *Psychon. Sci., 3*, 189–190.

Krugelis-Macrae, E., 1956. The occurrence of porphyrin in the planarian. *Biol. Bull., 110*, 69.

Lacey, D. J., 1971. Temporal effects of RNase and DNase in disrupting acquired escape behavior in regenerated planaria. *Psychon. Sci., 22*, 139–140.

Lecamp, M., 1942. Influence des acides amines sur la régénération. *Compt. Rend. Acad. Sci., 214*, 330–332.

Lee, R. M., 1963. Conditioning of a free operant response in planaria. *Science, 139*, 1048–1049.

Lender, Th., and Gabriel, A., 1960a. Sur la répartition des néoblasts de *Dugesia lugubris* (Turbellarie Triclade) avant et pendant la régénération. *Compt. Rend. Acad. Sci., 250*, 2465–2467.

Lender, Th., and Gabriel, A., 1960b. Etude histochimique des néoblasts de *Dugesia lugubris* (Turbellarie Triclade) avant et pendant la régénération. *Bull. Soc. Zool. Fr., 85*, 100–110.

Lender, Th., and Klein, N., 1961. Mis en évidence des cellules sécrétrices dans la cerveau de la Planaire *Polycelis nigra*. Variation de leur nombre dans le cours de la régénération postérieure. *Compt. Rend. Acad. Sci., 253*, 331–334.

McConnell, J. V., 1962. Memory transfer through cannibalism in planarians. *J. Neuropsychiat., 3*, Suppl. 1, s42.

McConnell, J. V., 1965. Cannibals, chemicals, and contiguity. *Anim. Behav., 13*, Suppl. 1, 61–68.

McConnell, J. V., 1966a. Comparative physiology: Learning in invertebrates. *Ann. Rev. Physiol., 28*, 107–136.

McConnell, J. V., 1966b. New evidence for the "transfer of training" effect in planarians. Paper presented at International Congress of Psychology, Moscow.

McConnell, J. V. (ed.), 1967a. *A Manual of Psychological Experimentation on Planarians*, 2nd ed., *J. Biol. Psychol.*, Ann Arbor, Mich.

McConnell, J. V., 1967b. Specific factors influencing planarian behavior. In Corning, W. C., and Ratner, S. C. (eds.), *Chemistry of Learning: Invertebrate Research*, Plenum Press, New York, pp. 217–233.

McConnell, J. V., 1967c. The biochemistry of memory. In Corning, W. C., and Ratner, S. C. (eds.), *Chemistry of Learning: Invertebrate Research*, Plenum Press, New York, pp. 310–322.

McConnell, J. V., 1968. In search of the engram. In Corning, W., and Balaban, M. (eds.), *The Mind: Biological Approaches to Its Function*, Wiley, New York, pp. 49–68.

McConnell, J. V., and Mpitsos, G., 1965. Effects of the presence or absence of slime on classical conditioning in planarians. *Am. Zoologist, 5,* 122.

McConnell, J. V., and Shelby, J., 1970. Memory transfer in invertebrates. In Ungar, G. (ed.), *Molecular Mechanisms in Memory and Learning,* Plenum Press, New York, pp. 71–101.

McConnell, J. V., Jacobson, R., and Maynard, D. M., 1958. Apparent retention of a conditioned response following total regeneration in the planarian. *Am. Psychologist, 14,* 410.

McConnell, J. V., Jacobson, A. L., and Kimble, D. P., 1959. The effects of regeneration upon retention of a conditioned response in the planarian. *J. Comp. Physiol. Psychol., 52,* 1–5.

McConnell, J. V., Cornwell, P., and Clay, M., 1960. An apparatus for conditioning planaria. *Am. J. Psychol., 73,* 618–622.

Meglitsch, P., 1967. *Invertebrate Zoology,* Oxford University Press, New York.

Morita, M., and Best, J. B., 1965. Electron microscopic studies on planaria. II. Fine structure of the neurosecretory system in the planarian *Dugesia dorotocephala. J. Ultrastruct. Res., 13,* 396–408.

Morita, M., and Best, J. B., 1966. Electron microscopic studies of planaria. III. Some observations on the fine structure of planarian nervous tissue. *J. Exptl. Zool., 161,* 391–395.

Owen, E. E., Weis, H. A., and Prince, L. H., 1938. Carcinogens and growth stimulation. *Science, 87,* 261–262.

Owen, E. E., Weis, H. A., and Prince, L. H., 1939. Carcinogens and planarian tissue regeneration. *Am. J. Cancer, 15,* 424–426.

Parker, G. H., and Burnett, F. L., 1900. The reactions of planarians with and without eyes to light. *Am. J. Physiol., 4,* 373–385.

Pearl, R., 1903. The movements and reactions of fresh-water planarians: A study in animal behavior. *Quart. J. Microscop. Sci., 46,* 509–714.

Pennak, R. W., 1953. *Fresh-Water Invertebrates of the United States,* Ronald Press, New York.

Pickett, J. B. E., Jennings, L. B., and Wells, P. H., 1964. Influence of RNA and victim training on maze learning by cannibal planarians. *Am. Zoologist, 4,* 158.

Ragland, R. S., and Ragland, J. B., 1965. Planaria: Interspecific transfer of a conditionability factor through cannibalism. *Psychon. Sci., 3,* 117–119.

Reynierse, J. H., 1967*a.* Aggregation formation in planaria, *Phagocata gracilis* and *Cura foremanii:* Species differentiation. *Anim. Behav., 15,* 270–272.

Reynierse, J. H., 1967*b.* Reactions to light in four species of planaria. *J. Comp. Physiol. Psychol., 63,* 336–368.

Riccio, D., and Corning, W. C., 1969. Slime and planarian behavior. *Psychol. Rec., 19,* 507–513.

Roe, K., 1963. In search of the locus of learning in planarians. *Worm Runner's Digest, 5,* 16–18.

Röhlich, P., 1968. Fine structural changes of photoreceptors induced by light and prolonged darkness. In Salanki, J. (ed.), *Invertebrate Neurobiology,* Plenum Press, New York, pp. 95–109.

Scott, J. F., Fraccastoro, A. P., and Taft, E. B., 1956. Studies in histochemistry. I. Determination of nucleic acids in microgram amounts of tissue. *J. Histochem. Cytochem., 4,* 1–10.

Shafer, J. N., and Corman, C. D., 1963. Response of planaria to shock. *J. Comp. Physiol. Psychol., 56,* 601–603.

Sengel, P., 1960. Culture *in vitro* de blastèmes de régénération de Planaires. *J. Embryol. Exptl. Morphol., 8,* 468–476.

Sengel, P., 1967. Aspects récents de la morphogenèse chez les Planaires. In Corning, W. C., and Ratner, S. C. (eds.), *Chemistry of Learning: Invertebrate Research*, Plenum Press, New York, pp. 73–115.

Soest, H., 1937. Dressurversuche mit Ciliaten und Rhabdocoelen Turbellarien. *Z. Vergl. Physiol., 24,* 720–748.

Taliaferro, W. H., 1920. Reactions to light in *Planaria maculata. J. Exptl. Zool., 31,* 59–116.

Thompson, R., and McConnell, J. V., 1955. Classical conditioning in the planarian, *Dugesia dorotocephala. J. Comp. Physiol. Psychol., 48,* 65–68.

Togrol, B. B., Ormanli, M., and Cantey, E. 1966a. Classical conditioning in the planaria *Polycelis tenuis* (Ijima), *Dugesia lugubris* (Schmidt) and the effects of regeneration upon retention of the conditioned response. *Rev. Fac. Sci. Istanbul., 31,* 147–166.

Togrol, B. B., Ormanli, M., and Cantey, E. 1966b. The effects of chemicals on the general behavior, regeneration and the learning capacity of planaria. *Polycelis tenuis* (Ijima) and *Dugesia lugubris* (Schmidt). *Rev. Fac. Sci. Istanbul., 31,* 167–181.

Tushmalova, N. A., 1967. Conditioned reflexes in the Baikal planarian *Podoplana olivacea* after injection of ribonuclease. *Zh. Vysshei Nervnoi Deyatel. I. P. Pavlova, 17,* 359–361.

Ungar, G., 1970. *Molecular Mechanisms in Memory and Learning,* Plenum Press, New York.

Van Deventer, J. M. (1963). Unpublished data.

Van Deventer, J. M., and Ratner, S. C., 1964. Variables affecting the frequency of response of planaria to light. *J. Comp. Physiol. Psychol., 57,* 407–411.

Van Oye, P., 1920. Over het geheugen bij fr flatwormen en andere biologische waarnemingen bji deze dieren. *Natuurwet. Tijdschr., 2,* 1.

Vattano, F. J., and Hullett, J. H., 1964. Learning in planarians as a function of interstimulus interval. *Psychon. Sci., 1,* 331–332.

Viaud, G., 1954. Etude quantitative de la force électro-motrice d'opposition produite par les planaires (Planaria = *Dugesia lugubris*) en réponse à une excitation électrique due à un courant continu. *Compt. Rend. Soc. Biol., 148,* 2068.

Walker, D. R., 1966. Memory transfer in planarians: an artifact of the experimental variables. *Psychon. Sci., 5,* 357–358.

Walker, D. R., and Milton, G. A., 1966. Memory transfer versus sensitization in cannibal planarians. *Psychon. Sci., 5,* 293–294.

Walter, H. E., 1908. The reactions of planaria to light. *J. Exptl. Zool., 5,* 35–163.

Wells, P. H., 1967. Training flatworms in a Van Oye maze. In Corning, W. C., and Ratner, S. C. (eds.), *Chemistry of Learning: Invertebrate Research*, Plenum Press, New York, pp. 251–254.

Wells, P. H., Jennings, L. B., and Davis, M., 1966. Conditioning planarian worms in a Van Oye type maze. *Am. Zoologist, 6,* 295.

Westerman, R. A., 1963. A study of the habituation of responses to light in the planarian *Dugesia dorotocephala. Worm Runner's Digest, 5,* 6–11.

Yaremko, R. M., and Kimmel, H. D., 1969. Two procedures for studying partial reinforcement effects in classical conditioning of the planarian. *Anim. Behav., 17,* 40–42.

Zelman, A., Kabat, L., Jacobson, R., and McConnell, J. V., 1963. Transfer of training through injection of "conditioned" RNA into untrained planarians. *Worm Runner's Digest, 5,* 14.

Chapter 5

BEHAVIOR MODIFICATION IN ANNELIDS[1]

JAMES A. DYAL

Department of Psychology
University of Waterloo
Waterloo, Ontario, Canada

Worms have played a more important part in the history of the world then most persons would at first suppose. (Charles Darwin, *The Formation of Vegetable Mould Through the Action of Worms: With Observations on Their Habits*, 1898, p. 309)

I. INTRODUCTION

While worms may be movers and moulders of mountains, they have managed to avoid the extended scrutiny of behavioral physiologists and psychologists. Nevertheless, as Bullock has noted, "Annelids offer some of the most promising material for study of learning mechanisms in simple systems . . . their availability, simplicity, phylogenetic position, learning capacity, and tolerance of mutilation suggest that they are well worth new attention" (Bullock and Quarton, 1966, p. 116).

The phylum Annelida is comprised of worms whose most distinguishing feature is an elongated body divided into highly similar segments or metameres.[2] Although they lack an internal skeleton, their musculature forms a tubular jacket filled with coelomic fluid, the combined action of which

[1]Preparation of this chapter was facilitated by Grant A 0351 from the National Research Council of Canada.
[2]The name of the phylum is taken from a French corruption of the Latin *annellus*, meaning *ring*.

results in a sort of "fluid skeleton" which aids in extension of the body and in burrowing. The segments are divided internally as well as externally by transverse septa. In the generalized annelid body plan, the basic structures of all of the vital body organs except the digestive tract are replicated in each segment. The circulatory system is typically closed and contains blood. The bilateral symmetry of the flatworm is further elaborated in the annelids, and the cephalocaudal axis gains greatly in functional import, the anterior end becoming quite specialized for sensory, neurointegrative, burrowing, and nutritive functions. Reproduction may be dioecious or hermaphroditic.

The basic annelid body plan has turned out to be extremely adaptable, and we thus find a great diversity of morphology, habitat, and mode of life within this phylum. Of the 6000 species of annelids, the bulk inhabit the seas, but some are freshwater animals and others are terrestrial. They range in length from a few millimeters to several feet (e.g., *Nereis brandti* of the North American Pacific Coast, which attains lengths of up to 3 feet, and the gigantic Australian earthworm, which may exceed 3 meters). The marine forms are divided into sedentary and free-moving types. While all of the sedentary forms build tube shelters, some of the errant forms also build tubes or occupy burrows between their excursions.

II. TAXONOMY AND PHYLOGENY

The phylum is most conveniently classified into three major classes: Oligochaeta (e.g., earthworms), Polychaeta (marine worms), and Hirudinea (leeches).[3] There is still considerable controversy about even the broad outlines of the evolutionary relationships of the annelid classes, and "the detailed phylogeny of the worms presents little more than a series of unsolved and probably insoluble problems" (Clark, 1969, p. 44).

Since it ill behooves an animal behaviorist to attempt to wrest taxonomic order from phylogenetic chaos, the curious reader is referred to the original sources (e.g., Clark, 1969) for both phylogenetic and taxonomic descriptions beyond the following general characterizations. Diagrammatic representations of several common annelids are presented in Fig. 1.

A. Class Polychaeta

Perhaps the general characteristics of the polychaetes can best be described by concentrating on a representative species of the errant and sedentary types, respectively. *Nereis virens* may be taken as a "typical" errant

[3]Some authorities prefer to designate three classes, Polychaeta, Myzostomaria, and Clitelata, with Oligochaeta and Hirudinea being subclasses of Clitelata (*cf.* Clark, 1969; Dales, 1963).

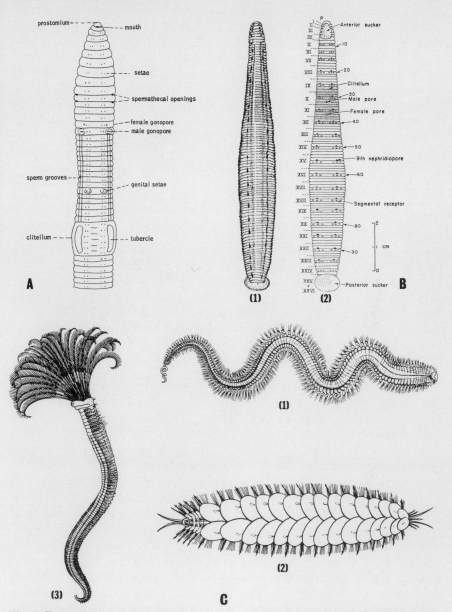

Fig. 1. External characteristics of a variety of annelids. A, Anteroventral surface of the earthworm *Lumbricus terrestris*. (Redrawn from Barnes.) B, (1) Dorsal view of *Hirudo medicinalis*; (2) diagrammatic ventral view of *H. medicinalis*. The segments are numbered in Roman and the annuli in Arabic. (Redrawn from Mann.) C, Three main polychaetes: (1) *Nephtys caeca* (Nephtyidae); (2) *Harmothoë imbricata* (Polynoidae); (3) *Megalomma*. (All redrawn from Light.)

polychate at this level of description. It has an elongated body which is slightly flattened dorsoventrally and which is strikingly segmented. It may range in length from 9 to 18 inches. It has a distinct, well-developed head consisting of a *prostomium,* which roofs over the mouth and on which are located four simple eyes, two antennae, and two palps, and a *peristomium,* which contains two pairs of long tentacular cirri. Each body segment except the last *(pygidium)* has prominent lateral appendages *(parapodia)* containing clusters of bristles *(setae).* The parapodia function in creeping and swimming movements. Its body is a handsome reddish brown with an iridescent greenish sheen. *N. virens* possesses a pair of powerful jaws which it uses to take prey (or nip experimenters' fingers). It is a raptorial feeder on dead clams and other bits of food from the mud or sand bottoms of the littoral zone.

A wide variety of Sabelliforma (e.g., *Sabella, Serpula,* and *Hydroides)* may be regarded as "typical" of sedentary polychaetes. These fanworms have developed elaborate crowns from the peristomium. These crowns consist of pinnate processes called *radioles* which function as structures for gathering food particles and also for respiration. The parapodia of these tube-dwelling sedentary animals are very much reduced and often are developed into hooks *(uncini)* which help maintain traction inside the tube.

B. Class Oligochaeta

As the class name implies, the oligochaetes are annelids with relatively few chaeta (setae); the comparison is with the abundance of setae which characterizes polychaetes. In oligochaetes, the setae are usually located in four bundles in each segment, and each bundle may contain one to 25 setae. The setae provide a grip on the substrate in locomotion and on another worm during copulation. All oligochaetes are hermaphroditic, and copulation involves contact of ventral anterior surfaces with the posterior of one worm directed toward the anterior of the other, permitting mutual cross-fertilization. The *clitellum* is a clearly identifiable external body characteristic of most earthworms, consisting of enlargement of a few segments of the anterior half of the worm. In most oligochaetes, it is a secondary sex characteristic present only during the reproductive season; however, in the two most familiar genera of earthworm, *Lumbricus* and *Allolobophora,* the clitellum is a permanent feature. Although copulation is the general rule among oligochaetes, asexual reproduction occurs in many aquatic oligochaetes.

C. Class Hirudinea

The body plan of the leeches is remarkably homogeneous. There are a prostomium and 32 body segments. The prostomium and the first two body segments bear an anterior sucker in the center of which is a mouth. The last seven segments bear the posterior sucker. The subesophageal ganglion is comprised of four fused ganglia and the posterior ganglion of seven; between them are 21 single ganglia, one per segment. The number of annuli per segment varies among the species. The complete systematics of marine, freshwater, and terrestrial leeches along with the identification keys may be found in Mann (1962).

III. SPECIAL ANNELID CHARACTERISTICS RELEVANT TO BEHAVIOR

In this section, we will consider the sensory capacities of annelids and important features of their nervous system. In the next section, we will discuss the complexifications of these sensory and neural processes as they manifest themselves in the basic behavioral repertoires of annelids.

A. Sensory Capacities

Mechanoception

Mechanical stimulation of the body surface readily initiates neural activity in all species of annelids. *Tactiception* is mediated by generalized sensory nerve cells and sense organs located in the epidermis. There is considerable species variation in structure and distribution of tactile receptors. It is possible that *proprioception,* independent of touch, is present in some species of nereid polychaetes, and several mechanoreceptor fibers have been found in all four roots of each ganglion in the ventral chain of *Hirudo* (Bullock and Horridge, 1965, p. 746).

Photoception

Photoception occurs in all annelids and is mediated by structures ranging in complexity from solitary photoreceptor cells to photoreceptor clusters to simple ocellar eyes to complex eyes equipped with cuticular lenses imbedded in deep and narrow pigmented cups. Simple light receptors are widely distributed over the bodies of annelids both in the epidermis and in specialized enlargements along the nerves of the head and caudal segments. These more primitive light receptors respond to light intensity. Generally, sedentary polychaetes do not have complex eyes but are equipped with photoreceptor cells in the epidermis of the branchial crown. Photoreception in Hirudinea, like in errant polychaetes, is mediated by a variety of structures ranging from a simple cluster of a few photoreceptors in a shallow cup (e.g., *Piscicola*) to a deep, narrow pigmented cup with a large number of receptors as in

Hirudo. With several such complex eyes pointing in various directions, it may be possible for *Hirudo* to detect form and movement at a crude level.

Chemoception

Chemoreceptors are housed in nuchal organs which typically consist of a pair of ciliated pits located in the head region. As with other sensory receptors, the specific form, and complexity, ranges from simple sensory pits to elaborate structures involving several segments. The maximal development occurs in some burrowing polychaetes. In leeches, the chemoreceptors seem to be confined entirely to the head region; apparently, the anterior sucker contains receptors sensitive to the chemical composition both of substances in the water and of substances contained in the blood of the hosts.

Statoception

Sensory organs mediating orientation of the body to gravity are found primarily among sedentary polychaetes which live in burrows or tubes (e.g., Arenicolidae, Terribellidae, and Sabellidae). In the natural environment, the statocysts function to tell the worm "which end is up" and thus serve to maintain normal burrowing activity. Leeches do not have statocysts but are able to orient properly by virtue of dorsal light reactions and presumably the effect of gravity on other body organs.

B. Nervous System

The basic plan of the nervous system in annelids is characterized by an anterior dorsal ganglionic mass (supraesophageal ganglion or brain) and some connectives which circle the pharynx and connect the brain with the enlarged ganglionic terminus of the ventral cord (subesophageal ganglion). The ventral cord extends the length of the body and contains segmental ganglia from which two pair of motor nerves and one pair of sensory nerves innervate each segment. The topographical layouts of the anterior part of the nervous system of a polychaete, a leech, and an earthworm are represented in Fig. 2.

The Brain

Polychaetes. The level of complexity of the annelid brain varies considerably among the various classes, roughly in proportion to the level of development of specialized sense organs. The brain achieves its highest level of complexification, as indicated by degree of differentiation, in errant polychaetes such as *Nereis*. In the case of these more complex brains, topographical features indicate differentiation into forebrain, midbrain, and hindbrain. However, it should be emphasized that these taxonomic categories do not

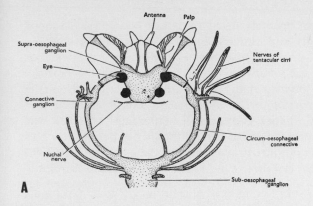

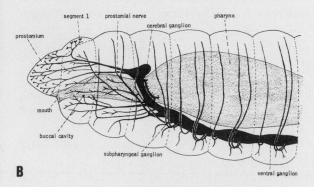

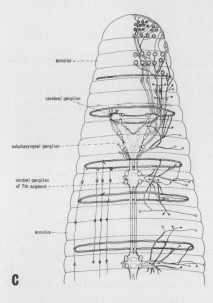

Fig. 2. A, Semidiagrammatic representation of the anterior nervous system and sense organs of the marine polychaete *Nereis*. (From Flint, 1965.) B, Anterior nervous system of the earthworm *Lumbricus* (lateral view). (From Barnes, 1966.) C, Anterior nervous system of the leech *Erpobdella lateralis*. (From Barnes, 1966.)

imply homologies to higher invertebrate or vertebrate brains. Furthermore, they may have no necessary evolutionary significance beyond reflecting the relative development of cephalic sense organs. Although the taxonomic categories of fore-, mid-, and hindbrain should not be regarded as homologous to arthropods, the specific centers may be homologized with greater justification. Detailed discussion of these internal structures and their connections may be found in Bullock and Horridge (1965, pp. 712–722), who also present an extensive comparison of several families of polychaetes with regard to development of brain structures. Some of the major aspects of the comparison which might be relevant to future behavioral work are as follows:

1. "From the standpoint of comparative neurology there could hardly have been a better choice of a polychaete as the common classroom type than *Nereis*. Well provided with sense organs, its brain and hence the rest of the nervous system is well differentiated and exhibits virtually all the features of advanced polychaetes" (Bullock and Horridge, 1965, p. 735).
2. The behavior of several closely related families (Aphroditidae, Polynoidae, Sigalionidae, and Hesionidae) "is not sufficiently well known to permit a correlation with or an explanation of the high development of the brain" (p. 735).
3. Glyceridae have also lost eyes and palps and have the simplest brain among the polychaetes studied.
4. Nephtyidae lack palps and eyes and have a low level of differentiation of the brain; "it may be expected that the behavior will be shown to be less complex than, say, *Nereis*" (p. 732).
5. In general, the sedentary families have quite simple, poorly differentiated brains.

Oligochaeta and Hirudinea. The brain of oligochaetes is far simpler and less differentiated than that of errant polychaetes or sabellids/serpulids. It is more comparable to that of the typical sedentary polychaete, and the brain of leeches is simpler still. The simplicity of brain structure in both classes is manifested in lack of lobes or specialized neuropile regions and by the diversity and simplicity of cell types. Although earthworm and leech brains are relatively simple compared to those of polychaetes, there is an important distinction between them. In the earthworm, the cerebral ganglion is a simple ganglion pair located in the prostomium with all the ganglia of the body segments represented below the esophagus. In the leech, one pair of segmental ganglia has migrated around the esophagus into the prostomium, an arrangement which foreshadows the arthropod device of compounding several segmental ganglia into the brain (Mann, 1962).

Detailed specification of the internal connections of the brains of oligochaetes and leeches may be found in Bullock and Horridge (1965).

The Ventral Cord

Polychaetes. Details of the gross anatomy, histological characteristics of cell types, and neuronal pathways may be obtained from Bullock and Horridge 1965, (pp. 673–680). For our purposes, it suffices to note that the ventral cord contains the intersegmental through-conducting giant fiber system which is present in most errant polychaetes and is exceptionally well developed in several sedentary polychaetes. In errant polychaetes, the number of fibers ranges from one (e.g., Eunicidae) to two or four (Polynoidae) or five (Nereidae) on to 20 or 30 (Lysaretidae). Sedentary polychaetes typically have one or two. The topographical arrangements are quite varied and involve many patterns of independence and fusion (see Nicol, 1948*a*).

Oligochaetes. In *Lumbricus terrestris,* the topographical arrangement involves a dorsal giant fiber system and a pair of ventral giant fibers; little is known of the structure and function of the ventral pair. The dorsal system consists of a large median fiber and two large lateral fibers which extend most of the distance of the cord. The median fiber appears to terminate in the subesophageal ganglion, while the laterals may decussate at this point and enter the circumesophageal connectives, although they have not been adequately traced into the brain. The two lateral giant fibers anastomose at frequent intervals, but there appears to be no structural or functional connection between the median giant and the laterals.

Hirudinea. The ventral cord of leeches does not contain high-speed giant fibers such as found in earthworms and polychaetes; however, their reflex activity suggests that there is a fast-conducting system. There are 21 paired ganglia spaced along the ventral cord and seven more fused into the anal ganglionic mass associated with the posterior sucker. Each ventral ganglion in the cord of the leech contains six cell capsules which contain internuncial nerve cells, giant cells, and sensory and motor cells. They give off processes which pass through the capsule and join the longitudinal fibers of the cord. The internuncial neurons often connect from one ventral ganglion to the next and join peripheral sensory and motor fibers.

The Peripheral Nervous System

There are typically three or four peripheral nerves leaving the ventral cord ganglion on each side of each body segment. In many errant polychaetes, the cell bodies which supply the nerves to the parapodia are not located in the ventral ganglia but in a podial ganglion on the second segmental nerve near the parapodial muscles. The pedal ganglia are also joined to the ventral ganglia by transverse fibers, and in some more primitive arrangements (e.g., Nereidae) there are lateral longitudinal nerves connecting one podial ganglion to the next. They run the length of the body and enter the brain independently of the circumesophageal connectives.

IV. REFLEXES, COORDINATIONS, AND ORIENTATIONS

A complete account of the many factors which influence the normal behavior of annelids would have to include a discussion of ecological factors beyond the scope of the present chapter. The interested reader will find an excellent treatment of the physicochemical, biological, and behavioral factors which are operative in habitat selection and maintenance in Newell (1970). A brief discussion of ecological factors relevant to behavior may be found in Dyal (1972).

It should be obvious that the most successful attempts to modify behavior will be based on a clear recognition of the intricacies of the innate response repertoire of the organism of choice. Historically it has been the case that psychologists concerned with animal learning have assumed that "a response is a response is a response." This unfortunate attitude was fostered by American interpretations of Russian reflexology and Loeb's tropism theory; it was bolstered and extended by Watsonian behaviorism and by the cumulative record of Watson's more radical descendants. But it is now clear that the rate, stability, and, indeed, the possibility of conditioning of a particular response depend intimately, and in ways probably as yet unknown, on the innate response hierarchy (*cf.* Breland and Breland, 1961; Staddon and Simmelhag, 1971). Any animal behaviorist who would modify behavior should know his animal well. This is especially true when his animal of choice is an invertebrate, since, for most psychologists, these curious creatures constitute "alien species."

Unfortunately, the present context does not permit any but the most cursory statement regarding reflexes, coordinations, and orientations in annelids. A more extended treatment may be found in Dyal (1972). Several reflexes are present in annelids, including a rapid withdrawal reflex (polychaetes and oligochaetes), parapodial pointing (polychaetes), shadow reflex (leeches), reflex arrest of creeping (oligochaetes), and luminance reflex (polychaetes). The rapid withdrawal reflex has been subjected to the most extensive neurophysiological and behavioral analysis over the past century. The reflex is mediated by the giant fiber system and is elicited by a variety of stimuli. Nicol (1948a) provides an extensive review of the function of the giant axons in the withdrawal reflex. Basic references include Bullock (1948), Gwilliam (1969), Hargitt (1912), Krasne (1965), Roberts (1962a), and Ruston (1945).

Coordinations of various types of creeping, crawling, and swimming movements reflect the integrative capacity of the nervous system, especially the ventral cord. The type of control ranges from a closed chain of reflexes which seems to be involved in the crawling locomotion of leeches to centrally controlled intrinsic rhythms as in earthworms and especially in the swimming movements of polychaetes and leeches. In the latter case, the intrinsic rhythm is under the control of pacemakers in the ventral core with a master pacemaker in the anal ganglion. A more detailed discussion of coordinations may be found in Bullock and Horridge (1965).

At the most general level, annelids may be described as exhibiting negative photoorientations, variable orientations to tactile stimuli, positive thigmotaxis, positive rheotaxis, positive or negative geotaxis, and adient *vs.* abient orientations to various chemical substances and temperature gradients. However, a gross generalization such as "annelids are photonegative" requires elaboration and qualification in order to reveal the complexities of behavior which lie hidden behind the blanket statement. Examination of the variety of stimulus, organismic and response parameters which influence annelid orientations may be found in Dyal (1972).

As we have seen in Chapter 1, a variety of behavior taxonomies relevant to behavior plasticity have been formulated. The classification system which will be utilized in the present context is a modification of Razran's taxonomy. We will thus examine annelid behavioral plasticities which may be classified as habituation, sensitization, aversive inhibitory conditioning (punishment), classical conditioning, and instrumental conditioning.

V. HABITUATION

The decrement of an innate response as a result of repeated stimulation is the prime characteristic of habituation, one of the simplest forms of behavioral plasticity. Parametric characteristics of behavioral and neurophysiological habituation processes in mammalian preparations as described by Thompson and Spencer (1966) have been discussed in Chapter 1. These considerations will not be reiterated here; rather, we will proceed directly to discuss demonstrations of habituation in the various classes of Annelida.

A. Class Oligochaeta

Basic Demonstrations

Kuenzer (1958) demonstrated habituation of examined "twitch" reactions of *Allolobophora, Eisenia,* and *Lumbricus* to mechanical, thermal, and electrical stimulation. It is apparent from Fig. 3 that the course of habituation varied with the type of stimulus. Gardner (1968) has demonstrated habituation of withdrawal responses to vibration in *L. terrestris.*

Influencing Variables

Stimulus Variables. Stimulus Intensity: Kuenzer (1958) examined the rate of habituation to electrical shock applied every 10 sec at intensities of

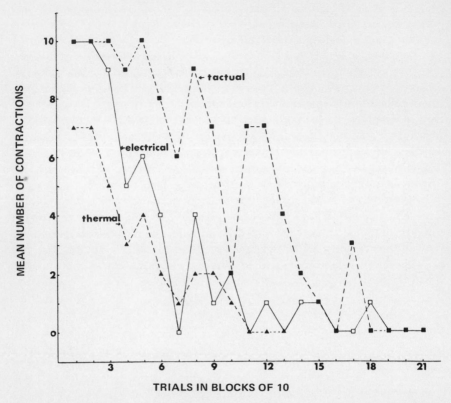

Fig. 3. Mean number of contractions elicited by tactile, electrical, and thermal stimuli applied every 5 sec to the fourth or fifth segment (Kuenzer, 1958).

150, 200, 250, and 300 mv. The course of habituation is presented in Fig. 4, where it may be seen that the number of trials required for complete habituation increased as a direct function of the intensity of the stimulation. Similar stimulus intensity effects have been shown for mechanical shock (Kuenzer, 1958) and GSR to light (Morgan *et al.*, 1965).

Interstimulus Interval: Kuenzer reports that within the range of 5–30 sec interstimulus interval does not influence the habituation of head twitching to any of the three types of stimuli tested (mechanical, galvanic, and thermal). Although Ratner and Stein (1965) have reported that the absolute number of responses elicited by a high-intensity light is greater if the stimuli are spaced (88 sec) rather than massed (6 sec), there is as yet no evidence that interstimulus interval significantly affects the *rate* of habituation in oligochaetes. Ratner (1972) reports significant effects on both habituation and retention.

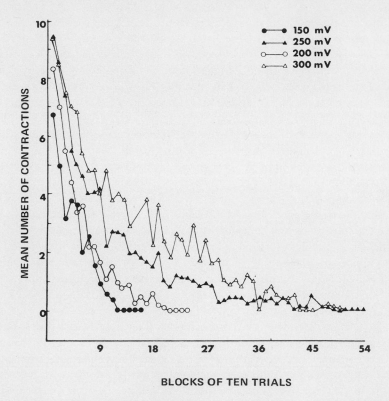

Fig. 4. Mean number of contractions as a function of the intensity of electrical stimulation applied every 5 sec (Kuenzer, 1958).

Response Characteristics. Gardner (1968) described the reflexive response of the earthworm to vibration as consisting of two components: a *withdrawal component,* which involves the rapid contraction of the anterior portion of the worm, and a *hooking component,* which consists of rapid contraction, flattening, and hooking of the posterior segments. Furthermore, these two components habituate at different rates, with the hooking component requiring twice as many trials for complete habituation. Gardner suggests that the anterior withdrawal response competes with feeding behavior and that this factor may be related to its more rapid habituation.

Related Phenomena

Retention. As was noted in Chapter 1, retention of the effects of an experience may be measured by simple retention, retroaction, and proaction paradigms.

Simple retention: This simple retention procedure was used by Kuenzer (1958) to test for spontaneous recovery from the effects of a single previous habituation as a function of the intensity of the habituating stimulus in microvolts. He found that following a 30 min recovery period the percent recovery was inversely related to the stimulus intensity. Gardner (1968) found that simple retention of habituation of the two components of the earthworm's response to vibration was retained (as measured by savings) for as long as 96 hr.

Repeated Habituations with Constant Recovery Time: A common characteristic of the habituation process in both lower and higher vertebrates has been called *potentiation of habituation* (Thompson and Spencer, 1966). This characteristic is a type of proactive facilitation and is manifested by habituation becoming successively more rapid as repeated series of habituation and spontaneous recovery are administered. Kuenzer has reported two extensive experiments which demonstrated this phenomenon in earthworms.

Stimulus Specificity: Thompson and Spencer (1966) have noted that, at least in mammals, habituation of response to a given stimulus tends to generalize to other similar stimuli; i.e., there is proactive or retroactive facilitation. Kuenzer attempted to determine if habituation of earthworms to mechanical, thermal, and electrical stimulation was stimulus specific or tended to generalize. He first matched the subjective stimulus intensities of the three stimuli by adjusting intensities so as to equate the number of responses to habituation criterion. He then habituated the worms to criterion with one stimulus (e.g., mechanical) immediately followed by habituation to an electrical stimulus, followed in turn by habituation to a thermal stimulus. Kuenzer's data support the conclusion that in the earthworm habituation tends to generalize very little, if at all; rather, it tends to be stimulus specific. There are no proactive effects, and because of the short intervals between series it is not reasonable to assume that this is due to lack of retention. Similar stimulus-specific habituation effects have been reported in other invertebrates (e.g., Drees, 1952, with Salticidae), amphibians (e.g., Eikmanns, 1955, with *Bufo bufo* L.), and birds (e.g., Hinde, 1954, with chaffinches). On the other hand, Kuenzer applied a 100 mv stimulus to the twelfth segment, habituated to criterion, and obtained a decreasing spatial generalization up to ten segments away from the point of original habituation.

B. Class Polychaeta (Sedentary)

Basic Demonstrations

Numerous experiments over the past 70 years have demonstrated habituation of the rapid withdrawal reflex which is elicited by light decrease and touch in sabellids and serpulids (Hargitt, 1906; Hess, 1914; Hesse, 1899; Krasne, 1965; Nicol, 1950; Rullier, 1948; Yerkes, 1906).

Krasne (1965) has described the habituation process in *Branchiomma vesiculosum* as follows:

> Gentle brushing of a fresh worm's protruded branchial crown causes a rapid and complete withdrawal of the worm into its tube. When the worm's crown again emerges, it can again be stimulated to withdraw, but as the cycle of stimulation, withdrawal, and re-emergence is repeated a stronger stimulus becomes necessary to prevent the waning of the response.
>
> The waning of the response to a constant stimulus shows a rather uniform course of symptoms leading to complete failure to react in any visible way to the stimulus, but proceeding at a rate which varies widely among individuals. Generally, delayed withdrawal replaces immediate withdrawal, slow withdrawal replaces fast withdrawal, rapid re-emergence replaces slow re-emergence, and incomplete withdrawal replaces complete withdrawal. (p. 309)

Influencing Variables

Stimulus Variables. Modality/Intensity: Yerkes (1906) obtained habituation of the withdrawal response of the serpulid *Hydroidus dianthus* to a shadow but not to a tactile stimulus. Similar results were reported by Nicol (1950) for habituation of the rapid withdrawal response of the sabellid *Branchiomma vesiculosum*. Krasne (1965), on the other hand, obtained substantial habituation of withdrawal to tactile stimulation of the branchial crown of *B. vesiculosum*. The serpulid *Mercierella enigmatica* habituates to light diminution and mechanical shock and requires longer to habituate to a pair of stimuli presented simultaneously than to either of the two stimuli alone (Rullier, 1948). It is not known whether differences in rate of habituation to various stimuli are due to the modality involved or to the intensity of the stimulus used, since no efforts have been made to equate the subjective intensity of the stimuli by either behavioral or neurophysiological techniques.

Interstimulus Interval: When moving shadow is used as the habituating stimulus, there is suggestive evidence from the early work of Hargitt (1906) and of Yerkes (1906) that habituation is faster with a short interstimulus interval (e.g., 0.25 sec) than with a longer interval (e.g., 1.0 sec). This variable requires more systematic study over several levels of interstimulus interval with a variety of species before firm conclusions may be drawn.

Related Phenomena

Retention. Simple Recovery: Krasne (1965) has obtained negatively accelerated increasing functions describing recovery of responsiveness to branchial touch in *B. vesiculosum*. Recovery is about 50% complete after 12 hr and still only 75% complete after 24 hr.

Stimulus Specificity: If habituation to a "light off" is followed by habituation to a moving shadow, no transfer occurs; i.e., the two habituation processes are independent (Nicol, 1950).

C. Class Polychaeta (Errant)

Basic Demonstrations

The literature on habituation in errant polychaetes is considerably richer than that on sedentary polychaetes, primarily due to the research efforts of Clarke and his students at the University of Bristol. In their first extensive study of habituation in errant polychaetes, they showed that in *Nereis pelagica* habituation of the withdrawal response to mechanical shock, a moving shadow, or a decrease in light intensity was essentially complete after 20 stimulus presentations (Clark, 1960*a,b*). Evans (1969*a*) replicated this basic demonstration and also extended the effect to related nereids (*Nereis diversicolor* and *Platyneris dumerillii*). Dyal and Hetherington (1968) demonstrated habituation in the polynoid *Hesperonoë adventor*, a scaleworm commensal with the echiuroid *Urechis caupo*.

Influencing Variables

The experiments of Clark and of Evans have shown that the habituation process in nereids is strongly influenced by a variety of stimulus variables which interact in their effects with the species and response characteristics.

Stimulus Variables. Modality/Intensity: Clark (1960*a*) found that habituation rate in *N. pelagica* varied with the type of habituating stimulus. In the most extensive study on nereid habituation, Evans (1969*b*) compared seven stimuli (light on, light off, moving shadow, mechanical shock, light off and mechanical shock combined, tactile stimulation anterior, and tactile stimulation posterior) in three species (*N. diversicolor, P. dumerillii,* and *N. pelagica*) with subjects that were either intact or decerebrate. Species differences and the effects of decerebration will be considered subsequently. In terms of habituation rates, the stimuli which were consistently the most potent in eliciting a response regardless of species were anterior and posterior tactile stimulation and the combination of "light-off" and mechanical shock. On the other hand, the most rapid habituation tended to occur to "light-off," moving shadow, and mechanical shock. Whether these differences are due to the modality stimulated, the intensity of the stimulus, or the "sign value" of the stimulus as a predator detector cannot be determined from these data.

Interstimulus Interval: Clark (1960*a*) has reported that the trials necessary for habituation of *N. pelagica* to a "light-on" increases within the range from 30 sec to 5 min interstimulus intervals.

Organismic Variables. Species: Although members of the family Nereidae show substantial similarity in the general characteristics of habituation, a more detailed analysis of the data reveals numerous species differences. When "light-on" is the habituating stimulus, *N. diversicolor* takes significantly longer to habituate than does *N. pelagica,* which in turn requires more trials

than does *P. dumerillii,* in which habituation is essentially complete after one
trial. Similarly, both *N. diversicolor* and *N. pelagica* require significantly
longer to habituate to mechanical shock than does *P. dumerillii*. That these
differences are not simply due to *P. dumerillii* being generally sluggish or
unresponsive to stimuli is indicated by the fact that this species is substantially
more responsive than *N. pelagica* to both tactile stimuli and to the combination of "light-off" and mechanical shock. In fact, both *P. dumerillii* and
N. diversicolor exhibit significant increases in responsiveness to the combined
stimulus compared to either of the two components alone. *N. pelagica,* on the
other hand, shows no tendency to respond more frequently to the combined
stimulus (Evans, 1969*b*).

Response Characteristics. In an attempt to characterize an inferred
process such as habituation it is important to remember that the observed
behavior may result from complex interactions of several response systems
and that different measures of the same response system may yield quite
different perspectives. As a case in point, it may be recalled that the withdrawal response of *N. diversicolor* habituated more rapidly (as measured by
frequency of response) to "light-off" than to "light-on." However, one should
not infer from this fact that the contraction itself was less vigorous to "light-
on" than "light-off." On the contrary, contractions to "light-off" were much
more rapid and of shorter latency than those to "light-on." Further analysis
revealed that different response systems were being activated by the two
stimuli; almost all (96%) of the contractions to "light-off" were withdrawals
of the anterior segments of the worm, whereas most of the responses to
"light-on" were posterior withdrawals (Evans, 1969*a*). Since different response systems are involved, it would clearly be improper to infer, as does
Clark from comparable data on *N. pelagica,* that it "habituates less rapidly
to a sudden decrease in light intensity and most slowly to a sudden increase
in light intensity" (Clark, 1960*a*, p. 90).

To complicate the matter somewhat further, stimulation by "light-off"
not only causes a fast, short-latency contraction of the anterior segments but
also results in the inhibition of the dorsoventral undulatory movements so
typical of tube-dwelling worms. The habituation of these two responses
proceeds at quite different rates, even though both responses involve the
longitudinal muscles of the body wall (Clark, 1965*b*).

Related Phenomena

Retention. Simple Recovery: Clark (1960*a,b*) has reported several experiments with *N. pelagica* which imply that recovery from the effects of
habituation may vary with the nature of the stimulus, the number of habituating trials, and the intertrial interval. Habituation of locomotion in

Hesperonoë adventor appears to be retained longer when trials are spaced rather than massed (Dyal and Hetherington, 1968). However, the results of both of these studies must be regarded as only suggestive, since the experimental effects may have been due to other confounding variables.

Sequential Interaction Effects. Retroaction and Retention: It will be recalled from Chapter 1 that the standard retroactive interference paradigm may be used to study the effects of an interpolated experience on the retention of prior training. Clark (1960*a*) has applied this paradigm to test for interaction effects between successive habituations to different types of stimuli. He found that if the two stimuli involve different modalities the habituation processes tend to be independent, whereas if they involve the same modality complex sequential interactions occur. In an experiment illustrating the first point, *N. pelagica* individuals were habituated to a moving shadow and then to a mechanical shock followed by a second habituation to moving shadow.

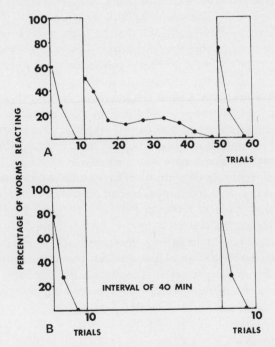

Fig. 5. Independence of the processes involved in habituation to a moving shadow followed by habituation to mechanical shock. In A, the first phase is habituation to a moving shadow causing a decrease in light intensity from 33.9 to 0.08 ft-c. Light is obscured for 9 sec on each trial. The second phase involves habituation to a mechanical shock (ISI 60 sec); the third phase involves ten test trials with the moving shadow. In B, the two periods of habituation to moving shadow are separated by a 40 min rest interval (Clark, 1960*a*).

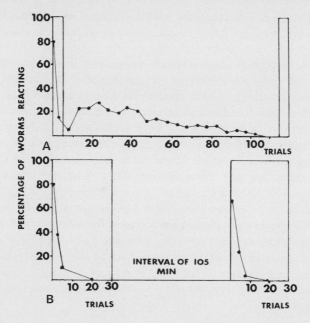

Fig. 6. Interactions between habituation to a moving shadow (phases 1 and 3) and to a
sudden decrease in light intensity (phase 2 in A) (Clark, 1960a).

The results are presented in Fig. 5, where it may be seen that the retention of
the experimental group did not differ from that of a control group which did
not receive the interpolated habituation. The two habituation processes are
independent; they are stimulus specific; no retroaction effects are present.
Furthermore, it should be noted that the level of retention after 40 min is
essentially zero.

A second series of experiments illustrated the complex sequential interactions which occur when the two habituating stimuli are similar. Figure 6
represents the results of an experiment in which habituation to a light decrease
was interpolated between two sessions of habituation to a moving shadow.
In this case, the interpolated habituation experience resulted in a complete
lack of responsiveness to the moving shadow on the second test. Clark
states that "Exposure to a sudden decrease in light intensity at regular intervals maintains an already established state of habituation to a moving shadow" (1960a, p. 88).

Comparison of the retention tests of experiments 1 and 2 (Figs. 5 and 6,
respectively) indicates that the effect of order of habituation is important.
Interpolation of "light decrease" following "moving shadow" "maintains"

habituation to the moving shadow, whereas interpolation of moving shadow following habituation to light decrease does not serve to "maintain" the response.

Proactive Effects: Proactive effects designate the influence of the prior learning of task A on the subsequent learning of task B. The effects may be either facilitory or inhibitory. Comparison of Figs. 5 and 6 reveals that prior habituation to either stimulus prolongs responsiveness during the subsequent habituation to the other stimulus. This interference with the normal progression of habituation is thus a case of proactive interference.

Proactive facilitation effects in the form of potentiation of habituation were reported by Dyal and Hetherington (1968). A special case of proactive facilitation is represented in the so-called latent habituation effect. The effect is illustrated in an experiment by Clark (1960a) in which *N. pelagica* individuals were presented with a light increase from 0.08 to 4.7 ft-c for 60 trials. Such an increase is a very weak stimulus and as such *did not elicit a withdrawal response* in any of the worms. During the next 60 trials, a much stronger light measurement (0.08–33.9 ft-c) was presented. These worms habituated to the strong stimulus much more rapidly than worms which had not received the "priming" with the weak stimulus.

D. Class Hirudinea

Basic Demonstrations

There have been only two demonstrations of habituation in leeches. In

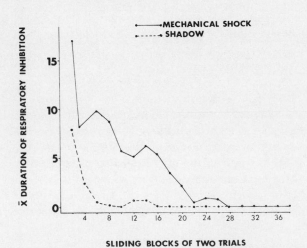

Fig. 7. Mean duration of respiratory inhibition by the leech *Dina microstoma* over 20 presentations of a habituating stimulus (Gee, 1913).

his classic study on the behavior of leeches, Gee (1913) included a series of experiments on what he called acclimatization using *Dina microstoma* (Moore) as subjects. Gee presented slight jars or a shadow for 20 trials and measured the frequency and duration of the cessation of ventilatory movements. The results for the duration measure are shown in Fig. 7. Both of the measures reveal substantial habituation of response cessation during the course of the 20 stimulus presentations. Habituation to the shadow was essentially complete within the first five trials. Habituation of this shadow reflex has also been reported in *Hirudo medicinalis, Haemopsis sanguisauga,* and *Herpobdella octoculata* (Kaiser, 1954).

Gee also presented data on repeated tactile stimulation of *D. microstoma.* The response measured was the duration of swimming following stimulation. There was no indication of any habituation of this response over a series of 40 stimulations. This result is consistent with Clark's data on nereids in which he finds little or no habituation of the rapid withdrawal response to tactile stimulation.

Influencing Variables

Stimulus Variables. Stimulus Intensity: Kaiser (1954) found that the rate of habituation of the shadow reflex (respiratory inhibition) was an inverse function of the percent decrease in light intensity. Slower habituation to stronger stimuli is consistent with the data from oligochaetes and seems to be a rather general characteristic of the habituation process from protozoans to man.

Interstimulus Interval: Gee (1913) compared the effects of short (15–30 min) and long (120 min) interstimulus intervals on habituation to shadow ($n = 2$) and mechanical shock ($n = 1$). Short intervals resulted in faster habituation. This evidence, while consistent, is not compelling and requires further research with larger samples and more species in order to be firmly established.

Related Phenomena

Retention. The only data available are based on repeated habituation of one leech and indicate that recovery from habituation of respiratory inhibition is essentially complete after 60 min (Gee, 1913). It is apparent that the problem of retention in the leech needs to have considerably more experimental attention devoted to it.

E. Neurophysiological Factors in Habituation of Reflexive Responses in Annelids

The Role of Giant Fibers

As we have seen, the giant fiber systems mediate the rapid "escape" contraction typical of many annelids. The behavioral habituation of this response could be mediated by neural transmission failure at (a) sensory receptors, (b) peripheral afferent synaptic junctions, (c) central-sensory-to-giant internuncial junctions, (d) central-giant-to-motor internuncial junctions, (e) peripheral motor neuron junctions, or (f) neuromuscular junctions. In addition, an observed waning of response could be strictly due to fatigue of the longitudinal muscle fibers and thus involve no failure of neural transmission at all.

Roberts (1962a) provided a thorough and systematic study of the possible sites of transmission failure in *Lumbricus terrestris* (L.). Using a series of preparations which permitted the logical elimination of various sites, he concluded that habituation to repetitive tactile stimulation is localized at internuncial sensory-to-giant and giant-to-motor junctions. Similar accommodation at sensory-to-giant junctions has been demonstrated in tubiculous sedentary polychaetes of the family Sabellidae *(Myxicola infundibulum)* by Roberts (1962b), in *Branchiomma vesiculosum* by Krasne (1965), and in errant polychaetes *Nereis virens* and *Harmothoë imbricata* by Horridge (1959). In the case of *Nereis,* electrical stimulation of the anal cirri induces accommodation in paramedial giant fibers and the lateral giants. It is identified most clearly in the lateral giants of *Harmothoë.* The sensory-to-giant junction fails to fire after a maximum of three stimulations of a single anal cirrus. *Nereis* and *Harmothoë,* like *Lumbricus,* show rapid accommodation of giant-to-motor junctions. Although the existence of giant-to-motor junctions in *Lumbricus* is a reasonable inference on the basis of the electrophysiological data, these junctions have not as yet been clearly identified by histological analysis. However, in the case of *Harmothoë* and *Nereis,* "An axon–axon synapse seen histologically between the lateral giant fibers and the large motor neurone to the longitudinal muscle has been identified with the rapidly accommodating physiological junction between these elements" (Horridge, 1959, p. 245). While the rapid withdrawal response can be mediated by all three types of giant fibers, backward parapodial pointing is mediated by the paramedials and forward parapodial pointing by the dorsal median. These aspects of neural mediation are reflected in independence of these systems during habituation.

A somewhat different giant-to-motor picture is presented in sabellids, which also apparently differ substantially across species. For example, accommodation does not occur at giant-to-motor junctions in *Myxicola,* and this is probably related to the fact that the longitudinal muscle fibers are directly innervated by motor branches from the giant fiber. Although direct stimulation of the giant fiber every 30 sec does result in accommodation in *B. vesiculosum,* such stimulation would be either impossible or exceedingly rare in the natural environment, since the worm would ordinarily stay in its tube longer than 30 sec following an "escape" response. Thus it seems doubtful that the giant-to-motor habituation plays a very significant role in the escape reflex of sabellids (Krasne, 1965).

Extirpation of Supraesophageal Ganglion and Transection of the Chord

The giant fibers of annelids tend to originate in or are represented in the brain (supraesophageal ganglion). In the earthworm, removal of the brain results in an animal which feeds and burrows more slowly and awkwardly. Although it locomotes normally, it tends to be more active (Peeke *et al.*, 1965). Similarly, Hirudinea are especially excitable and active following ganglionectomy, and *Nereis* persists in tube irrigation unceasingly. The total picture strongly suggests that the brain normally plays an inhibitory role in whole-body activities. Kuenzer (1958) removed the supraesophageal ganglion of *Lumbricus* and found that for a week postoperatively the animal's sensitivity to electric shock applied to the third segment increased, as indicated by increases in number of trials required to complete habituation and by a lowering of the threshold for withdrawal. These results are consistent with a hypothesized inhibitory role of the supraesophageal ganglion. In an intact worm, the ganglion may serve to dampen the tendency to react to "threat" stimuli and may thus facilitate habituation. While it is tempting to speculate that the brain might play a role in inhibition of the sensory-to-giant transmission, there is no direct evidence that it does so. Furthermore, it would appear that a general inhibitory function can be served by the portion of the cord anterior to the clitellum.

Evans (1969b) has studied the effects of decerebration on habituation of withdrawal in *N. diversicolor, N. pelagica,* and *P. dumerillii.* Although there tended to be some slight differences in responsiveness by some species to some stimuli, the general pattern of results was such that Evans concluded that the supraesophageal ganglion has little or no effect on the normal process of habituation in *Nereis.* Still, a word of caution is in order:

> Although much of the habituation process takes place in the segmental ganglia, we can perhaps be less confident that the supra-oesophageal ganglion is not involved when we consider the entire behavioural response of the animal to repeated, complex and varied stimuli such as it meets outside the laboratory. . . . this may involve a much more complicated neurophysiological process than any that has been envisaged so far. (Clark, 1965c, p. 365)

F. Biological Significance of Habituation of the Rapid Withdrawal Reflex

Consider the behavior and environment of a burrowing or tubiculous polychaete living in a mudflat. Most of its normal activity is devoted to feeding and respiration. However, in the process of these maintenance activities it exposes itself to predation by highly mobile hunters that move through its living space from time to time (e.g., bottom-feeding fish, shore birds, and large crustaceans). Since the evolution of annelids preceded the development of effective distance receptors, selection pressures would seem to capitalize on proximodetection of predators and rapid escape. The withdrawal response to a novel and potentially significant stimulus must thus be fast and integrated and require little "cerebration." The fast conduction of the through-conducting giant fiber system makes giant fibers ideal mediators of this rapid withdrawal reflex. Sedentary, tubiculous forms such as Sabellidae rely on the activation of the contraction by a single impulse from the giant fiber, a mechanism which makes adaptive sense in that the withdrawal reflex is their only line of defense. The free-moving oligochaetes and polychaetes are able to locomote away from a predator, and thus a graded response system is advantageous; such a system is achieved by providing that the full-blown reflex requires repetitive neural impulses for its activation. The graded response permits a "finer tuning" of the reciprocal relationships between maintenance activities and escape. On the other hand, a worm which failed to habituate its reflexive withdrawal response would run the risk of starvation and (in the case of sabellids and other fanworms) suffocation. Habituation of the withdrawal response is thus a biologically adaptive technique to reconcile the need to eat against the threat of being eaten.

G. Conclusions Regarding Habituation in Annelida

1. Behavioral habituation of several response systems to a variety of stimuli has been demonstrated to occur in representatives of all three classes of Annelida.
2. The rate of behavioral habituation in oligochaetes has been shown to be an inverse function of the intensity of the habituating stimulus. There is suggestive evidence that this relationship is also typical of polychaetes and leeches.
3. The rate of behavioral habituation in errant polychaetes is a direct function of the rate of presentation of the habituating stimuli over the range from 30 sec ISI to 5 min ISI. What meager evidence is available suggests that this relationship also characterizes the habituation of oligochaetes, sedentary polychaetes, and leeches.
4. Different stimulus modalities tend to habituate at different rates, and it would appear that responses to tactile stimuli habituate at a slower rate than those to light stimuli. It is not known whether this difference reflects a fundamental difference in the significance of the two modalities or is primarily associated with intensity differences.
5. Comparison of habituation rates in various species of errant polychaetes reveals species differences which can occasionally be related

to adaptation in the natural environment. Data for species comparisons in the other annelid classes are totally lacking.

6. Two or more relatively independent response systems may be activated by the same stimulus (e.g., anterior withdrawal and posterior hooking in oligochaetes or rapid contraction and parapodial pointing in polychaetes). These response systems are mediated by different giant fibers which habituate at different rates.

7. The response which has been most extensively investigated at both the behavioral and the neurophysiological levels is the rapid withdrawal reflex exhibited by most oligochaetes and polychaetes. (a) Habituation of the rapid withdrawal response is mediated by increased refractiveness at the sensory-to-motor junctions (sedentary polychaetes) and by failure of transmission at giant-to-motor junctions in oligochaetes and errant polychaetes. (b) Under laboratory conditions, the supraesophageal ganglion does not seem to play a role in determining the habituation rate.

8. Measurable retention of the effects of habituation has been shown in oligochaetes for periods as long as 96 hr. The maximum retention observed in sedentary polychaetes is somewhat over 24 hr and in errant polychaetes around 17 hr. Retention in leeches has not been shown to persist beyond 1 hr.

9. When the same behavioral response is habituated to different stimuli in successive sessions, the habituation processes may be independent or they may interact. The likelihood of interaction of a retroactive facilitation sort may increase with the similarity of the two stimuli. The nature of the interaction also depends on the order in which the habituating stimuli are presented.

10. It is apparent that all of these conclusions may be tempered by future research. While research is badly needed in all areas, the following are especially critical: (a) more adequate determination of the effects of interstimulus interval on rate of habituation in oligochaetes, sedentary polychaetes, and leeches; (b) extensive comparisons of habituation rates for several representative species within each annelid class, using a standard set of visual and tactile stimuli; (c) extensive studies of retention using both simple retention and transfer designs.

VI. SENSITIZATION

It is important to distinguish sensitization as a special and legitimate category of behavioral plasticity which is evolutionarily more advanced than habituation but less advanced than classical conditioning. Razran (1971) and Wells (1967) may be consulted for further elaboration of this point of view.

A. Class Oligochaeta

While modest sensitization effects may be involved in successful demonstrations of classical conditioning in oligochaetes, a clear demonstration of persistent and robust sensitization effects is not available. It may be seen in Fig. 10 that the presentation of a vibratory stimuli (CS) alone initially elicits a high frequency of withdrawal responses. This response wanes (habituates) over 100 training trials, but over all trials it is elicited significantly more frequently by the CS alone than by the other control conditions. However, the fact that the responsivity *decreases* over trials rather than increases indicates that the sensitization effect is a weak one.

Other evidence of sensitization processes in oligochaetes involves instrumental conditioning in straight alleys or mazes and is also rather tenuous (Krivanek, 1956).

B. Class Polychaeta (Sedentary)

Early experiments by Hargitt (1906, 1909) and by Yerkes (1906) seemed to imply the occurrence of classical conditioning. Yerkes first habituated the withdrawal response of *Harmothoë dianthus* to shadow and then gave a series of trials in which the shadow was followed 5–10 sec later by anterior touch (which also elicited the same withdrawal response). The experimental question was whether or not the pairing of "shadow" (conceived as the CS by Yerkes) with the tactile stimulus (UCS) would reactivate and maintain the withdrawal to the shadow. The expected withdrawal to the shadow did occur, but further training apparently resulted in habituation to both stimuli. This latter fact, along with the difficulty of demonstrating true classical conditioning in a paradigm in which the same response is elicited by both the CS and the UCS, requires that the obtained effect be interpreted as either a case of alpha conditioning or an example of dishabituation.

A more recent pilot study by Krasne[4] used a procedure similar to that of Yerkes to determine if habituation of the withdrawal reflex of *Eudistylia* could be slowed or prevented by pairing a light reduction (CS) with vigorous shaking of the tube. A sensitization control group was trained with the same stimuli but the CS–UCS interval was 5 min, the assumption being that 5 min trace conditioning would not be likely in annelids. The results indicated no reliable differences between the two groups in their rate of habituation. While this preliminary study does not permit strong inferences based on null

[4]Personal communication from F. Krasne (1971).

results, it is consistent with "sensitization" as opposed to "conditioning" effects.

C. Class Polychaeta (Errant)

Basic Demonstrations

Evans (1966a,b) has conducted three studies which support the existence of sensitization in *Nereis*. These three experiments could legitimately be considered to demonstrate conditioned sensitization effects which represent pseudoconditioned precursors or analogues of (a) instrumental conditioning based on appetitive reinforcement, (b) classical aversive conditioning, and (c) aversive inhibitory conditioning (Razran, 1971, p. 62).

Evans modeled his procedure in the appetitive reinforcement analogue after an experiment by Copeland (1930) which had been adduced to have demonstrated classical conditioning. The conditioning group received 48 training trials, six trials per day, with an ITI of 30 min. A light increase (1–9 ft-c) was followed "immediately" by presentation of food (wheat germ extract) at the end of the home tube. A test trial was interpolated every sixth trial in which the light-only was presented. In addition to this "classical conditioning" group, Evans included several sensitization/pseudoconditioning groups. In one of these groups, food was presented 5 *min after* the offset of the light CS. In another, the light-only was presented throughout training. The second pair of control treatments actually constituted a miniexperiment demonstrating the sensitization effects of food *per se*. One group was presented the light for 2 min followed immediately by the food. Thirty minutes later, the light was presented again as a probe for sensitization. The second group was not fed between the two presentations of the light. On the first test trial following training, all groups which had received food had significantly shorter mean latencies than either of the two light-only control groups. The mean latencies for all trials revealed a similar picture, with all groups which received food being significantly faster than the groups which did not; neither of these food *vs.* no-food clusters differed among themselves. The clear implication of these results is that "Food presentation sensitizes a worm's responses to sudden increases in illumination" (Evans, 1966b, p. 109).

In the classical aversive conditioning "precursor" experiment, Evans utilized a 30 sec light diminution as the CS (30 to 5 ft-c) and a 0.5 sec electric shock as the UCS; ITI was 90 sec. Three groups were trained: forward conditioning (CS–UCS interval 2.5 sec), backward conditioning (UCS–CS interval 40 sec), and CS only. During the course of conditioning, the forward and backward groups were not significantly different from each other but both were reliably superior to the CS–only group (Fig. 8).

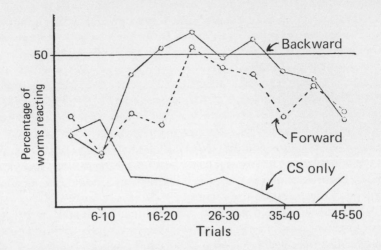

Fig. 8. Percentage of worms reacting to sudden decreases in illumination as the CS and electric shock as the US (Evans, 1966b).

How are these results to be interpreted? They could be regarded as valid demonstrations of forward and backward conditioning, or they could be viewed as examples of sensitization. The evidence from both oligochaetes (Herz *et al.,* 1967) and mammals (Razran, 1971) indicates that while backward conditioning is a valid phenomenon it is more labile and less strong than forward conditioning. It thus seems reasonable to interpret the strong, equivalent effects obtained for "forward" and "backward" conditioning in the present study as "almost certainly due to the sensitization effects of the unconditioned stimulus" (p. 118). More precisely, these effects are examples of alpha conditioning, since "in these experiments the responses to sudden decreases in illumination in all groups of worms were normally typical anterior-end withdrawals which *never became modified* to resemble the simultaneous anterior and posterior end withdrawals elicited by electric shocks" (p. 117) (italics mine).

Pseudoconditioning of response inhibition, a precursor of aversive inhibitory conditioning, was investigated by Evans (1966a) using a black Perspex straight alley equipped with electrodes for delivering shock at the terminal end. Five groups of *N. diversicolor* were trained in this "Bristol channel": group A was shocked at the exit of the channel, group B received no shock, group C was shocked 5–10 sec *after* leaving the channel, group D was shocked 5–10 sec before entering the apparatus, and group E was given pretraining shocks followed by the same treatment as Group A.

The time scores and percent refusal scores were almost identical for groups A, C, and D, and all groups showed more inhibition of response than group B, which was not shocked. These results suggest that the apparent inhibitory aversive conditioning exhibited by group A may in fact be pseudo-conditioning based on the effects of shock *per se*. Similarly, previous experiments reporting aversive inhibitory conditioning in *N. diversicolor, P. cultrifera,* and *N. virens* (Evans, 1963*a*) may be better interpreted as reflecting a pseudoconditioning precursor of true conditioning.

Influencing Variables

Stimulus Variables. Intertrial Interval: No direct evidence is available regarding the effects of intertrial interval or trial block spacing on any of the varieties of sensitization.

Interstimulus Interval: Apparently, the interval between warning stimuli and the delivery of punishment does not influence the rate of response decrement in pseudoconditioning based on aversive stimulation. Evans (1966*a*) found no differences between worms shocked immediately at the end of the runway, 5–10 sec after leaving the runway, or 5–10 sec before entering the runway. Indeed, it is the lack of acquisition differences attributable to this variable which carries the force of the argument for a pseudoconditioning interpretation as opposed to inhibitory aversive conditioning. Similarly, the argument for a sensitization interpretation of the conditioning experiments with food is based on the observation that worms given food immediately after the CS did not differ in their response latency from those to which the food was not presented until 5 min after the CS onset (Evans, 1966*b*). Likewise, sensitization, as opposed to conditioning, is inferred in the "classical aversive conditioning" experiment because the forward conditioning group did not differ from the backward conditioning group (Evans, 1966*b*).

Pretraining Shocks: Evans (1966*a*) gave one group of *N. virens* 40 pretraining shocks outside of the Bristol channel apparatus followed by 40 trials in the apparatus. This pretraining group made significantly more refusals and crawled more slowly along the channel than did a group without such pretraining. These results suggest that level of sensitization is some positive function of the number of presentations of the sensitizing stimulus. An experiment making number of pretraining shocks the dependent variable is needed to clarify the shape of that relationship.

Organismic Variables. Species: *N. virens* becomes sensitized to shock in the Bristol channel substantially faster than does either *N. diversicolor* or *P. cultrifera* (20 trials as opposed to 70–100 trials).

Other Variables. Neither response characteristics nor any other variables which might influence sensitization have been studied. Several problems which merit research come easily to mind: What is the time-dependent character of sensitization by pretraining shocks? What is the time-dependent character of sensitization by a single presentation of food? How do sensitizations based on food and shocks interact; i.e., what are the salient characteristics of this precursor of approach–avoidance conflicts? Does decerebration differentially affect approach/avoidance sensitizations?

Related Phenomena

Retention. Copeland and Brown (1934) trained worms to approach the end of their tubes for food when presented with a tactile stimulus. They discontinued training for 21–35 days and found substantial retention of the approach response to touch. Unfortunately, they continued to feed the animals on alternate days. In the light of the strong sensitization effects which food induces in polychaetes, the relevance of these retention data is restricted to the interval between the last feeding and the first test trial (48 hr). But even with this qualification, substantial retention of sensitization was demonstrated over this 48 hr period.

Combining all of the available evidence (Figs. 6 and 7; Evans, 1963*a*) on retention of sensitized response inhibition by *N. virens,* it would appear that retention decreases in a fairly linear fashion over the interval from 1 hr to 24 hr after original training. After 1 hr, retention is about 75%, whereas after 24 hr it has dropped to about 25%. There is clearly some retention of sensitizations over 24 hr; furthermore, its effects tend to be cumulative, since training to criterion every 24 hr resulted in a steady decrease in the trials necessary to reach a 40 sec inhibition criterion.

D. Class Hirudinea

The only demonstration of sensitization effects in leeches was reported by Gee (1913). It was previously noted that when a glass rod is brought into contact with the extreme anterior of *Dina microstoma* the animal may make approach, avoidance, or indifferent responses. There is a tendency for the number of negative responses to be decreased by the introduction of snail juice into the medium; there is concurrent increase in number of indifferent or positive reactions. This latter effect seems to be quite comparable to that obtained by Evans' (1966*b*) experiment from which he concluded that the presentation of food sensitizes approach responses.

E. Neurophysiological Basis of Sensitization

Facilitation in polysynaptic systems has been reported in *Harmothoë* and *Nereis* (Horridge, 1959) and *L. terrestris* (L.) (Roberts, 1966). Roberts used several preparations to demonstrate facilitation of anterior withdrawal response (giant fiber reflex) as a result of temporal summation of barely suprathreshold shocks to any portion of the worm's body. This series of careful experiments permitted the elimination of afferent fibers, longitudinal muscle, neuromuscular junctions, and synapses of the motor neuron as possible sites of the facilitation and left the site in the CNS, specifically the giant-to-motor junction. Roberts noted that "Facilitation is most pronounced in preparations whose 'giant-to-motor' junctions are accommodated. In such cases a single impulse in the median giant fiber is ineffective, two or more being required to produce a rapid response throughout the length of the animal" (p. 150). The neurophysiological data on facilitation are consistent with the role that simple incremental sensitization plays as a counterinfluence to habituation. While it is perhaps premature to identify these analogues as constituting the neurophysiological basis of sensitization, it is quite clear that future neurophysiological investigations of temporal summation and post-tetanic potentiation are likely to be helpful in elucidation of sensitization effects. A more extensive discussion of facilitation may be found in Bullock and Horridge (1965):

> Little studied, but perhaps of importance, is the phenomenon of *facilitation of facilitation*, where the course and degree of augmentation is itself enhanced by repetition. While some forms of facilitation last only milliseconds others are reported in simple preparations lasting hours and in one case seven days (Wilson, 1959). An upper limit for neurons cannot be given, for preparations and responses grade into complex and behavioral ones and hence into learning and memory. (pp. 269–270)

At this point, the role of the supraesophageal ganglion in sensitization must be inferred from the following meager behavioral data: (a) if ganglion-ectomy occurs immediately after sensitization training, retention is normal for 6 hr but is impaired at 24 hr; (b) if ganglionectomy occurs prior to training, sensitization is not significantly retarded, *but* retention time is dramatically shortened so that no retention occurs after 1 hr. These results imply that the supraesophageal ganglion is necessary for memory to be retained on a short-term basis up to 1 hr and that this short-term memory (STM) is necessary for even modest long-term memory (24 hr) to occur. Moreover, the presence of STM is not sufficient for LTM to occur; the supraesophageal ganglion apparently continues to play some role *after* the learning has been completed. It does not appear to be a critical memory storage site itself but "does appear to be implicated in some way in the initial process of learning and the transfer of the acquired behavior to a memory storage centre, presumably elsewhere in the nervous system" (Evans, 1963a, p. 177).

Speculation regarding the possible role of the supra- and subesophageal ganglia in nonassociative learning has been advanced by Evans (1966c) who has noted that nereids need to have mechanisms which will produce graduated commands and which will serve to integrate forward-going, food-seeking behavior with withdrawal–avoidance behavior. He suggests that the model proposed by Young (1963, 1964) to account for similar behavior requirements in *Octopus* may be applicable to nereids.

F. Biological Significance of Sensitization

Sensitization may be seen as a step in the evolutionary progression from sensation to symbolization. In contrast to habituation, it has the effect of keeping the animal active and responsive to salient stimuli in its environment. As an activator of stimulus-relevant response hierarchies, it permits the animal to match his behavior more closely to the environmental contingencies. For example, a tubiculous polychaete such as *Nereis* spends much of its time irrigating its tube and in "approach or avoidance" behavior. Its adaptive problem is to maximize the match between forward-going approach behavior and the appropriate stimuli, such as food, in the environment or to match withdrawal with the presence of predators. A mechanism which sensitizes approach and food-seeking activity to stimuli temporally associated with food would be highly adaptive. The mechanism would not have to be an associative one; indeed, a nonassociative mechanism would have the advantage of faster activation of the relevant response system. Furthermore, the mechanism would not need to be very selective (a broad-band stimulus filter would suffice), since if it were used to potentiate such fundamental activities as feeding and "flight" it would be adaptive for these activities to take precedence over the remaining limited response repertoire. The rapid sensitization of many stimuli to the elicitation of forward-going and "search" behavior by the presence of food is of course an example of such an adaptive mechanism. Similarly, the adaptive function of sensitization of the withdrawal response to a strong elicitor such as shock is apparent. What is also apparent is that it is necessary for interplay to occur between these approach and avoidance systems and between the processes of habituation, dishabituation, sensitization, and retention. Despite the fact that Darwin (1898) noted that when earthworms were copulating they were less responsive to shadows which would normally elicit withdrawal, few precise data are available regarding these important interactions.

Although annelids are capable of true associative conditioning, the relevance of this capacity to real-life problems of adaptation is questionable, since it requires a relatively large number of trials to become effective. Sensitization of approach or withdrawal tendencies, on the other hand, occurs much more rapidly and would seem capable of solving most of the "adjustive" problems of an animal with a relatively limited response repertoire.

VII. AVERSIVE INHIBITORY CONDITIONING (PUNISHMENT)

It will be recalled that, on logical and empirical grounds, Razran (1971) considered response inhibition as a result of punishment to be a primitive type of true associative conditioning. Examples of conditioned inhibition of both innate and learned reactions are available in the experimental literature on oligochaetes and errant polychaetes.

A. Class Oligochaeta

Basic Demonstrations

If earthworms are tested in a single-choice T-maze in which one arm is brightly lighted and the other is darkened, they will exhibit a strong natural preference for the dark arm, i.e., a negative phototaxis. If they are punished by electric shock for entering the darkened arm, they will learn to inhibit this response (Krivanek, 1956; Wherry and Sanders, 1941). Furthermore, if the arms of the T-maze are equally illuminated and entry into an arm is punished, two effects typically occur after 30–50 training trials: (a) an increase in frequency of correct turns (instrumental avoidance conditioning?) and (2) a nonspecific inhibitory effect characterized by vacillation and reversals in the choice area (Robinson, 1953; Schmidt, 1955). None of these experiments contained important sensitization control groups; nonetheless, because the response inhibition required substantially more trials to appear than would be expected if the effect were primarily sensitization it seems reasonable to infer that they represent *bona fide* examples of aversive inhibitory conditioning. It is also apparent that both "generalized avoidance" and conditioned inhibition to specific exteroceptive stimuli can occur in the same experiment (Robinson, 1953; Schmidt, 1955).

Smith and Dinkes (1971) have recently reported two experiments which they interpret as demonstrating passive avoidance conditioning in *L. terrestris* and which in the present context we interpret as instances of classical inhibitory conditioning. In the first experiment, a single group of worms were first given 20 trials (one per day) in which they were permitted to traverse a runway tube which exited into their home container of moist earth. In the second phase, a 25% sodium chloride solution was substituted for the moist earth and the worms were given ten more daily trials. The authors reported that "upon contact with the saline solution, the typical *S* recoiled markedly and usually backed out of the tube." In phase 3, six more daily trials were given to the moist earth. The results may be seen in Fig. 9.

In phase 2, the worms' response time (time from start point to *contact* with the goal box substrate with the prostomium) increased, suggesting a *conditioned* response inhibition. The results were replicated in a second ex-

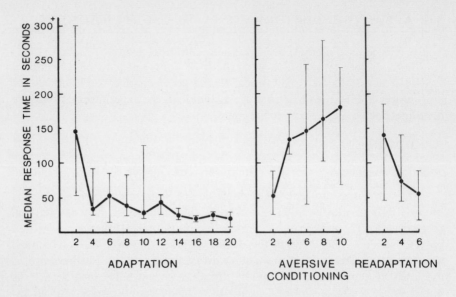

Fig. 9. Median response time in seconds during adaptation (moist earth in goal box), aversive conditioning (NaCl in goal box), and readaptation (moist earth) (Smith and Dinkes, 1971).

periment with the addition of some control groups. However, the size of the effect was much smaller than in the first experiment, and other aspects of the data suggest that the group differences could have been mediated by detectable airborne cues which were available prior to contact with the substrate. The Smith and Dinkes experiment is an important one and needs to be replicated with appropriate controls to eliminate possible confounding effects of airborne cues.

Influencing Variables

Stimulus Variables. Razran (1971) summarized much of the literature on punishment in both invertebrate and vertebrate forms. He indicated that the behavioral effects of punishment are dependent on several characteristics of punishing stimulus: absolute and relative intensity, the time interval between the response and the punishing stimulus, and the schedule of punishment. None of these variables have been investigated using oligochaetes as subjects.

Organismic Variables. Species: Schmidt (1955) compared *L. terrestris* and the manure worm *Eisenia foetida* in avoidance conditioning in a single-unit T-maze. He found that after 30–50 trials *L. terrestris* tended to exhibit generalized avoidance response to the stem and choice area, thus confirming

Robinson's (1953) observation on the same species. *E. foetida,* on the other hand, tended to confine its avoidance responses to the specific cue (sand-paper) associated with the shock. It may be that *L. terrestris* tends to generalize more and is less effective in utilizing specific discriminative stimuli as warning stimuli.

Related Phenomena

Retention. Although it is commonly held that the effects of punishment are quite persistent (as compared to the effects of reward), practically no evidence is available concerning retention of conditioned response inhibition in oligochaetes. The one experiment which measured retention obtained no evidence of retention over a period of 10 days (Smith and Dinkes, 1971).

B. Class Polychaeta (Errant)

Basic Demonstrations

Evans (1963b) trained three species of *Nereis* (*N. virens, N. diversicolor,* and *P. cultrifera*) in a single T-maze in which each arm could be electrified. An "incorrect" choice led to shock, while a correct choice led out of the maze without shock. The worms were trained in a single session terminated by the animals' refusal to crawl along the maze. The typical performance curve is characterized by an initial period of random choice and then a significant increase in correct choices; this is followed in turn by a decline in correct choices and a refusal to traverse the maze. It is the "decline and refusal" phase of the data which can be taken as an indication of aversive inhibitory conditioning.

The performance decrement and final refusal which typify performance in all these experiments may be related to the presence of potentially aversive stimulation in the "correct" arm. When the worm emerged from the maze, it "was maneuvered with a paint brush to the entrance in preparation for the next trial . . . . A maze in which severe punishment is given in one arm and mild punishment in the other presents a complex situation which may account for the refusal of worms to undergo an unlimited number of trials" (Evans, 1963b, p. 391).

Although it is quite possible that the results of Evans' simple T-maze experiments provide examples of aversive inhibitory conditioning, the validity of this interpretation must remain in question until the experiments are repeated with sensitization controls.

Influencing Variables and Related Phenomena

No data are available concerning the effects of a variety of punishment parameters (*cf.* Chapter 1), nor is there any evidence concerning such related phenomena as retention, intermittent punishment effects, and generalization.

C. Neurophysiological Basis of Aversive Inhibitory Conditioning

Evidence relevant to the neurophysiological basis of punishment is meager at any phylogenetic level and practically nonexistent in the case of annelids. "That the oldest, simplest, most universal and most effective level of associative learning should have thus been neglected is probably related to its confused conceptual status" (Razran, 1971, p. 126). Razran postulated that nonsynaptic electronic transmission, which has been demonstrated in leeches (Eckert, 1963; Hagiwara and Morita, 1962), mollusks (Tauc, 1959), lobsters (Wanatabe and Bullock, 1960), and cats (Belenkov and Chirkov, 1969), is "the underlying mechanism of long lasting effects of the inhibitory conditioning of punishment" (p. 157). Although the hypothesis is an interesting one, it is clearly "in anticipation of empirical evidence."

D. Biological Significance of Aversive Inhibitory Conditioning

That there is "evil" in the world cannot be denied, and recognition of signs of potential threat is of clear adaptive advantage, as we have seen in our discussions of habituation and sensitization. Aversive inhibitory conditioning goes beyond either of those two processes in serving as a mechanism for the learning of "what not to do." It is thus a device whereby both innate and previously learned responses may be overridden by new experience. On the basis of the above experiments, it seems likely that annelids possess this rudimentary capacity to inhibit an innate or learned approach response. However, it should be emphasized that this capacity is quite limited; it takes many trials to develop and is not easily tied to specific external stimuli. The adaptive significance of this ability in the animal's normal environment is questionable, since sensitization processes would seem to accomplish the "stop" function more rapidly. Nonetheless, there are two sources of potential advantage to worms that are able to achieve true conditioned response inhibition: (a) they are able to be more precise and discriminating in the stimuli to which they inhibit response, and (b) inhibitory conditioning may have a longer effective retention period than sensitization. Evidence on this latter point is notably lacking, easily obtainable, and clearly important.

VIII. CLASSICAL CONDITIONING

The question of the appropriate controls to demonstrate classical conditioning as opposed to sensitization or pseudoconditioning has been considered in Chapter 1. These considerations will be quite relevant to the question at hand: Is classical conditioning demonstrable in annelids?

A. Class Oligochaeta

Basic Demonstrations

Raabe (1939) conducted the first experiments which reported classical conditioning of an oligochaete *(Lumbricus variegatus)*. Unfortunately, Raabe did not run control groups to exclude nonassociative processes such as sensitization and pseudoconditioning. Ratner and Miller's (1959*a*) experiment was the first demonstration of classical conditioning in the earthworm to include many of the appropriate control procedures. It has thus become *the* reference experiment in this area. The UCS was the onset of a bright light which elicited a slight rearing ($\frac{1}{8}$–$\frac{1}{4}$ inch) and a withdrawal ($\frac{1}{4}$–1 inch) of the anterior portion of the body. The CS was a mild vibration. CS duration was 6 sec, and UCS duration was 2 sec. CS–UCS interval was 4 sec. Any withdrawal response which occurred during the CS–UCS interval was counted as a UCR. One-hundred training trials were given followed by 30 extinction trials in which the CS only was presented (classical conditioning group). The results may be seen in Fig. 10.

Group E (the classical conditioning group) significantly increased the number of CRs as compared to various control groups. Group V was a sensitization control group which received only the vibration. Group L was a pseudoconditioning control group which received the light only for a total 70 trials and was tested with the vibration for five trials following each block of ten light-only trials. Group R was a control for the base rate of the withdrawal response without stimulation of any kind. Replications of these effects were reported by Ratner and Miller (1959*b*) and by Wyers *et al.*, (1964) Although these experiments are rather compelling demonstrations of classical conditioning, the experimental design is somewhat weakened by the lack of inclusion of a control group receiving both the CS and the UCS but in random temporal sequences. Fortunately, this defect has been corrected by an experiment by Peeke *et al.* (1967) which included this control condition and which obtained strong classical conditioning of anterior withdrawal.

Influencing Variables

Stimulus Variables. Interstimulus Interval: Although it would not be surprising to find considerable phylogenetic variation in the optimal CS–UCS intervals, the values obtained thus far for earthworms are well within the range of values obtained with mammals (*cf.* Chapter 1). Raabe (1939) obtained his best conditioning in the aquatic oligochaete *Lumbriculus* with a $\frac{1}{2}$ sec interval, some conditioning at 3 sec, and no conditioning at either 0 or

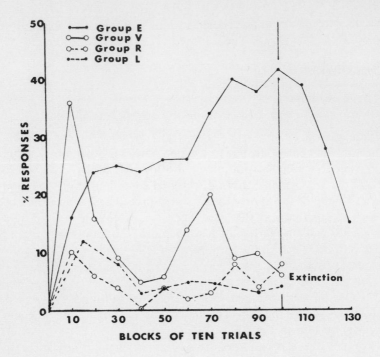

Fig. 10. Percentage of withdrawal responses as a function of experimental treatment (see text) (Ratner and Miller, 1959*a*).

4 sec. However, other experimenters have reported quite strong classical conditioning in *Lumbricus terrestris* with CS–UCS intervals of 4 sec (Ratner and Miller, 1959*a,b;* Wyers *et al.,* 1964; Herz *et al.,* 1967).

Intertrial Interval: The best evidence regarding intertrial interval effects comes from Ratner's laboratory. For the intact animal, the findings are consonant with the effects of the ITI variable in mammals; i.e., the shorter the ITI, the less adequate is the conditioning. The values which have been tested in various experiments range from 6 sec to 90 sec (Ratner, 1965). Ratner and Miller (1959*b*) found that a 90 sec ITI yielded significantly greater conditioning as measured by both percent CRs and CR latency than did an ITI of 10 sec. The effects of ITI are just the opposite for decerebrate worms; i.e., no learning occurs with spaced practice (90 sec), but substantial conditioning occurs under massed practice (6 sec) (Ratner, 1962; Ratner and Miller, 1959*b;* Ratner and Stein, 1965).

Organismic Variables. Species: There have as yet been no experiments devoted explicitly to species comparison. Indeed, the one comparison available is between *L. terrestris* (Ratner and Miller, 1959*a,b*) and *L. variegatus*

(Raabe, 1939), in which the former species was easily conditionable with a CS–US interval of 4 sec whereas no conditioning occurred in the latter. Considering the many other differences in the two experiments, it would be unwise to conjecture that this difference is a valid species difference.

Body Length: Ratner and Miller (1959b) reported that significant positive correlations exist between the length of the worm, measured in number of segments, and the number of CRs elicited during the training.

Response Characteristics. The limited response repertoire of annelids places substantial limitation on the possible responses which may be used as a UCR. Furthermore, the UCR to various stimuli is often a complex one, sometimes consisting of both forward movements and withdrawals to the same stimulus on different trials. For example, out of 120 presentations of a light-pulse UCS earthworms responded 70 times with withdrawal but 17 times with extension (Ratner, 1965). Herz *et al.* (1967) have shown that the response normally elicited by a vibratory stimulus depends on its intensity. A mild stimulus elicits a "sharp contraction which appears to habituate upon repeated presentations" (*cf.* group V, Fig. 10). A strong stimulus elicits an *extension* of the anterior portion of the worm. Herz *et al.* (1967) have conditioned this extension response and argue that it may offer some advantages over the use of withdrawal as the UCR in studies of the functional role of the supraesophageal ganglion. They point out that ganglionectomy often changes the response to photic stimulation from contraction to extension and thus makes comparison between operated-on and control subjects difficult. Ganglionectomy appears to have no similar untoward effects on the extension response to strong vibrations.

Related Phenomena

Extinction. One of the characteristic features of all conditioning is that the conditioned response declines in frequency when the CS is presented alone, unpaired with the UCS. The validity of the experiments of Herz *et al.* (1967), Ratner and Miller (1959a), Peeke *et al.* (1967), and Wyers *et al.* (1964) as demonstrations of classical conditioning is substantially strengthened by virtue of the fact that the classically conditioned subjects also exhibited extinction when the UCS was discontinued (*cf.* Figs. 10 and 11).

The Partial Reinforcement Extinction Effect (PREE). The partial reinforcement extinction effect is a robust and ubiquitous phenomenon which manifests itself in a wide variety of vertebrate species (*cf.* Chapter 1). Wyers *et al.* (1964) conditioned earthworms in the Ratner–Miller apparatus; one group had the CS paired with the UCS on every conditioning trial, while the partially reinforced group received the CS–UCS pairing on a random half of the 50 daily trials. The results are presented in Fig. 11. A strong PREE is revealed by group P making significantly more CRs in extinction than the

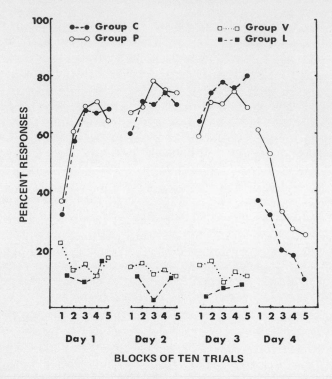

Fig. 11. Percentage of withdrawal responses as a function of experimental treatment (see text) (Wyers *et al.*, 1964).

continuously reinforced group. While these results are consistent with those from studies of the PREE in *instrumental* conditioning of vertebrates, it is not clear that a similar effect is obtained in classical conditioning of vertebrates (*cf.* Chapter 1).

Simple Retention. No experiment has been reported in which retention of a classically conditioned response was a prime variable. There are, however, two experiments which provide data relevant to retention for worms conditioned in the Ratner–Miller apparatus. Following 80 conditioning trials, Ratner and Miller (1959*b*) introduced a 20 min rest interval and then administered ten more conditioning trials. It was found that the performance of the massed practice group (ITI 10 sec) dropped precipitously after the rest from 58% to 24% (estimated from figures); the spaced practice group dropped from 70% to 56%. While additional research will be required before this effect can be regarded as well established, the Ratner and Miller data strongly suggest that at least short-term retention is positively related to the intertrial interval used during conditioning; the longer the interval (up to some unknown maximum), the better the retention.

Further evidence relevant to retention is available from the Wyers *et al.* (1964) experiment. Note in Fig. 11 that a 23 hr retention interval occurred between daily training sessions. Evaluation of the level of retention could be based on a rough "savings" estimate, from which it may be seen that by the end of the first ten trials following the 23 hr rest interval the continuously reinforced group was still about 10% below its performance at the end of the previous session. This value is roughly comparable to the 15% loss seen in the Ratner and Miller study for the 100% spaced group. It is unfortunate that these data were not analyzed to determine the relative amount retained in the first few trials during a daily session. It would not be surprising to find that there was very substantial forgetting between training sessions as judged by performance on the first one or two trials, since this sort of effect has been obtained both in habituation of polychaetes (Dyal and Hetherington, 1968) and in instrumental conditioning of earthworms (Datta, 1962). It would appear that the learning process in annelids may take the form of increasingly rapid response acquisition within a training session but with almost complete forgetting between daily sessions; this problem will be discussed further in the section on instrumental conditioning.

Backward Conditioning. Experiments by Peeke *et al.* (1967) and by Herz *et al.* (1967) have included backward conditioning groups in which UCS onset either preceded or was simultaneous with CS onset. In both experiments, the backward conditioning group showed a nonsignificant increase in responding during the conditioning phase and a marginally reliable decrease in responding during extinction. No tests have been conducted to determine if the CS acquired active inhibitory properties as a result of its negatively contingent association with the UCS (*cf.* Chapter 1).

B. Class Polychaeta (Sedentary and Errant)

Basic Demonstrations

As we have noted in the previous section, experiments by Hargitt (1906, 1909) and Yerkes (1906) were interpreted as demonstrations of classical conditioning in *H. dianthus*. However, closer examination of these experiments along with that of Krasne suggests that the results are more properly considered to be demonstrations of sensitization or perhaps simple dishabituation. Coppock and Bitterman (1955) reported that they had successfully conditioned the rapid withdrawal response of *Mercierella enigmatica* using light increase as the CS and electric shock as the UCS. Unfortunately, it is not clear from this abstract whether or not adequate sensitization controls were included in the experiment. However, it should be noted that light *increase* would seem to be the CS of choice in attempts to demonstrate classical conditioning in sedentary polychaetes, since the withdrawal response is not normally elicited by this stimulus. Although it is not possible at this time to feel confident that true classical conditioning has been demonstrated in sedentary polychaetes, the Coppock and Bitterman procedure seems promising and should be replicated.

Early experiments by Copeland (1930) and Copeland and Brown (1934) demonstrated that behavior of the errant polychaete *H. virens* could be modified so that it approached the end of its home tube to a light or touch stimulus which had preceded food. The experiments were interpreted by their authors (and by Jacobson, 1963) as demonstrating true classical conditioning despite the fact that neither of the experiments contained sensitization control groups. In the light of Evans' demonstration that Copeland's results could be obtained on the basis of the simple sensitization of approach responses by the presence of food (Evans, 1966*b,c*), we must regard Copeland's experiments as inconclusive.

The conclusion which must be drawn from the meager and methodologically inadequate research reported thus far is that classical conditioning has not been demonstrated to occur in polychaetes.

C. Hirudinea

Basic Demonstrations, Influencing Variables, and Related Phenomena

Henderson and Strong (1971) have recently reported classical conditioning in a leech. They used two levels of shock (UCS) and two intensities of light increase (CS) with paired and unpaired CS–UCS combinations. It is clear from Fig. 12 that substantial conditioning occurred in the paired groups but not in the unpaired control groups. Significant main effects were obtained for CS intensity, and UCS intensity interacted with training and extinction blocks. Both experimental extinction and spontaneous recovery after 24 hr were demonstrated. This experiment demonstrates the feasibility of exploring a wide variety of conditioning phenomena with the leech.

D. Neurophysiological Factors in Classical Conditioning

Ratner and Miller (1959*b*) removed the anterior five segments of earthworms prior to conditioning in their ring apparatus. They found that decorticate worms conditioned more slowly than normal worms and that this effect was strongest for worms trained under spaced conditions. Ratner (1962) replicated the effect and showed that there was a strong tendency for the decorticate worms trained with spaced trials to make fewer URs as well as fewer CRs. A third study confirmed this observation and showed that the normal negative response (anterior withdrawal) to light increase was substantially reduced and was replaced by a forward-going response (Ratner

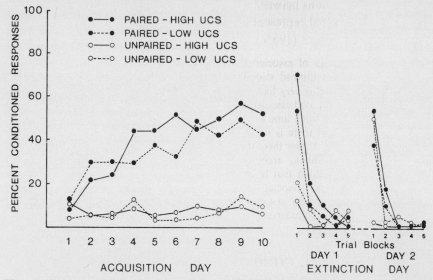

Fig. 12. Percent conditioned responses in acquisition and extinction as a function of pairing and UCS intensity (Henderson and Strong, 1971).

and Stein, 1965). Although explanation of the effects of ITI and decerebration on negative movements and forward-going tendencies is not available, it would seem that the data again implicate the supraesophageal ganglion as an inhibitor of approach responses. Furthermore, the effect is greater for worms in which five anterior segments are removed than for those subjected to more precise suprapharyngeal or subpharyngeal ganglionectomy. Thus, the debilitating effects of ganglionectomy on classical conditioning may reflect a loss of sensory equipment and a resulting changed responsiveness to light more than a loss of ability to form associations.

The investigation of neurophysiological factors in annelid conditioning has been restricted to the relatively gross ablation approach. Electrophysiological contributions to our understanding of neurophysiological activities which may be critical correlates of conditioning have been confined to higher invertebrates and vertebrates. While more complex preparations have provided conditioning analogues ranging from the trivial (Kandel, 1965, p. 687) to the compelling (Livingston, 1966), electrophysiological and neurochemical explorations of annelid conditioning processes have been notable by their absence.

E. Biological Significance of Conditioning

The adaptive advantages of being able to predict environmental contingencies are apparent and obvious in the case of complex organisms.

However, the effective utilization of this ability to form associations depends on complex interactions between sensory capacities and capacities for storage and retrieval of stored representations of previous experience. This point is well expressed by Clark (1965c) as follows:

> The great advantage of associative over non-associative learning seems to be that, since the conditioned stimulus immediately precedes the unconditioned stimulus, the animal is very likely to form associations between events that are causally related. In non-associative learning, any unconditioned stimulus received at about the same time as the conditioned stimulus may elicit the conditioned response and there is much less likelihood that the two stimuli will be related in any way. Under these circumstances there is no reason why the conditioned stimulus should accompany the unconditioned stimulus on future occasions and it follows that brief retention, which appears to be a feature of animals that learn non-associatively, is of considerable benefit. A good memory without advanced powers of discrimination and the ability of associative learning is positively disadvantageous. (p. 367)

IX. INSTRUMENTAL CONDITIONING

The instrumental conditioning paradigm involves the reinforcement of a response by a "reward" and/or the avoidance of punishment. The procedure which has been most frequently used to condition annelids involves a single-choice T-maze or Y-maze in which an aversive stimulus is encountered in one maze arm and a "neutral" or "positive" stimulus in the other arm. It will be recalled that many of the experiments previously considered under the *punishment* paradigm involved precisely the same procedures (e.g., Robinson, 1953). In the case of instrumental conditioning of annelids, the primary differences in the two paradigms are the dependent variable being measured and often the phase of training being considered. When a particular measure, such as time to traverse the maze, indicates a general inhibition of response, we are obviously dealing with a generalized effect of punishment; on the other hand, to the extent that the choice behavior of the worm is modified, we *may* be dealing with a true conditioned response selection.

A. Class Oligochaeta

Basic Demonstrations

The classic study of instrumental conditioning in oligochaetes was conducted in 1911–1912 on a single specimen of the manure worm *Allolobophora foetida* (now *Eisenia foetida*) by Robert Yerkes (1912). The task to be learned was a turning response in a T-maze. Entrance into the negative arm resulted in contact with a sandpaper strip which preceded electric shock (or contact with saline); the "positive" arm permitted entry into a wooden-tube "artificial burrow." Although the behavior was quite variable, Yerkes did observe a reduction in number of errors and running times during the more than 800 training trials. Yerkes interpreted the results to indicate that the worm was able to associate the sandpaper cue with aversive stimulation and to learn a turning response. Heck (1920) reported a series of successful replications of Yerkes' results using several species of oligochaetes (*Lumbricus castaneus, Eisenia foetida, Allolobophora caliginosa, A. longa,* and *A. chlorotica*). In Heck's experiments, no specific explicit cue preceded the aversive stimulation; thus the turning was probably based on kinesthetic cues rather than exteroceptive cues.

Swartz (1929) trained *Helodrilus caliginosis, H. foetidus,* and *H. parvus* in a Y-maze. Following a free-choice baseline period which was used to establish preference shock was introduced in the preferred arm of the maze. According to Swartz, *H. caliginosis* reduced its percentage of responses to the shocked side by a mean of 14.2%, whereas *H. foetidus* and *H. parvus* slightly increased their preference for the shocked side (9.3%). Swartz interprets the results on *H. caliginosis* as reflecting "habit formation," but this is quite in error since she failed to note that the nonshocked control animals also showed a comparable shift.[5]

The most extensive experiments on instrumental conditioning in the earthworm were reported by Datta (1962). She trained *L. terrestris* in a T-maze with shock (15 ma) as the aversive stimulus in the negative arm; the positive arm led to the home containers. If the worms did not leave the stem, they were activated by bright light, heat, and shock. The correction method was used. Worms were trained five trials per day for 15 days with an ITI of 5 min. Three salient features of the results are represented in Figs. 13, 14, and 15. Figure 13 presents the percent of times that the worm's initial choice on a given trial was correct (i.e., the percent of errorless trials). This measure approaches an asymptotic value of around 70% errorless trials by day 5. While such a curve would ordinarily be taken to reflect improved performance over the training days, more detailed analyses presented in Figs. 14 and 15 reveal that such an interpretation would be incorrect. Figure 14 shows that "the between-sessions improvement may be accounted for entirely in terms of change of rate of within-session improvements" (Datta, 1962, p. 538). The absence of *between-session* improvement in number of correct choices is further demonstrated by the lack of improvement in correct choice on the *first* daily trial (Fig. 15). The number of repetitive errors per

[5]Matching experimental and control subjects on the bases of total number of trials and dichotomizing the distribution for the control animals at the point when shock was introduced for the experimental animals reveal that the unshocked control animals reduced their preference by 16.6%.

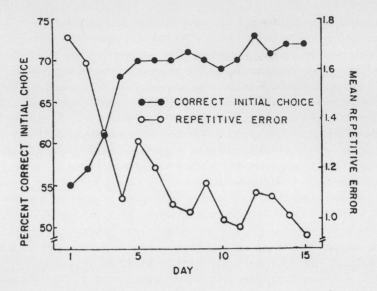

Fig. 13. Mean probability of correct initial choice and mean repetitive error as a function of days of training (Datta, 1962).

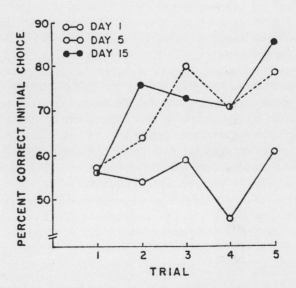

Fig. 14. Mean probability of correct initial choice as a function of trials on the first, fifth, and fifteenth days of training (Datta, 1962).

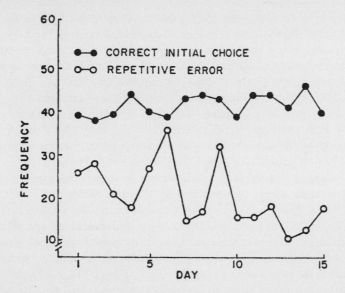

Fig. 15. Frequency of correct initial choice and of repetitive error on trial 1 as a function of days of training (Datta, 1962).

trial is reduced by half over the course of training both in the case of all trials (Fig. 13) and when the analysis is restricted to the first daily trial (Fig. 15). This latter fact represents *between-session* improvement in the reduction of repetitive errors and suggests that the two response measures (percent correct and number of repetitive errors) may reflect the operation of two different learning processes. Specifically, the reduction of repetitive errors may reflect an aversive inhibitory conditioning process, whereas the increase in percent initial correct responses may indicate the development of choice learning based on aversive and/or appetitive reinforcement.

Other experiments demonstrating choice maze learning by earthworms have been reported by Fraser (1958), Schmidt (1955), and Zellner (1966).

Ratner (1964) has criticized most of the experiments which have utilized single-choice mazes in that the experimenter has often used a weak tactile stimulus as a motivator to leave the stem and has discouraged reversing by "the use of a long probe applied to the point of maximum effectiveness" (Robinson, 1953). Ratner contends that in many of these experiments "the results that were interpreted as learning the maze by worms may be accounted for as *learning by the experimenter to guide the movements of a flexible object, the worm*" (p. 31). While Ratner is correct that it *is possible* to guide a worm through a maze, the implication that all or even most of the results obtained in a T-maze may be accounted for by such experimenter bias seems to be rather gratuitous, Rosenthal's (1966) cautions to the contrary notwithstanding. Nonetheless, being persuaded that the T-maze is not the apparatus of choice, Ratner and his students have conducted several experiments in straight-alley runways (Ratner, 1964; Reynierse and Ratner, 1964). They obtained a "home-cage-reinforcer effect," but three later attempts to replicate this effect have been unsuccessful (Kirk and Thompson, 1967; Reynierse *et al.*, 1968). Furthermore, any reliable effects which might have been present were probably highly confounded with nonassociative factors such as stimulus change and substratum height (Reynierse *et al.*, 1968).

In addition to instrumental conditioning of approach responses in T-mazes and straight alleys, two experiments have provided the opportunity for escape learning and active avoidance conditioning. Kirk and Thompson (1967) trained *L. terrestris* in a straight alley and evaluated the effects of runway lighting conditions, plain or moss-filled goal boxes, and shock or no shock in the start box. In the escape/avoidance condition, shock was administered if the worm had not left the starting area 30 sec after the opening of the start-box door. Since no significant decreases in locomotor speeds occurred over trials for the shock groups, Kirk and Thompson concluded that "there is no evidence for escape or avoidance learning in any of the

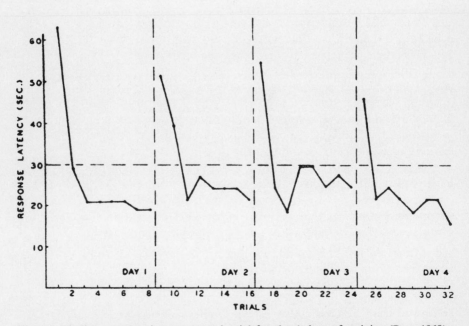

Fig. 16. Median response latency on each trial for the 4 days of training (Ray, 1968).

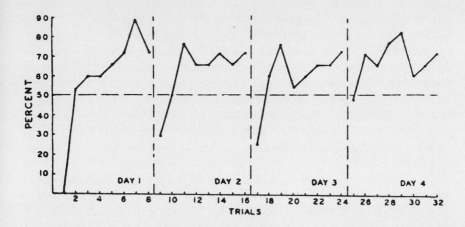

Fig. 17. Percent avoidance responses on each trial for the 4 days of training (Ray, 1968).

shocked groups." On the other hand, Ray (1968) has reported active avoidance conditioning in *L. terrestris*. The CS was a gentle vibration, and the aversive stimulus was the onset of a 3200 ft-c light. The CS–UCS interval was 30 sec; eight trials were given every 24 hr with an ITI of 3 min. The worm could avoid the light by flowing underneath the ledge to a moist darkened area within the 30 sec CS–UCS interval; a similar response after the light onset permitted escape from the light. The median response latency and the percent avoidance responses are presented in Figs. 16 and 17. Within each daily training session, there were statistically reliable decreases in response latency and significant increases in percent avoidance responses to a level of about 70–75% successful avoidances. While Ray's results appear to warrant the assertion that successful active avoidance conditioning has been demonstrated, it is unfortunate that he did not include a sensitization control group which received vibration and the light but in a random unpaired series in order to clearly exclude the possibility that the light was simply sensitizing responses to the vibration.[6] It is clear that Ray's active avoidance procedures deserve further experimental efforts.

Influencing Variables

Stimulus Variables. Distinctive Visual/Tactile Stimuli: Although Yerkes had interpreted his data to indicate that his worm was able to make use of sandpaper prior to shock as an effective warning stimulus, subsequent

[6]Ray included control groups which received CS only and UCS only in early pilot studies; he reported that the worms quickly habituated to CS only but did not indicate what happened to the UCS-only group (personal communication).

research has failed to support that contention in the case of *L. terrestris* (Datta, 1962; Robinson, 1953; Schmidt, 1955; Zellner, 1966).

Intertrial Interval: The only study which has systematically investigated ITI revealed that instrumental conditioning of *L. terrestris* is strongly influenced by this parameter (Datta, 1962). Both 1 min and 5 min ITIs resulted in significant increase in percent correct on choice, but this between-session increase was entirely due to increases in the within-session rate. ITIs of 25 min or 125 min resulted in no significant increase in performance either between sessions or within sessions.

Trials per Day: Yerkes (1912) compared 5, 10, 15, and 20 trials per day and from inspection of the data concluded that "The results of training on the basis of 5 trials per day, or every other day, are more satisfactory than those obtained with 10, 15 or 20 trials per day" (p. 351). It is probable that giving a large number of trials per day involves both fatigue and habituation effects.

Temperature and Temperature Change: Reynierse (1968) studied the effect of maintenance (intersession) temperature, intertrial interval temperature, and temperature during the training trial on locomotion in a straight alley. His results are consistent with those obtained by Herz *et al.* (1964) in classical conditioning of the earthworm, and together they suggest that the most frequently used procedure (maintaining the subjects under refrigeration and testing them at room temperature) is probably the least satisfactory of the possible procedures and may account for low levels of conditioning in many of these studies.

Organismic Variables. Diurnal Cycle: Earthworms exhibit a definite diurnal activity cycle, with the most active period occurring between 6 PM and midnight (Baldwin, 1917). Arbit (1957) showed that activity cycle significantly influenced the speed with which earthworms learned to avoid the shocked arm of a T-maze; worms trained during the peak activity period (8 PM to 12 PM) learned significantly more rapidly than those trained from 8 AM to 12 AM. Similarly, Ray reported that worms trained in the morning never learned to criterion in his active avoidance task (personal communication, 1971).

Species: It will be recalled that Yerkes (1912) had found that *Eisenia foetida* seemed to react to a sandpaper strip designed to serve as a warning stimulus immediately preceding electric shock, whereas Robinson (1953) found no such tendency in *L. terrestris*. Schmidt (1955) showed that although these two species did not differ in the number of trials required to learn a T-maze to criterion, there were important species differences in behavior in the maze. *L. terrestris* gave generalized avoidance responses to the maze, whereas *E. foetida* was quite specific in its avoidance of the sandpaper. These results by Schmidt should serve to caution against overgeneralization of results across species.

Response Characteristics. Even in a relatively simple task such as a single-choice maze, the behavior exhibited is quite complex and is not necessarily unitary in either its manifestation or its mechanism. In Zellner's (1966) experiment, he measured seven aspects of the worm's behavior on each trial: correct choice, error, number of shocks, negative movements before the choice points, negative movements at the choice points, number of contacts with the sandpaper, and number of withdrawals from the sandpaper. Of these measures, only the number of correct choices consistently and reliably reflected the effects of training over all three of the experiments. However, in one of the experiments there was a significant decrease in negative movements at the choice point. Both of these measures presumably reflect some process of instrumental conditioning. The fact that the measures which most directly reflected the effect of the noxious stimulus (number of shocks, number of contacts with the "warning" sandpaper, and number of withdrawals from the sandpaper) did *not* reliably decrease with training suggests that the learning of the correct choice was *not* based on learning to *avoid* the noxious arm. The experiment probably represents a demonstration of instrumental escape conditioning based on escape from a bright light, since the stem and negative arm were illuminated to 80 ft-c while the correct goal box was shaded.

Related Phenomena

Latent Learning. Thorpe (1956) has held latent learning (*cf.* Chapter 1) to be a type of insight learning present in higher invertebrates (Hymenoptera) and other animals "on a relatively high behavioral level" (p. 397). Bharucha-Reid (1956, 1961) reported that he had demonstrated latent learning in *L. terrestris.* However, his experimental design did not include a proper control group to exclude the possibility that nonassociative factors accounted for the differences in performance. Such a control group was run by Arbit (1961). Although his procedures did not replicate those of Bharucha-Reid, they did provide a conceptually valid test of the existence of latent learning in the earthworm, and no evidence of latent learning was obtained. James and Woodruff (1965) also failed in their attempt to demonstrate latent learning of a T-maze. We may conclude that latent learning in earthworms has not been convincingly demonstrated as a reliable phenomenon.

Extinction. Zellner (1966) demonstrated both acquisition and extinction of a turning response in a T-maze. The extinction process was quite rapid, being essentially complete for a normal (not operated on) control group at the end of the first ten extinction trials. Extinction of a locomotor response has also been studied in a straight-alley runway (Ratner, 1964; Reynierse and Ratner, 1964; Reynierse *et al.,* 1968). The experiment by Reynierse *et al.* strongly suggests that the apparent extinction effects obtained by Reynierse and Ratner were due primarily to nonassociative factors (stimulus change and lowered goal box substrate) rather than to elimination of the "reinforcement" characteristics of a moss-filled goal box.

Darwin (1898) made extensive observations on the tendency of worms to drag leaves into their burrows to plug the entrance of the burrow. Malek (1927) describes an aspect of this process which may be analogous to the extinction of an instrumental response. He reports that an earthworm *(L. herculeus)* gave up on its attempts to pull an immovable leaf into its burrow after about a dozen such attempts.

Retention. The most extensive research on the retention of an instrumental response has been reported by Datta (1962). Her investigation of the effects of intertrial interval on acquisition of a turning response is of course directly relevant to the problem of retention. It will be recalled that with intertrial intervals of 1 or 5 min there occurred significant increase in the percent correct choice; however, with intertrial intervals of 25 or 125 min there was no evidence of correct response acquisition. As Datta observes, "the change in the earthworm which is responsible for the increase and the probability of correct initial choice is quite unstable, lasting no longer than 25 min" (p. 543). On the other hand, number of repetitive errors is not influenced by increasing ITIs up to 2 hr. It would appear that the mechanism which mediates learning not to repeat errors (memory of shock?) persists longer than the mechanism which mediates correct choice (memory of reward?).

In addition to the ITI experiment, Datta conducted two other experiments focused on the problem of retention. The first of these involved a 2 × 3 factorial design consisting of two levels of original training (5 and 15 days) and three retention intervals (1, 5, or 15 days). Worms were given five trials per day with an ITI of 5 min during original training. Following the retention period, they were tested for retention by giving 5 days more of training. Unfortunately, no savings scores were reported, but some indication of relative retention may be obtained from Fig. 18. In this figure, each of the points on the relearning curve represents the mean percent of correct initial choices over the 25 relearning trials (five trials per day). It may be seen that both simple retention and proactive inhibition (discrimination reversal) tests indicate considerable retention up to 5 days but zero retention after 15 days. These results are generally consistent with those of Krivanek (1956), who found some evidence of retention of aversive inhibitory conditioning after 3 days (his longest in retention interval). Yerkes, on the other hand, reported that his worm exhibited perfect retention over a 3 week period following over 500 training trials. A more detailed discussion of these retention data may be found in Dyal (1972).

Learning to Learn and Overlearning Reversal Effects. Datta obtained an apparent improvement in discrimination reversal over a series of reversals.

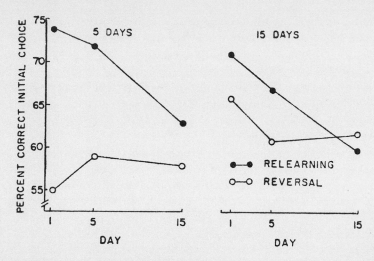

Fig. 18. Retention as measured by relearning and reversal after 5 and 15 days of training (Datta, 1962).

In view of the difficulty in obtaining this effect in other species (*cf.* Chapter 1), Datta quite rightly viewed this surprising effect with a degree of suspicion and conducted a further experiment to determine if the worms were actually learning to reverse *as a result of experience with reversals* or simply as a result of *more training per se*. She trained a control group of animals for 40 days and then reversed them for the first time. The performance of this group was essentially the same as that of the reversal animals after four reversals. It may thus be concluded that the apparent "learning to learn" exhibited by the worms is in fact a manifestation of the number of training trials *per se* and not a true learning set phenomenon.

Another point of interest in the data represented in Fig. 18 is the reversal performance of the 5 and 15 day training groups after 1 day of retention. The fact that the group trained for 15 days performed better than the group trained for 5 days is suggestive of an *overlearning reversal effect* (*cf.* Chapter 1). Unfortunately, Datta did not indicate whether those two points are significantly different.

Spontaneous Alternation. Spontaneous alternation of choices at a choice point has been commonly reported for rats and other higher vertebrates (Dember and Fowler, 1958). Similar effects have been reported by Wayner and Zellner (1958) for *L. terrestris*. Iwahara and Fujita (1965) repeated Wayner and Zellner's experiment using the common Japanese earthworm *Pheretima communissima* and varying the intertrial interval at 0 min, 10 min, and 24 hr. They, like Arbit and McLean (1959), found no significant alternation in any of the subjects. Fraser (1958) reported that *A. terrestris longa* alternated reliably more than chance, whereas *L. rubellus* alternated significantly less often than chance. Further research on species differences in alternation tendencies would seem to be indicated in that these tendencies may be relevant to possible species differences in instrumental conditioning.

B. Class Polychaeta (Errant)

Basic Demonstrations

Fischel (1933) trained *N. virens* in a Y-maze with a darkened area as the positive goal and an aversive sea anemone as the negative goal. The worms learned the discrimination quite rapidly. Evans (1963*b*) has trained *N. virens, N. diversicolor,* and *P. cultrifera* in a T-maze with large darkened chambers and obtained evidence of instrumental conditioning of a turning response.

On a given training trial, the worm was shocked if it made an incorrect choice and immediately removed from the maze with a small paintbrush. A correct choice was not punished, and the worm was allowed to enter the darkened chamber, where it remained for at least 5 min. All worms reached a criterion of learning of ten successive correct trials within 70–85 trials. (*N. virens* mean 68, *P. cultrifera* and *N. diversicolor* means 85). The performance of a typical worm involved three distinct phases (Fig. 19). Phase 1 was characterized by random choice and lasted about 50 trials. Phase 2 was the incremental learning phase, while phase 3 was the stable asymptotic performance during which very few errors were made. Flint (1965), using the

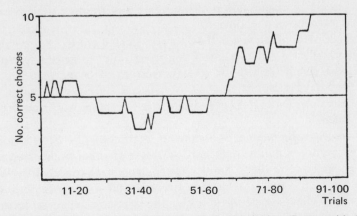

Fig. 19. The performance of an individual *Nereis diversicolor* in a T-maze with darkened chambers. The data points represent sliding means of the previous ten trials (Evans, 1963*b*).

same apparatus and procedure as Evans, confirmed this three-phase curve with 33 specimens of *N. diversicolor*.

Influencing Variables

Stimulus Variables. Presence or Absence of a Darkened Chamber: It would appear that whether or not the worm is given an opportunity to enter and remain in a darkened chamber at the end of a correct run is critical in determining whether stable instrumental choice behavior will be obtained. We noted in our previous discussion of aversive inhibitory conditioning that if such a chamber is not available the worms tend to refuse to run following a series of trials in which correct choices first increase and then decrease. On the other hand, if such a chamber is available, the worms reach a high and stable level of correct choice (Fig. 19).

Related Phenomena

Retention. No definitive statements can as yet be made regarding retention of instrumental conditioning in nereid polychaetes. However, Evans (1963b) presents some intriguing data on one specimen of *N. virens* which suggest that rest periods do not change performance when they are introduced prior to learning (phase 1) or after the learning criterion has been attained (phase 3). However, introduction of a rest period during phase 2 results in substantial performance decrement. These results could be interpreted to say that once a stable habit is formed it is relatively resistant to disruption by the rest interval but during the labile learning period (phase 2) a retention interval results in considerable forgetting. These provocative results demand replication and further elaboration.

C. Sedentary Polychaetes and Hirudinea

The tubiculous habits of sedentary polychaetes tend to contraindicate their use as subjects in instrumental conditioning. Leeches, however, would seem to be ideal subjects for research in this area, but thus far no experiments have been reported.

D. Neurophysiological Factors in Instrumental Conditioning

Effects of Ganglionectomy on Subsequent Learning and Performance

Heck (1920) removed the supraesophageal ganglion from two specimens of *L. castaneus* and then trained them in a turning response in a Yerkes-type T-maze. He concluded that ganglionectomy did not interfere with learning of the response. Zellner (1966) confirmed these results in the case of *L. terrestris*. The conclusion is quite clear: *the supraesophageal ganglion is not necessary for learning to occur in oligochaetes*.

The evidence is quite the opposite in the case of nereid polychaetes. Evans (1963*b*) found that decerebrate nereids *(N. virens)* were unable to learn in his T-maze with darkened chambers. Flint (1965) also found that decerebrate nereids *(N. diversicolor)* were unable to learn in Evans' T-maze. However, before it can be concluded that ganglionectomy prevents instrumental conditioning of polychaetes the possible confounding effects of two other factors should be noted. First, it is apparently necessary to use *spaced* trials (5 min or more in the darkened area) in order to get learning by *intact* worms in this situation (Clark, 1965*b*), yet we know that in other learning situations *massed* training is necessary in order for decerebrate worms to learn either an inhibitory avoidance response (Evans, 1963*a*) or classical conditioning (Ratner and Miller, 1959*a*). Second, since ganglionectomy reduces the negative phototaxis of earthworms (Prosser, 1934), it may be that entry into a darkened chamber no longer serves as an effective positive reinforcer (indeed, Evans noted that some of the decerebrate worms crawled out of the darkened chamber into the lighted arm). The proper interpretation of the observed differences between ganglionectomized polychaetes and oligochaetes awaits the development of a suitable massed training technique which does not rely on negative phototaxis to motivate locomotion.[7]

It will be shown later that the effect of decerebration on the *retention* of an *already learned* turning response may be at least in part due to loss of information from anterior sense organs rather than inability to retrieve information which was stored in the supraesophageal ganglion. However, such an explanation is *not* applicable to the present problem of the inability of decerebrate worms to learn the maze, the reason being that Flint (1965) has clearly demonstrated that a *N. diversicolor* which has had *all of its anterior. sense organs removed but has its supraesophageal ganglion intact* can learn the maze actually somewhat *better* than normal, fully intact animals!

Effects of Ganglionectomy on Performance of a Previously Learned Habit

Heck (1920) used a series of nonreinforced (extinction) trials to test for the effects of ganglionectomy on two specimens of *A. chlorotica*. He found them to be highly persistent in extinction (e.g., one of them made 60 consecutive correct turns). Zellner (1966) also utilized an extinction test to measure retention following ganglionectomy. He confirmed Heck's finding that decerebrate animals were significantly more persistent in correct turns than sham-operated controls. While increased persistence is apparently a reliable behavioral effect, it should not be interpreted to imply that ganglionectomy necessarily increases retention in earthworms. Several nonassociative factors

[7]It would seem that a T-maze having a weak activating shock in the stem plus a favored food incentive in the positive arm might provide a suitable behavioral assay.

are probably involved; for example, if the supraesophageal ganglion is an inhibitory center, its removal would be expected to prolong behavioral effects which require response inhibition for their manifestation (e.g., extinction). Furthermore, Zellner found that the decerebration resulted in significantly more withdrawals from the sandpaper and more correct choices following contact with the sandpaper. These results suggest that decerebrate worms tend to react more negatively to the sandpaper, a nonassociative factor which would have contributed to the increased persistence of correct responses.

In all three experiments (learning, extinction, and regeneration), Zellner found that decerebration resulted in a dramatic and reliable reduction in the number of negative movements in the stem of the maze but, surprisingly, no change in the number of negative movements in the choice area. Similar reductions of negative movements in the stem have been observed for ganglionectomized worms in tests of alternation behavior (Wayner and Zellner, 1958). "Evidently negative movements in the maze stem represent a kind of behavior that is mediated primarily by the suprapharyngeal ganglion, while negative movements at the choice point are a different kind of behavior in which the ganglion plays little or no part" (Zellner, 1966, p. 15). The above evidence seems to support the conclusion that in the case of *L. terrestris* retention of a previously learned turning response is not dependent on the supraesophageal ganglion.

Again, a quite different picture is apparent in the case of nereid polychaetes. Evans (1963b) trained a group of *N. virens* in his dark-chambered T-maze and then extirpated the ganglion. He retested the worms 2 hr later and found no evidence of any retention of the previously learned turning response. This effect could be due to several factors, including (a) inability to retrieve previously stored information (i.e., a real memory loss) or (b) a sensory deficit which eliminated cues that had been conditioned to the correct response. Evans pointed out that his operation involved cutting the nerves from the sensory receptors in the prostomium, including the antennae, palps, and tentacular cirri.

Flint (1965) attempted to resolve these alternative interpretations in a series of experiments. She found that the removal of a single set of sense organs such as the cirri, the palps, or the antennae (Fig. 2) had no effect on the performance of trained worms. However, removal of all of the anterior sense organs resulted in completely random choice behavior. Thus it is clear that deprivation of sensory input can disrupt a well-learned response, and these results support the view that previous ganglionectomy effects have been in some substantial way confounded by a sensory loss. An even more convincing demonstration was performed by Flint in which she cut the circumesophageal connectives just anterior to the nerves of the tentacular cirri.

This operation disconnects the supraesophageal ganglion and information from the eyes, palps, and antennae but leaves intact information input from the cirri. Half the worms showed high retention (75% or more correct choices), whereas the other half showed completely random behavior.[8] The conclusion is obvious: *the supraesophageal ganglion is not necessary for continued correct performance of a turning habit as long as information from the cirri is available.* The performance decrement often associated with decerebration in nereids is probably due, in large part, to the loss of sensory information rather than to loss of the critical memory store.

We are again confronted with possible differences in the role of the supraesophageal ganglion between oligochaetes and polychaetes. Ganglionectomy (or deafferentiation) disrupts performance of polychaetes but does not disrupt performance of oligochaetes. However, several other questions would need to be answered before we could conclude that the "brain" plays a different role in the two classes of annelids. For example, perhaps the difference is due to differential roles played by sense organs in the learning of the maze; maybe oligochaetes rely more on kinesthetic cues and polychaetes more on information from tentacular cirri. Furthermore, some of the behavioral responses may be quite species specific, and thus the differential results do not necessarily imply a differential role for the supraesophageal ganglion. For example, Iwahara and Fugita (1965) observed no identifiable negative responses in the stem of a T-maze in *P. communissima,* whereas for *L. terrestris* such negative movements are a dominant part of the behavior (Wayner and Zellner, 1958; Zellner, 1965); likewise, *L. terrestris* shows generalized negative reactions to the runway, whereas *E. foetida* does not (Yerkes, 1912; Schmidt, 1955).

E. Biological Significance of Instrumental Conditioning

It is clear that annelids can learn to "choose" the least aversive of two arms of a T-maze. This fact is of considerable importance to psychologists, since it permits them to engage in quandaries as to the most meaningful conceptualization of this "fact." (Does it represent "classical aversive conditioning" or "instrumental avoidance conditioning" or both?) Regardless of the eventual conceptual resolution, the existence of such learning may prove to be of much more significance to the animal behaviorist than to the worm. Indeed, there is no evidence available that any worm has actually used this capacity to effect a more adequate adaptation to his environment. While it may well be that worms *can* learn associatively, the demands of their normal environment seldom, if ever, require that they do so.

---

[8]It is interesting to note that the animals which showed the best retention also required significantly longer to learn the original task.

X. CONCLUSIONS AND DIRECTIONS FOR FUTURE RESEARCH

A detailed summary of the conclusions regarding *habituation* in annelids is given in the appropriate section of this chapter. Suffice it to say at this point that behavioral habituation occurs in all three classes of annelids but there are many gaps in our knowledge within each of the classes regarding the fundamental parameters which define habituation. For example: Does the rate of habituation vary with rate of stimulus presentation and how does this vary among the major classes? Species? Does retention of habituation really vary as widely between leeches and oligochaetes as present data would suggest (1 hr *vs.* 96 hr)? How do various habituation processes interact? Do proactive and retroactive effects interact with retention interval? What is the neurophysiological basis of habituation and of potentiation of habituation?

Sensitization processes have been studied most in nereid polychaetes; little information is available for oligochaetes, sedentary polychaetes, and leeches. Perhaps the recognition of sensitization as an important behavioral modification in its own right will encourage more intensive exploration of this important phenomenon. These efforts could profitably be directed to a comparative analysis of the four types of sensitization in "representative" species of each annelid class. In the light of the proposed importance of sensitization processes for the behavioral adaptation of the annelids, it would seem that concentrated research on this phenomenon by future worm runners would result in a high payoff.

The use of aversive stimuli to explore *aversive inhibitory conditioning* is likely to provide a most powerful paradigm for the investigation of rudimentary associative learning processes. It is difficult to accept that the capacity of annelids to modify their behavior via such learning is as limited as the existing data might suggest. Although response inhibition based on punishment may require many more trials to learn than sensitization processes, the possibility that inhibitory conditioning may have a longer retention period makes the process of potential importance to worm and worm runner alike. Future research devoted to studies of retention of classical inhibitory conditioning is clearly necessary.

Classical excitatory conditioning has been adequately demonstrated in oligochaetes and leeches, but a convincing demonstration in sedentary or errant polychaetes is still lacking.

The availability of a reliable conditioning procedure opens up the exciting possibility of intensive exploration of a variety of phenomena related to classical conditioning. Some related phenomena such as extinction and the PREE have been demonstrated, but the many parameters which control these effects in mammals remain uninvestigated in annelids. Similarly, no experiment devoted explicitly to the study of retention of classical conditioning has yet been reported, and the host of associated phenomena such as latent inhibition provide tempting areas for future exploration.

The large number of *single choice learning* experiments utilizing aversive conditioning that have been conducted with oligochaetes yield the conclusion that earthworms are capable of discrimination learning but that the primary effect may be a within-session as opposed to a between-session variable. Differential conditioning based on aversive stimulation (*cf.* Table IV, Chapter 1) has been reported in only one experiment, and escape conditioning has not yet been demonstrated. A few acquisition parameters (ITI, trial block length, temperature) have begun to be investigated, and a start has been made on the description of related extinction and retention phenomena.

With the exception of the Bristol group, the investigation of instrumental learning in polychaetes has been largely ignored, and research on instrumental learning in Hirudinea has been notable by its absence. It is also striking that no successful techniques of instrumental learning based on appetitive reinforcement have been forthcoming. It would seem that both polychaetes and leeches would be likely candidates for successful attempts in this critically important area.

Investigation of *neurophysiological processes* in behavior modification of annelids has been primarily limited to the use of ganglionectomy. From these studies, certain conclusions seem warranted regarding the role of the supraesophageal ganglion in learning and retention processes in oligochaetes and polychaetes:

1. The supraesophageal ganglion is not necessary for habituation to occur.
2. The ganglion is not necessary for sensitization but is required for retention (polychaetes).
3. No evidence is available concerning its role in learning or retention of aversive inhibitory conditioning.
4. The apparent debilitating effects of ganglionectomy on classical conditioning may reflect a loss of sensory equipment and a changed responsiveness to light rather than a loss of ability to form association (oligochaetes).
5. An intact ganglion is not necessary for learning or retention of choice learning in oligochaetes, but the situation is less clear-cut in the case of polychaetes, where the ganglion may be necessary for choice learning to occur.

In contrast to other invertebrate preparations, electrophysiological and psychopharmacological investigations of learning and retention in annelids have been meager or totally absent. Future research will most certainly involve the use of annelid preparations for the exploration of the effects of protein synthesis inhibitors and facilitators. Furthermore, the special ability of some annelids to regenerate their brains makes them attractive subjects for brain transplants and collateral attempts at interanimal transfer of acquired information.

Habituation and sensitization processes are probably quite significant in the many adjustments which annelids must make in their normal habitats, but the degree to which their environments demand the utilization of higher forms of associative learning is still unknown. Future research and theoretical speculation might profitably be directed to this question.

REFERENCES

Arbit, J., 1957. Diurnal cycles and learning in earthworms. *Science, 126,* 654–655.
Arbit, J., 1961. A failure to confirm "latent learning in earthworms." *Worm Runner's Digest, 3,* 129–134.
Arbit, J., and McLean, J. P., 1959. The spatial gradient of alternation and reactive inhibition in the earthworm. Paper read at the annual meeting of the Illinois Academy of Science, Chicago.
Baldwin, F. M., 1917. Diurnal activity of the earthworm. *J. Anim. Behav., 7,* 187–190.
Barnes, R. D., 1966. *Invertebrate Zoology,* Saunders, Philadelphia.
Belenkov, N. Yu., and Chirkov, V. D., 1969. Nonsynaptic (ephatic) factors in the function of cortical neurones. *Zh. Vyssh. Nerv. Deyatel. I. P. Pavlova, 19,* 1033–1043.
Bharucha-Reid, R. P., 1956. Latent learning in earthworms. *Science, 123,* 222.
Bharucha-Reid, R. P., 1961. Confirmation or refutation of latent learning in earthworms? *Worm Runner's Digest, 3,* 179–183.
*Bohn, G., 1902. Contribution à la psychologie des annelides. *Bull. Inst. Gen. Psychol., 2,* 317–325.
Breland, K., and Breland, M., 1961. The misbehavior of organisms. *Am. Psychologist, 16,* 681–684.
Bullock, T. H., 1948. Physiological mapping of giant nerve fibres systems in polychaete annelids. *Physiol. Comp. Occol., 1,* 1–14.
Bullock, T. H., and Horridge, G. A., 1965. *Structure and Function in the Nervous Systems of Invertebrates,* Freeman, San Francisco.
Bullock, T. H., and Quarton, G. C., 1966. Simple systems for the study of learning mechanisms. *Neurosci. Res. Progr. Bull., 4,* 105–233.
Caldwell, W. E., and Kailan, H., 1955. An investigation of the role of exteroceptive motivation in the behaviour of the earthworm. *J. Psychol., 40,* 133–144.
*Clark, R. B., 1959a. The neurosecretory system of the supraoesophageal ganglion of *Nephtys* (Annelida: Polychaeta). *Zool. Jb. (Physiol), 68,* 395–424.
*Clark, R. B., 1959b. The tubiculous habit and the fighting reactions of the polychaete *Nereis pelagica. Anim. Behav., 7,* 85–90.
Clark, R. B., 1960a. Habituation of the polychaete *Nereis* to sudden stimuli. I. General properties of the habituation process. *Anim. Behav., 8.* 82–91.
Clark, R. B., 1960b. Habituation of the polychaete *Nereis* to sudden stimuli. II. The biological significance of habituation. *Anim. Behav., 8,* 92–103.

---

*While these references are not cited in this chapter, they are quite relevant to the more general topic of the behavior of annelids and are discussed in my more extended treatment of the topic (Dyal, J. A., Behavior of Annelids, *University of Waterloo Research Reports in Psychology,* Report No. 41, December 15, 1972).

*Clark. R. B., 1965a. Endocrinology and the reproductive biology of polychaetes. In Barnes, H. (ed.), *Oceanography and Marine Biology, an Annual Review,* Vol. 3, Allen and Unwin, London, pp. 211–255.

Clark, R. B., 1965b. The learning abilities of nereid polychaetes and the role of the supraoesophageal ganglion. *Anim. Behav. Suppl., 1,* 89–100.

Clark, R. B., 1965c. The integrative action of a worm's brain. *Symp. Soc. Exptl. Biol., 20,* 345–379.

Clark, R. B., 1969. Systematics and phylogeny: Annelida, Echiura and Sipuncula. In *Chemical Zoology,* Vol. 4, Academic Press, New York.

Copeland, M., 1930. An apparent conditioned response in *Nereis virens. J. Comp. Psychol., 10,* 339–354.

Copeland, M., and Brown, F. A., Jr., 1934. Modification of behavior in *Nereis virens. Biol. Bull. 67,* 356–364.

*Copeland, M., and Wieman, H. L., 1924. The chemical sense and feeding behaviour of *Nereis virens* Sars. *Biol. Bull. Woods Hole, 157,* 231–238.

Coppock, W., and Bitterman, M. E., 1955. Learning in two marine annelids. *Am. Psychologist, 10,* 501 (Abst.).

Darwin, C., 1898. *The Formation of Vegetable Mould Through the Action of Worms: With Observations on Their Habits,* D. Appleton and Co., New York.

Datta, L. -E., 1962. Some experiments on learning in the earthworm, *Lumbricus terrestris. Am. J. Psychol., 75,* 531–553.

Dember, N. D., and Fowler, H., 1958. Spontaneous alternation of behavior. *Psychol. Bull., 55,* 412–428.

Drees, O., 1952. Untersuchungen über die angeborenen Verhaltensweisen bei Springspinnen (Salticidae). *Z. Tierpsychol., 9,* 169–207.

*Durchon, M., 1962. Neurosecretion and hormonal control of reproduction in Annelida. *Gen. Comp. Endocrinol., 1,* 227–240 (Suppl.).

Dyal, J. A., 1972. Behavior of annelids. *University of Waterloo Research Reports in Psychology,* Report No. 41, December 15, 1972.

Dyal, J. A., and Hetherington, K., 1968. Habituation in the polychaete, *Hesperonoë adventor. Psychon. Sci., 13,* 263–264.

Eckert, R., 1963. Electrical interaction of paired ganglion cells in the leech. *J. Gen. Psychol., 46,* 573–587.

Eikmanns, E. K., 1955. Verhaltensphysiologische Untersuchungen über den Beutefang und das Bewegungssehen der Erdkröte (*Bufo bufo L.*). *Z. Tierpsychol., 12,* 229–253.

Evans, S. M., 1963a. The effect of brain extirpation on learning and retention in nereid polychaetes. *Anim. Behav., 11,* 172–178.

Evans, S. M., 1963b. Behaviour of the polychaete *Nereis* in T-mazes. *Anim. Behav., 11,* 379–392.

Evans, S. M., 1965a. Learning in the polychaete *Nereis. Nature, 207,* 1420.

Evans, S. M., 1965b. Learning in nereid polychaetes. Thesis, University of Bristol.

Evans, S. M., 1966a. Non-associative avoidance learning in nereid polychaetes. *Anim. Behav., 14,* 102–106.

Evans, S. M., 1966b. Non-associative behavioural modifications in the polychaete *Nereis diversicolor. Anim. Behav., 14,* 107–112.

Evans, S. M., 1966c. Non-associative behavioural modifications in nereid polychaetes. *Nature, 211,* 945–948.

Evans, S. M., 1969a. Habituation of the withdrawal response in nereid polychaetes. I. The habituation process in *Nereis diversicolor. Biol. Bull., 137,* 95–104.

Evans, S. M., 1969b. Habituation of the withdrawal response in nereid polychaetes. 2. Rates of habituation in intact and decerebrate worms. *Biol. Bull., 137,* 105–117.

Fischel, W., 1933. Über bewahrende und wirkende Gedachtnisleistung. *Biol. Zbl., 53,* 449–471.

Flint, P., 1965. The effect of sensory deprivation on the behaviour of the polychaete *Nereis* in T-mazes. *Anim. Behav., 13,* 187–193.

Fraser, C. H. T., 1958. Maze-learning in earthworms. Unpublished master's thesis, University of Aberdeen.

Gardner, L. E., 1968. Retention and overhabituation of a dual-component response in *Lumbricus terrestris sp. J. Comp. Physiol. Psychol., 66,* 315–318.

Gee, W., 1913. The behavior of leeches with especial reference to its modifiability. *Univ. Calif. Publ. Zool., 11,* 197–305.

*Golding, D. W., 1965a. Endocrinology and morphogenesis in *Nereis diversicolor. Gen. Comp. Endocrinol., 5,* 681.

Gwilliam, G. F., 1969. Electrical responses to photic stimulation in the eyes and nervous system of nereid polychaetes. *Biol. Bull., 136,* 385–397.

Hagiwara, S., and Morita, H., 1962. Electrotonic transmission between two nerve cells in a leech. *J. Neurophysiol., 25,* 721–731.

Hagiwara, S., Morita, H., and Naka, K., 1964. Transmission through distributed synapses between two giant axons of a sabellid worm. *J. Comp. Biochem. Physiol., 13,* 453–460.

Hargitt, C. W., 1906. Experiments on the behavior of tubiculous annelids. *J. Exptl. Zool., 3,* 295–320.

Hargitt, C. W., 1909. Further observations on the behavior of tubiculous annelids. *J. Exptl. Ecol., 7,* 157–187.

Hargitt, C. W., 1912. Observations on the behavior of tubiculous annelids. *Biol. Bull., 2,* 67–94.

*Harper, E. H., 1905. Reactions to light and mechanical stimuli in the earthworm, *Perichaeta bermudensis* (Beddard). *Biol. Bull., 10,* 17–34.

*Harris, J. D., 1943. Studies on nonassociative factors inherent in conditioning. *Comp. Psychol. Monogr., 18,* 1 (Whole No. 93).

*Hauenschild, C., 1960. Lunar periodicity. *Cold Spring Harbor Symp. Quant Biol., 25,* 491–497.

Heck, L., 1920. Über die Bildung einer Assoziation beim Regenwurm auf Grund von Dressurversuchen. *Lotos, 68,* 168–189.

Henderson, T. B., and Strong, P. N., 1972. Classical conditioning in the leech *Macrobdella ditetra* as a function of CS and UCS intensity. *Cond. Reflex, 7,* 210–215.

*Herter, K., 1926. Versuche über die phototaxis von *Nereis diversicolor* O. F. Muller. *Z. Vergl. Physiol., 4,* 103–141.

Herz, M. J., Peeke, H. V. S., and Wyers, E. J., 1964. Temperature and conditioning in the earthworm *Lumbricus terrestris. Anim. Behav., 12,* 502–507.

Herz, M. J., Peeke, H. V. S., and Wyers, E. J., 1967. Classical conditioning of the extension response in the earthworm. *Physiol. Behav., 2,* 409–411.

Hess, C., 1914. Untersuchungen über den Lichtsinn mariner Würmer und Krebse. *Pflugers. Arch. Ges. Physiol., 155,* 421–435.

*Hess, W. N., 1924. Reactions to light in the earthworm *L. terrestris. J. Morphol. Physiol., 39,* 515–542.

Hesse, R., 1899. Untersuchungen über die Organe der Lichtempfindungen bei niederen Thieren: V. Die Augen der polychaten Anneliden. *Z. Wiss. Zool., 65,* 446–516.

Hinde, R., 1954. Factors governing the changes in strength of a partially inborn response, as shown by the mobbing behaviour of the chaffinch *(Fringitla coelebs)*. I. The nature of the response, and an examination of its course. II. The waning of the response. *Proc. Roy. Soc. Ser. B, 142,* 306–358.

288   James A. Dyal

*Holmes, S. J., 1905. The selection of random movements as a factor in phototaxis. *J. Comp. Neurol., 15,* 98–112.

Holmes, S. J., 1911. *The Evolution of Animal Intelligence,* Holt, New York.

Horridge, G. A., 1959. Analysis of the rapid responses of *Nereis* and *Harmothoe* (Annelida). *Proc. Roy. Soc. Ser. B, 150,* 245–262.

*Howell, C. D., 1939. The responses to light in the earthworm, *Pheretima agrestris* (Goto and Hatai), with special reference to the function of the nervous system. *J. Exptl. Zool., 81,* 231–259.

Iwahara, S., and Fujita, O., 1965. Effect of intertrial interval and removal of the suprapharyngeal ganglion upon spontaneous alternation in the earthworm, *Pheretima communissima. Jap. Psychol. Res., 7,* 1–14.

Jacobson, A. L., 1963. Learning in flatworms and annelids. *Psychol. Bull., 60,* 74–94.

James, J. P., and Woodruff, A. B., 1965. Latent learning in earthworms. *Psychol. Rep., 16,* 406.

*Jennings, H. S., 1904. *Contributions to the Study of the Behavior of Lower Organisms,* Carnegie Institute of Washington Publication No. 16, 256 pp.

*Jennings, H. S., 1905. The method of regulation in behavior and other fields. *J. Exptl. Zool, 2,* 448–494.

Kaiser, F., 1954. Beiträge zur Bewegungsphysiologie der Hirudeen. *Zool. Jb. (Allg. Zool.), 65,* 59–90.

Kandel, E. R., 1965. Cellular studies of learning. In Quarton, G. G., Melnechuk, T., and Schmitt, F. O. (eds.), *The Neurosciences,* Rockefeller University Press, New York, pp. 666–689.

Kasper, P., 1961. Maze learning and spontaneous alternation in the earthworm. Paper read at meeting of Midwestern Psychological Association, Chicago.

Kirk, W. E., and Thompson, R. W., 1967. Effects of light, shock, and goal box conditions on runway performance of the earthworm, *Lumbricus terrestris. Psychol. Rec., 17,* 49–54.

Krasne, F. B., 1965. Escape from recurring tactile stimulation in *Branchiomma vesiculosum. J. Exptl. Biol., 42,* 307–322.

Krivanek, J. O., 1956. Habit formation in the earthworm, *Lumbricus terrestris. Physiol. Zool., 29,* 241–250.

Kuenzer, P. P., 1958. Verhaltenphysiologische Untersuchungen über das Zucken des Regenwürms. *Z. Tierpsychol., 15,* 31–49.

*Laverack, M. S., 1963. *The Physiology of Earthworms,* Macmillan, New York.

Livingston, R. B., 1966. Brain mechanisms in conditioning and learning. *Neurosci. Res. Progr. Bull., 4,* 235–347.

Malek, R., 1927. Assoziatives Gedächtnis bei den Regenwürmern. *Bio. Gen., 3,* 317–328.

Mann, K. H., 1962. *Leeches (Hirudinea): Their Structure, Physiology, Ecology and Embryology,* Pergamon, Oxford.

*Mast, S. O., 1938. Factors involved in the process of orientation of lower organisms to light. *Biol. Rev., 13,* 186–224.

*Moore, A. R., and Kellogg, F. M., 1916. Note on the galvanotropic response of the earthworm. *Biol. Bull. Woods Hole, 30,* 131–134.

Morgan, R. F., Ratner, S. F., and Denny, M. R., 1965. Response of earthworms to light as measured by the GSR. *Psychon. Sci., 3,* 27–28.

Newell, R. C. (1970). *The Biology of Intertidal Animals.* London, Logos Press.

Nicol, J. A. C., 1948a. Giant axons of annelids. *Quart. Rev. Biol., 23,* 291–323.

*Nicol, J. A. C., 1948b. The giant nerve fibres in the central nervous system of *Myxicola* (Polychaeta: Sabellidae). *Quart. J. Microscop. Sci., 89,* 1–45.

Nicol, J. A. C., 1950. The responses of *Branchiomma vesiculosum* (Montagu) to photic stimulation. *J. Mar. Biol. Ass. U.K., 29,* 303–320.

*Nicol, J. A. C., 1951. Giant axons and synergic contractions in *Branchiomma vesiculosum.* *J. Exptl. Biol., 28,* 22–31.

Peeke, H. V. S., Herz, M. J., and Wyers, E. J., 1965. Ganglia removal and photically driven activity in the earthworm *(Lumbricus terrestris).* *Psychon. Sci., 3,* 187–188.

Peeke, H. V. S., Herz, M. J., and Wyers, E. G., 1967. Forward conditioning, backward conditioning, and pseudoconditioning sensitization in the earthworm *(Lumbricus terrestris).* *J. Comp. Physiol. Psychol., 53,* 534–536.

Prosser, C. L., 1934. Effect of the central nervous system in response to light in *Eisenia foetida* Sav. *J. Comp. Neurol., 59,* 71–96.

Rabbe, S., 1939. Zur Analyse der Assoziationbildung bei *Lumbriculus variegatus* Mull. *Z. Vergl. Physiol., 26,* 611–643.

Ratner, S. C., 1962. Conditioning of decerebrate worms, *Lumbricus terrestris.* *J. Comp. Physiol. Psychol., 55,* 174–177.

Ratner, S. C., 1964. Worms in a straight alley: Acquisition and extinction of phototaxis. *Psychol. Rec., 14,* 31–36.

Ratner, S. C., 1965. Research and theory on conditioning of annelids. In Davenport, D., and Thorpe, W. H. (eds.), *Learning and Associated Phenomena in Invertebrates. Anim. Behav. Suppl., 1,* 101–108.

Ratner, S. C. The cerebral ganglion of earthworms in habituation and retention. Paper presented at Psychonomic Society Meetings, St. Louis, Nov. 3, 1972.

Ratner, S. C., and Miller, K. R., 1959a. Classical conditioning in earthworms, *Lumbricus terrestris.* *J. Comp. Physiol. Psychol., 52,* 102–105.

Ratner, S. C., and Miller, K. R., 1959b. Effects of spacing of training and ganglia removal on conditioning in earthworms. *J. Comp. Physiol. Psychol., 52,* 667–672.

Ratner, S. C., and Stein, D. G., 1965. Responses of worms to light as a function of an intertrial interval and ganglion removal. *J. Comp. Physiol. Psychol., 59,* 301–304.

Ray, A. J., 1968. Instrumental light avoidance by the earthworm. *Commun. Behav. Biol., 1,* 205–208.

Razran, G., 1971. *Mind in Evolution: An East–West Synthesis of Learned Behavior and Cognition,* Houghton-Mifflin, Boston.

Reynierse, J. H., 1968. Effects of temperature and temperature change on earthworm locomotor behavior. *Anim. Behav., 16,* 480–484.

Reynierse, J. H., and Ratner, S. C., 1964. Acquisition and extinction in the earthworm, *Lumbricus terrestris.* *Psychol. Rec., 14,* 383–387.

Reynierse, J. H., Halliday, R. A., and Nelson, M. R., 1968. Non-associative factors inhibiting earthworm straight-alley performance. *J. Comp. Physiol. Psychol., 65,* 160–163.

Roberts, M. B. V., 1962a. The giant fibre reflex of the earthworm, *Lumbricus terrestris* L. I. The rapid response. *J. Exptl. Biol., 39,* 219–227.

Roberts, M. B. V., 1962b. The rapid responses of *Myxicola infundibulum* (Grube). *J. Mar. Biol. Ass. U.K., 42,* 527–539.

Roberts, M. B. V., 1966. Facilitation in the rapid response of the earthworm, *Lumbricus terrestris* L. *J. Exptl. Biol., 45,* 141–150.

Robinson, J. S., 1953. Stimulus substitution and response learning in the earthworm. *J. Comp. Physiol. Psychol., 46,* 262–266.

Rosenthal, R., 1966. *Experimenter Effects in Behavioral Research,* Appleton-Century-Crofts, New York.

Rullier, F., 1948. La vision et l'habitude chez *Mercierella enigmatica* Fauvel. *Bull. Lab. Mar. Dinard, 30,* 21–27.

Rushton, W. A. H., 1945. Action potentials from the isolated nerve cord of the earthworm. *Proc. Roy. Soc. Ser. B , 132,* 423–437.

Schmidt, H., Jr., 1955. Behavior of two species of worms in the same maze. *Science, 121,* 341–342.

*Sherrington, C. S., 1906. *The Integrative Action of the Nervous System*, Yale University Press, New Haven.
*Smith, A. C., 1902. The influence of temperature, odors, light and contact in the movements of the earthworm. *Am. J. Physiol., 6*, 459–486.
Smith, G. E., and Dinkes, I., 1971. Passive avoidance learning by the earthworm. Paper delivered at the Eastern Psychological Association Meeting, New York City, April 15–17.
Staddon, J. E. R., and Simmelhag, V. L., 1971. The "superstition" experiment: A reexamination of its implications for the principles of adaptive behavior. *Psych. Rev., 78*, 3–43.
*Stammers, F. M. G., 1930. Observations in the behavior of land leeches. *Parasitology, 40*, 237–245.
Swartz, R. D., 1929. Modification of behavior in earthworms. *J. Comp. Psychol., 9*, 17–33.
Tauc, L. T., 1959. Interaction non-synaptic entre deux neurones adjacent du ganglion abdominal de l'aplysie. *Compt. Rend. Hebd. Seanc. Acad. Sci. Paris, 248*, 1857–1859.
Thompson, R. F., and Spencer, W. A., 1966. Habituation: A model phenomenon for the study of neuronal substrates of behavior. *Psychol. Rev., 73*, 16–43.
Thorpe, W. H., 1956. *Learning and Instinct in Animals*, Longmans Green, London.
Wanatabe, A., and Bullock, T. A., 1960. Modulation of the activity of one neurone by subthreshold slow potentials in another cardiac ganglion. *J. Gen. Physiol.,* 1031–1045.
Wayner, M. J., Jr., and Zellner, D. K., 1958. The role of the suprapharyngeal ganglion in spontaneous alternation and negative movements in *Lumbricus terrestris*. *J. Comp. Physiol. Psychol., 51*, 282–287.
*Wells, G. P., 1939. Intermittent activity in polychaete worms. *Nature, 144*, 940–941.
*Wells, G. P., 1949. The behavior of *A. marina* L. in sand, and the role of spontaneous activity cycles. *J. Mar. Biol. Ass. U.K., 28*, 465–478.
Wells, G. P., 1950. Spontaneous activity cycles in polychaete worms. *Symp. Soc. Exptl. Biol., 4*, 127–142.
*Wells, G. P., and Albrecht, E. B., 1951. The integration of activity cycles in the behavior of *A. marina* L. *J. Exptl. Biol., 28*, 41–50.
*Wells, G. P., and Dales, R. P., 1951. Spontaneous activity patterns in animal behavior: The irrigation of the burrow in the polychaetes *Chaetopterus variopedatus* Renier and *Nereis diversicolor* O. F. Muller. *J. Mar. Biol. Ass. U.K., 29*, 661–680.
Wells, M. J. Sensitization and evolution of associative learning. In Salanki, J. (ed.), *Neurobiology of Invertebrates*. Academic Kiado, Budapest, pp. 391–411.
Wherry, R. J., and Sanders, J. M., 1941. Modifications of a tropism in *Lumbricus terrestris*. *Trans. Ill. Acad. Sci., 34*, 237–238.
Wilson, D. M., 1959. Long-term facilitation in a swimming sea anemone. *J. Exptl. Biol., 36*, 526–532.
Wilson, D. M., 1960. Nervous control of movement in annelids. *J. Exptl. Biol., 37*, 46–56.
Wyers, E. J., Peeke, H. V. S., and Herz, M. J., 1964. Partial reinforcement and resistance to extinction in the earthworm. *J. Comp. Physiol. Psychol., 57*, 113–116.
Yerkes, A. W., 1906. Modifiability of *Hydroides dianthus*. V. *J. Comp. Neurol., 16*, 441–450.
Yerkes, R. M., 1912. The intelligence of earthworms. *J. Anim. Behav., 2*, 332–352.
Young, J. Z., 1963. Some essentials of neural memory systems. Paired centres that regulate and address the signals of the results of actions. *Nature, 198*, 626–630.
Young, J. Z., 1964. *A Model of the Brain*, Clarendon Press, Oxford.
Zellner, D. K., 1966. Effects of removal and regeneration of the suprapharyngeal ganglion on learning, retention, extinction and negative movements in the earthworm *Lumbricus terrestris* L. *Physiol. Behav., 1*, 151–159.

INDEX